D1277987

Shalaby
Biomedical Polymers

Biomedical Polymers

Designed-to-Degrade Systems

Edited by Shalaby W. Shalaby

With contributions by A.L. Allen, S. Amselem, S. Arcidiacono, D. Ball, T.H. Barrows, A.U. Daniels, A.J. Domb, R.A. Gross, J. Heller, R.A. Johnson, J. Keith, D.L. Kaplan, J. Kohn, R. Langer, S.J. Lombardi, M. Maniar, J.M. Mayer, A. Nathan, K. Park, A.G. Scopelianos, S.W. Shalaby, W.S.W. Shalaby, B.J. Wiley

Hanser Publishers, Munich Vienna New York

Hanser/Gardner Publications, Inc., Cincinnati

The Editor:
Dr. Shalaby W. Shalaby, Dept. of Bioengineering, 301 Rhodes Research Center, Clemson University, Clemson, South Carolina 29634-0905, USA

Distributed in the USA and in Canada by
Hanser/Gardner Publications, Inc.
6600 Clough Pike, Cincinnati, Ohio 45244-4090, USA
Fax: +1 (513) 527-8950

Distributed in all other countries by
Carl Hanser Verlag
Postfach 86 04 20, 81631 München, Germany
Fax: +49 (89) 98 48 09

Library of Congress Cataloging–in–Publication Data
Biomedical polymers : designed–to–degrade systems / edited by Shalaby
W. Shalaby
   p. cm.
Includes bibliographical references and index.
ISBN 1–56990–159–7 (Hanser/Gardner). - - ISBN 3–446–16531–2
(Hanser)
   1. Polymers in medicine. 2. Polyesters - - Biocompatibility,
I. Shalaby, Shalaby W.
   [DNLM: 1. Biocompatible Materials. 2. Polymers.
3. Biodegradation. QT 34 B641 1994]
R857. P6B566 1994
610'.28 - - dc20
DNLM/DLC
for Library of Congress   93–50660

Die Deutsche Bibliothek – CIP-Einheitsaufnahme
Biomedical polymers : designed–to–degrade systems / ed. by
Shalaby W. Shalaby. – Munich ; Vienna ; New York ; Hanser ;
Cincinnati : Hanser/Gardner Publ., 1994
   ISBN 3–446–16531–2 (Hanser)
   ISBN 1–56990–159–7 (Hanser/Gardner)
NE: Shalaby, Shalaby W. [Hrsg.]

© Carl Hanser Verlag, Munich Vienna New York, 1994
Typesetting in the USA by Agnew's Electronic Manuscript Processing Service,
   Grand Rapids, MI
Printed and bound in Italy by Editoriale Bortolazzi–Stei s.r.l., Verona

Preface

Interest in absorbable polymers, and particularly the synthetic ones, which are sometimes referred to as erodible, bioabsorbable, resorbable, or biodegradable, has grown considerably over the past two decades. This is primarily because of their transient nature when used as biomedical implants or drug carriers. Need to replace the tissue-reactive resorbable collagen-based sutures with synthetic chain molecules, which elicit milder tissue reaction led to the development of absorbable poly (α-hydroxyesters) such as polyglycolide (PG). These and other polyesters are discussed in the opening chapter of this book. In spite of the extensive R&D activities on absorbable polymers, polyesters remain as the dominant type of absorbable polymers. Since the early development of absorbable polyesters there have been four basic questions which presented a challenge to the polymer community. These questions include (a) other than polyesters, can new synthetic chain molecules be designed to provide absorbable materials? (b) can new absorbable molecules be made by biological processes? and (c) can the resorption of natural polymers be modulated to achieve desirable degradation profiles? A final, and perhaps the key, question to the three earlier questions is: Can one, synthetically or biologically, design to produce chain molecules with controlled degradation profiles? These questions prompted the development of this volume by a uniquely assembled team of authoritative authors, who either pioneered pertinent areas of research or made notable contributions to the field at large.

Thus, it is not surprising to have developed a book that provides an integrated update of one of the most promising forms of biomaterials by authors who are pioneering authorities or notable contributors to the field of polymeric biomaterials. Not only does the book portray past and present experience with invaluable critical notes by these authors, but it reflects their valuable views as to the future directions of the field. The nine chapters of the book cover major types of the designed-to-degrade polymeric systems of proven and projected biomedical significance. These include the traditional or less traditional polyesters, poly(orthoesters), poly(anhydrides), poly(esteramides), amino acid-derived polymers, poly(phosphazenes), bacterial polyesters, biosynthetic polysaccharides, and chemically modified proteins and polysaccharides.

Shalaby W. Shalaby

Contents

CHAPTER 1

Synthetic Absorbable Polyesters

Shalaby W. Shalaby and Russell A. Johnson

1.1 Scope

This chapter provides an overview of the chemical and physical aspects as well as key applications of three families of absorbable polyesters: lactone- and oxalate-based polymers and radiostabilized systems. In addition, the last section of the chapter focuses on the analysis of contemporary issues and emerging trends in the technology and clinical applications of synthetic absorbable systems. Although individual

Shalaby W. Shalaby and Russell A. Johnson, Department of Bioengineering, 301 Rhoses Research Center, Clemson University, Clemson, South Carolina 29634-0905, U.S.A.

characteristics and applications of these polymers have been the subject of a number of reviews, original articles, and patents [1–17], this chapter provides an integrated, broad coverage of the area.

1.2   Lactone-Based Polyesters

This family of polyesters encompasses primarily those made of glycolide (G), *l*-lactide (LL) and its isomers, ε-caprolactone (CL), trimethylene carbonates (TMC), *p*-dioxanone (PD), 2-methyl glycolide (MG), 2,2-dimethyl glycolide (DMG), 1,5-dioxapane-2-one (DOX-5), 1,4-dioxapane-2-one (DOX-4), 3,3-dimethyltrimethylene carbonate (DMTMC), glycosalicylate (GS), and morpholine-2,5-dione (MD). The structures of most of these monomers are shown in Figure 1.1.

1.2.1   Ring Opening Polymerization and Chain Reequilibration

A commonly accepted mechanism of ring opening [18–23] entails the activation of the ester-carbonyl of the cyclic monomer with an inorganic or organometallic

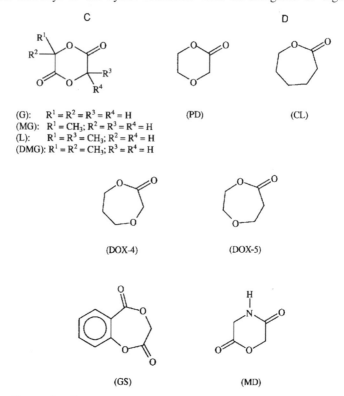

(G): $R^1 = R^2 = R^3 = R^4 = H$
(MG): $R^1 = CH_3; R^2 = R^3 = R^4 = H$
(L): $R^1 = R^3 = CH_3; R^2 = R^4 = H$
(DMG): $R^1 = R^2 = CH_3; R^3 = R^4 = H$

(PD)   (CL)

(DOX-4)   (DOX-5)

(GS)   (MD)

Figure 1.1   Types of cyclic monomers.

catalyst, based on Zn or Sn^{II}. Typical examples of these catalysts are diethyl or dibutyl zinc, stannous octoate, and stannous chloride. In the presence of a hydroxyl-bearing compound in the vicinity of the activated carbonyl group, a ring-opening reaction can take place to produce a new linear ester moiety with a hydroxyl end group. The hydroxyl-bearing compound can be in the form of a hydroxyacid (resulting from the hydrolysis of the ring structure in the presence of trace amounts of water existing as an impurity) or mono- or dihydric alcohol. Several authors refer to the hydroxyl compounds as the chain initiators, and rightfully so; their initial concentration is known to control the molecular weight of the polymeric chains, quite predictably.

Depending on the size of ring, the presence of heteroatoms, thermodynamic chain stability, and prevailing reaction temperature, the resulting polymers may coexist with about 1–15% of their monomers. This is attributed to the monomer/polymer equilibrium known to exist in ring-opening polymerization. Thus, a ring with highly nucleophilic ester-carbonyl and high strain energy usually provides a relatively fast rate of polymerization and low equilibrium monomer concentration. Steric requirements about the ester are expected to affect the reactivity and thermodynamic stability of the cyclic monomer. This can be exemplified best by comparing glycolide with lactide, where the latter monomer undergoes fast polymerization and results in a polymer that is in equilibrium with usually 1–3% of monomer (depending on polymerization temperature). On the other hand, lactide is a slow reacting monomer, and a monomer concentration of 2–7% can be found in equilibrium with resulting high molecular weight polymer.

During the advanced stages of most ring-opening polymerizations, additional reactions such as ester—ester interchange and chain unzipping may take place and can have a significant effect on the composition of the final product. Once again, the extents of those reactions are affected by the reactivity of the ester moieties. As a result of those reactions, not only monomer can be reformed, but also cyclic dimers, trimers, and to a lesser extent, cyclic oligomers. If the chain is made of more than one type of ester moiety, additional randomization will be a consequence of the ester—ester interchange reaction.

The presence of monomers and low molecular weight cyclic oligomers in absorbable polymers should be avoided, for they degrade more rapidly than the polymers and can lead to undesirable chemical and biological effects, as discussed later in the chapter. Hence, it has become prudent to remove these species from the polymers by distillation/sublimation under reduced pressure or extraction using nonreactive solvents. Although the issue of low molecular weight species in the polymerization products has been addressed reasonably well by many authors and manufacturers, the presence of monomers/oligomers in absorbable devices made by melt processing is practically ignored. Expectedly, a monomer-free polymer can undergo chain depolymerization to form monomers/oligomers on exposure to certain thermal treatments during melt processing or annealing. The extent of these reactions is dependent on the same factors noted previously relative to monomer/polymer equilibrium. A possible strategy to minimize the chain depolymerization is based on using alcohols or diols as chain initiators to control the catalytic effect of acid end groups. Having these impurities in an implant can affect its biological performance, particularly reaction of the surrounding tissue at the implant site. This topic is readdressed later in the chapter.

1.2.2   Physical Properties and Processing

1.2.2.1   General Considerations

With the exception of chain molecules of MG, DMG, and optically inactive *dl* forms of other lactones (e.g., *dl*-lactide), all homopolymers of glycolide and substituted glycolides can be obtained as highly crystalline polymers, having glass transitions and melting temperatures ranging from about 25°C to 65°C and from about 185°C to 225°C, respectively. Polymers of MG, DMG, and *dl*-lactide are amorphous materials. This is because such polymers are copolymeric in nature and the consistent alternation of their corepeat units (which can allow crystallization) is practically impossible. In addition to being amorphous due to lack of chain regularity and/or the steric requirements of its dimethyl groups, poly(2,2-dimethyl glycolide) (P-DMG) could not be produced as a high polymer [24]. The highest degree of polymerization attained was that of a sample having an inherent viscosity of about 0.3 in hexafluoroisopropyl alcohol (HFIP). This may be related to the tendency of P-DMG to undergo ester pyrolysis and formation of nonreactive end groups.

Almost all glycolide and substituted glycolide homopolymers can be melt processed by extrusion or molding. With the exception of polyglycolide (P-G), which has a limited or no solubility in ordinary organic solvents, all substituted glycolide polymers can also be processed in a solution form. In melt processing polylactides, care must be taken to avoid excessive heating of the polymer to minimize the extent of monomer formation by chain depolymerization.

Of all the glycolide and substituted glycolide polymers P-G, poly(*l*-lactide) (P-LL), poly(*dl*-lactide) (P-DLL), and poly(*p*-dioxanone) (P-PD) have attracted the attention of most investigators. Hence, specific data for these polymers are given in this section. Polyglycolide can be obtained from a highly pure monomer as fiber-forming or molding grade material with about 1–3% monomer/oligomer content. The polymer melts at approximately 225°C and displays about 50% crystallinity. As an absorbable material, its thermal stability is good and under dry extrusion or molding conditions, its melt characteristics are also good. On the other hand, lactide polymers are usually formed with about a 2–7% monomer/oligomer content. P-LL is less sensitive to hydrolysis than P-G and is a highly crystalline polymer with T_g and T_m of about 65°C at 185°C, respectively. It can be melt processed within a temperature range of about 200–250°C, depending on its molecular weight. Minimum residence time in the molten state is recommended. The poly(*dl*-lactide) is an amorphous polymer that can be melt processed below 250°C depending on its particular weight; it has a T_g of about 65°C.

P-PD is produced at a relatively low polymerization temperature, due to the high tendency of the chains to depolymerize. Monomer concentration present in the resulting high molecular weight polymer can vary between 4 and 15%, depending on prevailing polymerization conditions. The monomer can be removed by evaporation/sublimation under reduced pressure or extraction with nonreactive solvents. High molecular weight polymer (inherent viscosity of 1.5 to 2.2 in HFIP) displays a T_g and T_m about -10 to $0°C$ and $110-115°C$, respectively, and can be easily extruded or molded at temperatures of about $125-220°C$ [25]. The processing temperature depends, primarily, on the polymer melt viscosity (a function of molecular weight) and the residence time of the polymer in the molten state. Having an ether linkage adjacent to an active methylene group increases the susceptibility of the polymer to oxidative and photooxidative degradation. For this, photooxidative stabilizers are recommended for use in PD-based devices. The higher seven-membered ring homologues of PD, that is, DOX-4 and DOX-5, polymerize more readily than the six-membered ring PD. Compared with PD, polymeric DOX-4 and DOX-5 are usually viscous liquids at room temperature and their T_g values are near $-50°C$ [25–27]. High molecular weight P-DOX-5 can be obtained as a semicrystalline material with a with T_m near $25°C$, whereas lower molecular weight polymers are viscous fluids. Both P-DOX-4 and P-DOX-5 are much more thermally stable than P-PD due to the instability of their seven-membered ring monomers. On the other hand, these monomers tend to form crystalline cyclic dimers rather readily [28]. Sufficient data on the absorbability of P-DOX-5 and P-DOX-4 are unavailable; however, they are expected to be slow-absorbing.

Although high molecular weight P-CL is considered to be relatively nonabsorbable compared with glycolide, low molecular weight chains may undergo absorption. For this reason, and the fact that P-CL segments are key components of several absorbable copolymers, discussion of P-CL is included in this section. Caprolactone has comparable polymerizability to its isomorph, DOX-5, and their corresponding polymers show comparable thermal stability. The presence of an additional heteroatom in P-DOX-5 imparts slightly higher chain stability. The P-CL chains crystallize readily and display melting temperature near $70°C$. On the other hand, the T_g of P-CL is near that of P-DOX-5, approximately $-50°C$.

Different from the above mentioned lactones is GS, which is a cyclic codimer of glycolic acid and salicylic acids. Available data on GS indicate its high tendency to polymerize with or without stannous octoate as a catalyst [29]. This and the fact that polyglycosalicylate (P-GS) could be produced only in low to moderate molecular weights suggest the presence of impurities that not only initiate the ring-opening polymerization but also interfere with the chain propagation to a higher degree of polymerization. The nature of these impurities is yet to be determined. Although preliminary nuclear magnetic resonance (NMR) data of P-GS indicate alternating glycolate and salicylate sequences, the formation of crystalline fractions in P-GS could not be ascertained consistently. In most cases P-GS was an amorphous material with T_g of about $57°C$. The polymer was shown to absorb but its absorption rate was slower than that of P-G. P-GS could be easily melt processed with no discernible decrease in molecular weight [29].

The cyclic carbonate monomers TMC and DMTMC are quite similar to traditional lactones in terms of polymerizability and thermodynamic stability of their polymeric chains. Although both monomers polymerize readily, the resulting polymers have a high tendency to undergo thermal depolymerization. P-DMTMC is expected to be much less stable than its unsubstituted analog. Absorbability was implied for P-DMTMC in the patent literature [30], but convincing absorption data for both P-TMC and P-DMTMC could not be found. High molecular weight P-TMC is crystalline near room temperature but P-DMTMC melts near 70°C [30–32]. The major use of TMC and DMTMC is in the formation of copolymers with glycolide and/or lactide.

MD is the parent monomer of several lactam—lactone codimers, and its synthesis and polymerization were first reported by Shalaby and Koelmel [33]. It polymerizes quite readily to a highly degradable and readily solubilizable polymer. An interesting use of this monomer is its copolymerization with lactones such as P-D to produce absorbable melt-processable polymers with enhanced absorbability [33].

1.2.2.2   Specific Technological Considerations

P-G was the first absorbable polymer specifically made for the development [34,35] and introduction of a clinically successful new absorbable suture [36]. This was followed by several types of glycolide copolymers including those used in the production of a braided suture based on 10/90 poly(l-lactide–coglycolide) (10/90 P-LL/G) [37] and a monofilament suture made of 40/60 segmented copolymer of trimethylene carbonate and glycolide [38,39]. These are sold under the trade names of Vicryl and Maxon, respectively. It must be noted that the elegant technology used in handling the monomer glycolide and its conversion to high molecular weight, fiber-forming polymers has laid the foundation for a fast-growing segment of the biomedical industry. Hence, it is worthwhile to discuss some critical aspects of this technology in this section. First, monomeric precursors of absorbable polymers are handled and stored in moisture-free, hermetically sealed packages to avoid water intervention and, therefore, compromise of the demonstrated ability to convert glycolide to high molecular weight polymer. To minimize the effect of any unexpected traces of water during storage, it is customary to store the monomer at very low temperature. On the other hand, this practice may be reexamined at least in the case of glycolide. This monomer can exist in two polymorphic forms, α and β, which prevail below and above 42°C, respectively. Although the α-form is more thermodynamically stable and can form and persist until melting of the monomer at 83°C, it is less hydrolytically stable than the β-form. Such information about glycolide and glycolide-like monomers should be reviewed on designing the handling and storage protocol of these monomers. In addition to the usual precautions used in melt processing of polyesters, exceptional care to eliminate moisture intervention is taken during any thermal processing (including fiber orientation) of P-G and similar polymers.

Conversion of pure, moisture-free glycolide to high molecular weight polymer can be achieved in the presence of a tin halide, or, more commonly, stannous octoate. Detailed aspects of the polymerization mechanism and kinetics of glycolide and related monomers have been addressed by a number of authors [18–23,36,40–42]. In addition to the traditional bulk, ring-opening polymerization of glycolide-type monomers, Nieuwenhuis [43] noted the successful polymerization of these monomers in solution of aprotic solvents or suspensions with a nonsolvent liquid hydrocarbon as a medium (in the presence or absence of a small amount of silicone oil as phase stabilizer). Due the tendency of certain polylactones to depolymerization at high temperatures, small-scale polymerization is often conducted below the T_m of the resulting polymer. The process is denoted as solid-state polymerization [43–46].

1.2.3   Mode of Chain Dissociation

Early studies of Salthouse and Matlaga [47,48] on 10/90 poly(*l*-lactide–coglycolide) sutures indicated that the chain dissociation of this class of polymers entails chemical hydrolysis of the polyester chain to produce water-soluble, low molecular weight fragments. These fragments are then attacked by enzymes to produce lower molecular weight metabolites. Although it has been reported [48] that certain enzymes are not required for the early stages of chain dissociation, it is likely that under certain conditions the mere presence of enzyme can affect the type, and perhaps rate, of chain dissociation to some extent. This may be related to nonspecific oxidative degradation and/or base-catalyzed chemical hydrolysis. Such nonspecific effects could have been, in part, pertinent to the results obtained by Williams and co-workers [49–53], where the chain degradation was considered to occur through enzyme-catalyzed processes. Interest in the effect of free radicals on implants has prompted the investigators to initiate studies on the degradation of poly(*dl*-lactide acid) in the presence of hydroxyl radical [54].

A review by Vert [55] stated that biodegradation of absorbable polyester is far from simple and listed 20 factors that can contribute to such a process. Meanwhile, he emphasized his extraordinary results on the effect of mass on accelerated degradation in the center of a large amorphous implant [56,57] as well as the well accepted mode for semicrystalline polyester hydrolysis, where degradation of the amorphous regions precedes that of the crystalline component [55,58–61]. The latter process is expected to be associated with a gradual increase in polymer crystallinity as the degradation progresses, leading to continued loss of the polymer mass but preferentially of its amorphous component. Indeed, that was observed by Vert, Fredericks et al., and Chu [55,58,59] and more recently in this laboratory [62]. The latter studies [62] were intended to provide new insight into the degradation mechanisms taking into account (a) the type of polymer used, (b) correlation between in vivo and in vitro data (at 37°C in a phosphate buffer at 7.26 pH) in terms of both the loss of mechanical properties and mass of implant, and (c) use of relatively large mass, molded tubes, with minimum degree of orientation. Results of these studies suggest that molded, annealed short tubes of 10/90 P-LL/G copolymer, PolyDioxanone Suture (PDS®), and a copolymer of hexamethylene oxalate and hexamethylene terephthalate have in vivo degradation profiles that parallel those associated with in vitro aging at 37°C. This was reflected in the mass loss and compressive yield data, and supports the earlier thesis [47,48] that chemical hydrolysis is the prevailing mechanism for polymer degradation. In addition, percent crystallinity of the devices increased progressively as the degradation advanced, both in the in vitro and in vivo samples. Such data for essentially unoriented devices are consistent with those reported earlier for highly oriented sutures [58].

In a new in vitro test addressing the interaction of osteoclastic cells (responsible for bone resorption), PDS® (P-PD) was found to have a similar effect on these cells as an acceptable control in terms of inducing morphological changes [63]. However, this study was limited to periods where PDS® is not known to undergo noticeable degradation. Hence, the test was recommended for studying cytotoxicity and not absorption [63].

1.2.4   Applications and Evaluation of Commercial and Experimental Systems

This section addresses existing applications of commercial polyesters, and advanced and early evaluation studies of both commercial and experimental polymers. For simplicity, the discussion is subdivided to deal with specific polymers and their existing or potential products. Earlier reviews can be consulted for more details [2,16,17,60,61].

1.2.4.1   Polyglycolide

P-G was developed specifically for manufacturing suture products under the trade name Dexon [35,36]. It can be formed by the ring-opening polymerization of glycolide in presence of a tin catalyst and a suitable initiator (a hydroxylic compound). P-G is a crystalline material (40–55% crystallinity) that melts at about 225°C. A fiber grade polymer (η_{inh} = 1.2–1.6 dL/g in HFIP) can be spun to multifilament or monofilament yarns for the production of braided and monofilament sutures, respectively [35,36,64]. A typical suture braid has a straight tensile strength of 80–100 Kpsi and retains over 50% of its initial strength at 2 weeks postimplantation and absorbs in about 4 months [3,16,36,64,65]. The mechanical properties of P-G are shown in Table 1.1.

Table 1.1   Comparative Physical Properties of P-G (Dexon) and 10/90 P-LL/G (Vicryl) Braided

	P-G	10/90 P-LL/G
Straight tensile strength (Kpsi)	106	95
Knot strength (Kpsi)	65	63
Elongation (%)	24	25

The same yarn used to prepare suture braids has been used to manufacture knitted and woven fabrics. These were tested as absorbable vascular grafts [66] and meshes for repairing hernias. However, limited clinical success has been achieved with these meshes. Other attempts to utilize P-G in biomedical devices other than sutures such as ligating molded clips and staples were fruitless due to early loss of strength in unoriented forms.

There have been a number of questionable claims on the use of low molecular weight P-G in controlled delivery systems, such as drug-loaded microcapsules. The insolubility of the P-G in common organic solvents, which are traditionally used in microencapsulation, limits its use in this technological area; however, soluble copolymers are well-suited for such applications.

1.2.4.2   High Glycolide-Based Copolymers

These glycolide systems include random, segmented, and block copolymers with glycolate sequences constituting more than 50% of the chain mass. The most important member of this group of copolymers is the 10/90 P-LL/G, manufactured under the trade name of Vicryl sutures and meshes. With the exception of a few differences, Vicryl sutures and meshes are made and used similar to those composed of Dexon (Table 1.1). Compared with Dexon, Vicryl [37] displays (a) a lower T_m of around 205°C and, hence, a lower processing temperature; (b) lower crystallinity and crystallizability, which may be related to a preferred suture morphology and corresponding prolonged strength retention in the biological environment; and (c) near complete absorption in about 90 days or less (in fine size sutures) [37]. The absorption and strength retention profiles of P-G and 10/90 P-LL/G sutures are shown in Figure 1.2. Vicryl meshes found limited use as surgical aids for repairing hernias. However, knitted, woven, and multifilament yarns of Vicryl, and to a lesser extent Dexon, have been used (a) in manufacturing "baskets" for the protective wrapping of soft organs (such as the kidney and liver during abdominal surgery), (b) in producing different shapes of absorbable barrier to inhibit postoperative epidermal down-growth in the surgical procedure associated with the treatment of certain periodontal diseases [67], (c) in studying the in vivo performance of absorbable vascular grafts [68], and (d) as template or scaffolding for in vitro growth of tissue cultures such as those of liver and cartilage [69,70].

In a successful effort to have a monofilament absorbable suture with higher compliance than monofilament Dexon sutures and prolonged retention of breaking strength in typical surgical sites, a segmented copolymer of about 40/60 trimethylene carbonate and glycolide (40/60 P-TMC/G) was developed about 9 years ago [38]. The suture is marketed under the trade name Maxon and typical properties are given in Table 1.2. The suture tensile strength approaches that of Dexon braided sutures. It retains over 50% of its initial original breaking strength at 3 weeks postimplantation, and absorbs completely in less than 6 months [38,39]. The in vivo strength retention profile of Maxon (P-TMC/G) suture is compared with that of Dexon (P-G) in Figure 1.3.

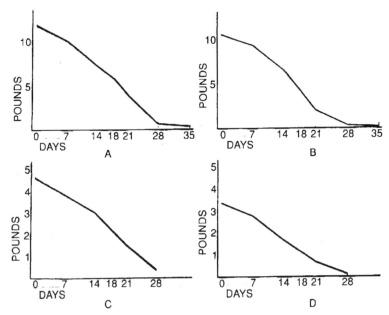

Figure 1.2   In vivo strength loss profiles of 10/90 P-LL/G and P-G sutures. (A) Size 0–0 10/90 P-LL/G sutures. (B) Size 0–0 P-G sutures. (C) Size 4–0 10/90 P-LL/G sutures. (D) Size 4–0 P-G sutures [37].

1.2.4.3   Poly(*p*-dioxanone)

P-PD was developed and marketed under the trade name PDS®, well before Maxon and practically for the same reasons the latter was sought [25,70]. The polymerization can be conducted in the melt or below the polymer T_m, which is about 115°C, using a tin catalyst and a hydroxylic initiator [25]. Due to its chain flexibility, the amorphous T_g of P-PD is in the range of about −10 to 0°C. P-PD crystallizes quite readily, displaying crystallinities of around 55%. Because the polymer tends to requilibrate with its cyclic monomer above its T_m, its melt processing is usually conducted at the lowest possible temperature, that is, between 120 and 220°C, depending on the molecular weight and residence time in the molten state. From a polymer having an inherent viscosity of about 2.0 dL/g (in HFIP) high-strength, low-modulus (high-compliance) monofilament sutures can be obtained by melt extrusion followed by drawing. Representative properties include a straight tensile strength of 70–90 Kpsi, a Young's modulus of 250–350 Kpsi, and an ultimate elongation of 40–70%, depending on the suture size and drawing conditions (Table 1.2). A typical suture retains over 50% of its initial breaking strength at 3 weeks postimplantation, which provides an

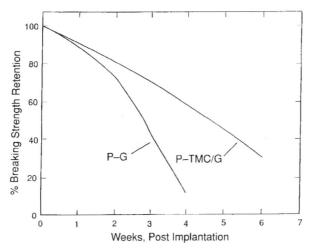

Figure 1.3 Comparison of in vivo breaking strength retention (BSR) of P-G braided sutures to P-TMC/G monofilament sutures [61].

advantage over the glycolide-based Dexon and Vicryl in slow healing wounds. A comparison of PDS® and those sutures is shown in Figure 1.4. The suture absorbs in about 6 months [25,71].

1.2.4.4   High PD-Based Copolymers

Toward improving the properties of PDS® sutures, random and segmented copolymers of PD with other monomers were made and converted to monofilament sutures having new attributes [33,74,75]. Segmented copolymers of PD with up to 20% glycolide-based sequences form sutures with comparable compliance to PDS® but exhibit an absorption profile comparable to those of Vicryl and Dexon [74]. On the other hand, copolymers with up to 15% *l*-lactide form sutures that are more compliant than PDS® and have absorption profiles that are similar to that of the parent suture

Table 1.2   Comparative Physical Properties of 40/60 P-TMC/G (Maxon), P-PD (PDS), and 5/95 P-MD/PD Monofilament Sutures

	40/60 P-TMC/G	P-PD	5/95 P-MD/PD
Straight tensile strength (Kpsi)	88	87	62
Knot strength (Kpsi)	57	55	45
Elongation (%)	38	30	58
Young's modulus (kpsi)	460	390	—

12   Shalaby W. Shalaby and Russell A. Johnson

Table 1.3 Comparison of In Vitro Absorption Data of P-PD to 5/95 P-MD/PD Size 4–0 Monofilament Sutures

| | Percent weight remaining | |
Time	P-PD	5/95 P-MD/PD
2 Weeks	70	61
4 Weeks	50	23
Complete absorption	180–210 Days	120 Days

[75]. Copolymerization of PD with 3–5% MD produces fiber-forming materials with unusual properties as monofilament sutures [33]. These were shown to absorb 10–25% faster than PDS® but maintained an in vivo strength retention profile comparable to that of PDS® monofilament sutures (Figures 1.5 and 1.6). Comparative properties of PDS® and 5/95 P-MD/PD monofilament sutures are shown in Tables 1.2 and 1.3. Although articles other than sutures were not described in sufficient detail for these copolymers, they are likely to be useful in several biomedical devices.

1.2.4.5   Polylactide and High Lactide-Based Copolymers

This section considers polylactide and copolymers with at least 50% lactide-based chains. The use of polylactides, particularly P-LL and its copolymers with glycolide, as suture materials has been claimed in the early patent literature [76]; however, none

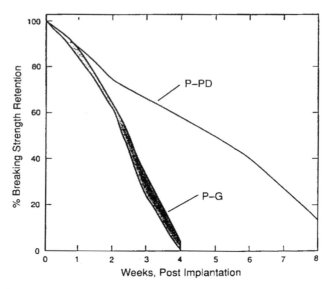

Figure 1.4 Comparison of in vivo breaking strength retention (BSR) of P-G braided sutures to P-PD monofilament sutures [61].

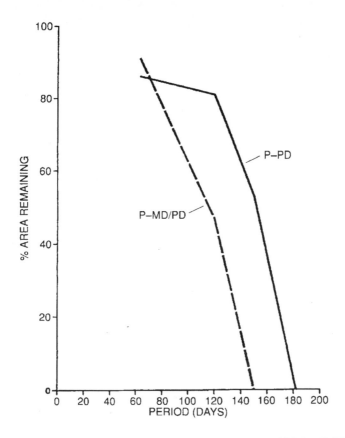

Figure 1.5 Comparison of in vivo absorption profiles of P-PD to 3/97 P-MD/PD monofilaments [33].

has been commercialized. This is likely to a consequence of the (a) slow absorption of the P-LL fibers and other fiber-forming crystalline copolymers containing at least 90% of *l*-lactide-based sequences; (b) insufficient crystallinity, or lack of it, in copolymers having more than 10% of hydrolytically more reactive sequences (i.e., glycolide); and to a lesser extent, (c) the thermodynamic instability of P-LL, which limits its melt processability. Copolymers of *l*-lactide with *dl*- or *d*-lactide are also subject to the usual compositional effect on crystallinity with modest improvements in in vivo absorption. Thus, it is not surprising to see that most of the work on these polymers over the past two decades has been directed toward orthopedic applications, where slow absorption is an advantage [44,77–99]. Applications of the high lactide-based polymers include those pertinent to suture coating [100], vascular grafts [68,101–103], controlled delivery of drugs [104–106], and meshes to facilitate wound healing following dental extraction [66,107].

P-LL and the high lactide-based copolymer can be prepared in a similar manner as polyglycolide. Due to the tendency of lactide polymers to depolymerize at high

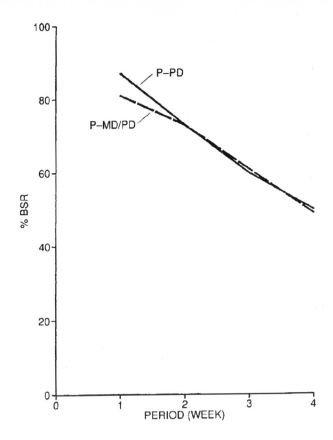

Figure 1.6   Comparsion of in vivo breaking strength retention (BSR) of P-PD to 3/97 P-MD/PD monofilaments [33].

temperatures, they are frequently prepared in the solid state [43–46] and melt processed at the lowest possible temperature, using extrusion [44,45] molding [76,108] techniques. Although difficult to pursue on a commercial scale, solution and suspension polymerization [43] and solution spinning [109] were recommended for these polymers.

Recommended inherent viscosity (in chloroform) for lactide-based polymers suitable for melt processing can be between about 1.5 and 5.0 dL/g [76,108,110–113]. Polymers with higher viscosities are more suited for solution processing [109,114,115]. The use of ultrahigh molecular weight polylactide to produce ultrahigh strength fiber by gel spinning is a special case of solution spinning [115,116]. In this study, polylactide having an intrinsic viscosity as high as 13 dL/g was used [116].

Melt spinning of P-LL having an $\overline{M}_v$ of 3.6 $\times 10^5$ resulted in fiber with an $\overline{M}_v$ of only 1.1 $\times 10^5$ [110]. The fibers lost 20% of their initial tensile strength after 9 months in a phosphate buffer solution at pH 7.4 and 37°C; no weight loss was recorded under these conditions. A more definitive study of melt spinning and drawing of P-LL and 5/95 P-G/LL was conducted by Benicewicz et al. [44,117] in a successful effort to produce low denier per filament, multifilament yarns for the development of prosthetic tendons and ligaments. It was noted that a fibrous, porous implant would provide a scaffolding for new tissue to form along the filaments of the prosthesis. It was also suggested that fine diameter filaments (between about 3 and 15 μm) are desirable to maximize tissue ingrowth. Model forms of such devices in the form of braided constructions were tested in vitro, in phosphate buffers (pH 7.27 at 50°C). The breaking strength retention profiles of the P-LL and 5/95 P-G/LL braids are given in Figure 1.7 and the weight loss data are shown in Table 1.4. The available data on these systems reflect the obvious advantage of the copolymeric system for its acceptable strength profile during the normal healing period for bones which showed a noticeable improvement over the homopolymer in terms of absorption.

A relatively new family of lactide-based copolymers are those with TMC and DMTMC [30–32,118,119]. A copolymer with 10/90 TMC/LL or DMTMC/LL was proposed as a potential absorbable orthopedic device. Copolymers with less crystallizable P-LL segments were converted to flexible conduits, which were evaluated for channel nerve regeneration [119]. More details on polycarbonates are given in the next section.

1.2.4.6 Aliphalic Polycarbonates

Aliphatic polycarbonates include polymers that are based primarily on DMTMC and made by ring-opening polymerization (30–32,118–125] as well as copolymers of

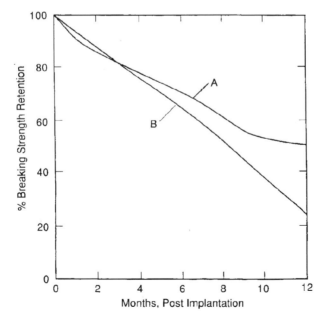

Figure 1.7 Comparison of in vivo breaking strength retention (BSR) of the homopolymer P-LL (A) to the copolymer 5/95 P-G/LL (B) braided constructions [60,61].

Table 1.4   Comparison of In Vitro Absorption Data of P-LL to 5/95 P-G/LL Braided Sutures at 50°C

	Percent weight remaining	
Time (Weeks)	P-LL	5/95 P-G/LL
0	100	100
4	100	99.8
12	99.5	95.3
26	95.9	58.3
39	91.9	20.5
52	58.5	2.1
65	23.3	1.2

AA–BB type prepolymers, made from *trans*-1,4-cyclohexanedimethanol [124,125]. A crystalline homopolymer of DMTMC, which is expected to have low thermodynamic stability, was reported to be an absorbable material that can be melt processed above its T_m to produce fibers with variable deniers [30,118]. The homopolymer was claimed to be biocompatible with living tissue, having a controllable rate of biodegradation [30,118]. Fibers made of this homopolymer as well as random copolymers DMTMC and CL and/or TMC (e.g., 97.5 DMTMC and 2.5 TMC) were described as being useful for the production of absorbable sutures, tendon prostheses, and vascular grafts [30–32,118]. Random and block copolymers of TMC, *dl*-lactide, and *l*-lactide were proposed as useful compositions for the production of absorbable coatings, tendon prostheses, and nerve channels [119–123].

The polycarbonate of *trans*-1,4-cyclohexanedimethanol was found to be a crystalline material with a T_m of about 150°C [124]. A high molecular weight polymer was spun and drawn into strong monofilament yarns which underwent no significant mass loss on incubation in a buffered solution at pH 7.26 and 37°C after about 1 year. Thus, it was considered to be a nonabsorbable polymer. On the other hand, low molecular weight copolymers of such a material with PD were patented as absorbable coatings for braided sutures [125].

1.2.4.7   Absorbable Orthopedic Implants

In spite of the successful application of absorbable polymers in devices for soft tissue repair, their use in orthopedic surgery (other than in the form of sutures) has been limited to the commercially available P-PD pins and a few similar devices for internal fracture fixation that are made of P-LL or P-G; some have been the subject of clinical studies [126,127]. In fact, for the internal fixation of fractured diaphyseal bones and the repair of skeletal damage such as joint fractures, metal implants still prevail, in spite of the documented disadvantages of certain metal implants. These disadvantages include the unnecessary stiffness of metal implants, leading to stress-shielding and inadequate bone remodelling and the sensitivity of many patients to nickel, chromium, and cobalt present in many metal alloys. Several years ago this provided strong incentive to explore the development of transient absorbable systems with tailored mechanical and biological properties to substitute (at least in part) the metal implants. In 1966, Kulkarni et al. [109] described the use of polylactic acid (PLA) for surgical implants. A few years later Cutright and co-workers showed that polylactide sutures, plates, and screws resulted in the successful healing of mandibular fractures but the evaluation of the tissue reactions was limited to sutures [128–130]. Over a decade later, interest in using polylactide, its copolymers, and other absorbable polymers for bone fracture fixation was revived and grew at a considerable rate [77–99]. This included the use of (a) P-PD implants in experimental osteotomies in animal models and human ankle fractures (85–89), (b) P-LL and copolymers as bone plates in sheep [90] or as plates and screws for bone fracture fixation [91], (c) self-reinforced P-G rods having over 60% fiber loading based on ordinary [92] or ultradrawn fibers [93] for bone fracture repair, and (d) injection-molded [94] or highly oriented extruded P-LL rods [87–95,96] for fracture fixation in animal models.

Reports on the first follow-up and clinical results in orthopedic devices made from polymers and composites reflected few complications [17,77,131–135]. Among those of most concern is the development of sterile sinuses, with a frequency of 4–8% [77,131]. The exact cause of the sterile sinus formation is yet to be determined, although a few factors have been suggested to contribute to this event. These include immunological and foreign body reactions [132], and sudden release of acids by degrading materials at a certain period and overtaxing the transportation capacity of surrounding tissue [133,134]. It is interesting to note that the sterile sinus formation was considered to have no negative consequences for the functional outcome of these implants [77,135]. In spite of the large amount of data available on the properties and biological performance of absorbable implants for orthopedic applications, there appears to be no clear consensus on the clinical relevance of most of these data. This may be, in part, attributed to (a) variability of the residual monomer/oligomer content, which leads to premature loss of in vivo strength and amplified tissue reactions; (b) nonuniformity in composite systems, in terms of fiber—matrix adhesion and fiber loading or mechanical properties; (c) dependence on the animal model used; (d) variability in mass of implant relative to the animal weight; (e) difference in the type of surgical implant site; (f) the effect of polymer or device morphology; and particularly orientation and the crystalline—amorphous interface; (g) variability in the device surface that is interfacing directly with the living tissues; and (h) the expected dependence on compositional variability for copolymers not only in the exact co-repeat ratio, but also their distribution along the copolymer chain. Toward addressing those issues, there have been a number of definitive, as well as far from definitive, studies over the past 8 years. These include studies on (a) differences in the strength retention of P-G rods implanted in the femoral medullary canal and the less degrading subcutis of rabbit [87]; (b) loss in shear strength of P-PD pins implanted in the subcutis, muscle, and femoral medullary canal of in rabbit [98]; (c) the effect of slow-absorbing coating on the strength retention of self-reinforced P-G composite implants [99]; and (d) the in vitro and in vivo performance of carefully purified, extruded, and oriented ultrahigh

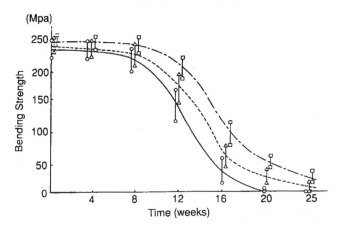

Figure 1.8   In vitro bending strength profiles of P-LL rods at three different diameters and 5 cm length (○), 3.2 mm; (△), 4 mm; (□) 5 mm diameter [96].

strength P-LL [96]. In the latter study, 3.2 mm, drawn rods (2.5 ×draw ratio) showed initial bending strength and modulus values of 240 MPa and 13 GPa, respectively. The bending strength retention of those rods implanted in the rabbit subcutis were almost equal to those aged in a buffered saline at 37°C and slightly higher than those of rods implanted in the medullary canal of rabbit. The effect of surgical implant site on those P-LL rods was consistent with that reported by Bhatia et al. [98]. Using larger diameter P-LL rods (4.0 and 5.0 mm), Matsusue et al. [96] have shown that the strength retention increases with the increase in diameter [Figures 1.8 and 1.9]. It is interesting to note that these P-LL rods [96] lost 22 and 70% of their mass at 52 and 78 weeks postimplantation, respectively, in the medullary canal of the rabbit femur and their bending

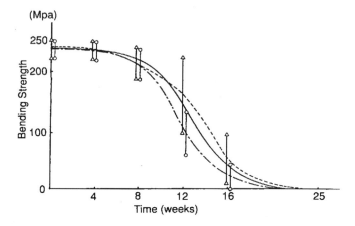

Figure 1.9   Bending strength profiles of P-LL rods of 3.2 mm diameter and 5 cm length in three different locations of implantation. (-), in vitro; (△), the subcutis; (○), the medullary cavity [96].

strength exceeded that of bone at 8 weeks. Based on these data and histological findings at 12, 16, 50, and 78 weeks, Matsusue et al. [96] advocated the clinical use of this P-LL system, in the repair of human bone fracture, in spite of its slow biodegradation.

1.3   Oxalate-Based Polymers

This section addresses primarily aliphatic oxalate-based polymers [136–143] and polyesteramides (or polyoxamates) [144,145]. The discussion entails (a) the development of this family of polymers as new absorbable materials [136,143], (b) properties of monofilament sutures and other devices made of those polymers [137,138,141,145], (c) the use of low molecular polyesters as absorbable coatings for braided sutures [136], and (d) use of polyalkylene oxalate in controlled drug delivery [139,140].

1.3.1   Poly(alkylene oxalates) and Their Isomorphic Copolymers

In the late 1970s, Shalaby and Jamiolkowski developed a family of oxalate polymers [136–139] as candidates for absorbable monofilament sutures and coatings to improve the characteristics of commercially available braided sutures. For uses as absorbable coatings, acyclic polyalkylene oxalates were derived from linear chains shown below, where the R group represented $(CH_2)_6$ or a mixture of $(CH_2)_4$ to $(CH_2)_{12}$. These were synthesized by the condensation of diethyl oxalate and n-alkanediols [136]. Their low melting temperatures were suggested to allow the formation of a liquid lubricant due to friction-induced heating during the tie-down and sliding of two strands of sutures. This, in turn, provided minimum friction at the contact area between the sliding sutures. Because of the high crystallinity of the polymer, the liquid interface was presumed to resolidify immediately as the sutures separated and the surface cooled. The tissue response, determined with coated braided sutures made from 10/90 P-LL/G (Vicryl) with the various poly(alkylene oxalates) and implanting them in the gluteal muscles of Long–Evans rats, was reported to be minimal. Polyoxalates were synthesized in various forms with carbon chains ranging from 3 to 16; those based on C3 and C6 moieties were later proposed as absorbable drug delivery systems. The absorbability of these polymers was shown to decrease primarily with increases in the methylene (or hydrocarbon) fraction of the polymer chain [136].

$$—ROOC—COO—$$
Alkylene oxalate unit

In other reports, Shalaby and Jamiolkowski [137,138] described monofilaments comprising extruded and oriented polymers of poly(alkylene oxalates) with the intent of producing absorbable sutures, surgical aids (such as in fabrics for hernia repair), and prostheses from these polymers. The filaments were described as soft, flexible, and absorbable in animal tissue with minimal adverse tissue reaction. The absorption

Table 1.5   Absorption Properties and Breaking Strength Retentions (BSR) of Poly(alkylene oxalates)[a]

			Absorption, [% remaining, (days)]		In vivo BSR
$-(CH_2)_n-$	T_m (°C)	Cryst. (%)	In vitro[b]	In vivo	[%, (days)]
n = 4	103	50	37(7)	50(9); 0(28)	35(3)
n = 6	70	47	17(31)	80(42); 0(121)	79(3)
n = 8	75	54	34(22); 4(199)	—	—
n = 10	78	—	89(44); 38(177)	—	—

[a] References [137,138].
[b] 37°C in phosphate buffer (pH 7.26).

rates were characterized as slow for higher alkylene oxalate polymers, and rapid for lower ones. Properties of these systems are shown in Table 1.5.

With the increase in hydrocarbon component of the polyester chains, melting temperatures of the alkylene oxalate polymers approached that of polyethylene (130°C), but those of useful systems were < 95°C. Such melting behavior was considered less than optimal for suture processing [143]. Subsequently, cyclic and aromatic diols were considered as polymer precursors for higher melting alkylene oxalate polymers. Polyesters based on 1,4-*trans*-cyclohexanedimethanol and 1,4-benzene and 1,3-benzene dimethanol (1,4-HDM, 1,4-Bz, and 1,3-Bz, respectively) were synthesized with high melting temperatures; however, polymers of the latter two diols could not be made as high molecular weight materials. This was attributed to the thermal instability of the monomers. Polymers consisting of 1,4-HDM had a higher hydrolytic stability than commercial absorbable polymers, and for this reason it was necessary to find polyesters with intermediate propensities for absorption, between those of the acyclic alkylene oxalates and the cyclic 1,4-HDM oxalates. The concept of isomorphic replacement was sought as a practical strategy to tailor-make cyclic/acyclic copolymeric chains and develop new absorbable crystalline materials with a broad T_m range (including those that melt above 100°C) [141]. More specifically, this entailed the copolymerization of 1,6-hexanediol and 1,4-HDM (Figure 1.10). After characterization, it was surprising to find that a substantial level of crystallinity was attained throughout the entire range of copolymer compositions. Furthermore, the T_m increased progressively with increase in the cyclic fraction and displayed only one major melting transition. In a key patent on this technology [141], it was also noted that this family of isomorphic polymers was unusual in that all copolymers through the entire composition range of from 5 to 95% of each isomorphic comonomer demonstrate

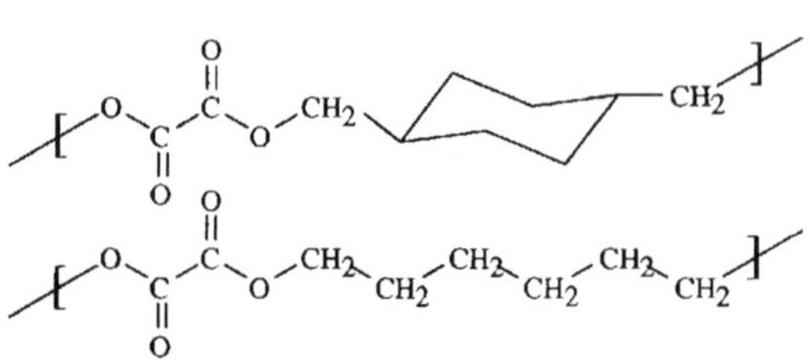

Figure 1.10   Isomorphic copolyoxalates [162].

Table 1.6   Absorption and Breaking Strength Retention (BSR) Data of Isomorphic Copolyoxalates[a]

HDM/C$_6$	T$_m$ (°C)	Cryst. (%)	Absorption, [% remaining, (days)]		In vivo BSR [%, (days)]
			In vitro[b]	In vivo	
100	216	38	89(42); —	—	—
95/5	210	—	79(42); 34(127)	—	—
80/20	193	37	61(42); 8(127)	56(180)	87(7); 17(14)
70/30	179	—	70(28); —	—	—
50/50	138	—	42(38); —	—	—
30/70	85	44	—; 0(141)	2(150)	—
5/95	69	—	7(42); —	—	—

[a] Reference [141].
[b] 37°C in phosphate buffer (pH 7.26).

levels of crystallinity comparable to those encountered in the parent homopolymers, namely between 30 and 50% depending on the thermal history. Constructed curves of the melting temperature versus composition did not reveal any positive "eutectic" composition in these systems. These copolyoxalates could be fabricated into pliable monofilaments that had good in vivo strength retention and were absorbed in animal tissue with minimal adverse reaction. As sutures, the isomorphic copolyoxalates were said to have good initial tensile and knot strength and a high order of softness and flexibility [141], although the time of complete absorption was not provided. Typical properties of representative isomorphic polymers are shown in Table 1.6.

1.3.2   Polyoxamates

This is a new family of polymers with ester–amide sequences. A typical scheme for the synthesis of these polymers [144,145] is shown in Figure 1.11. The condensation of the

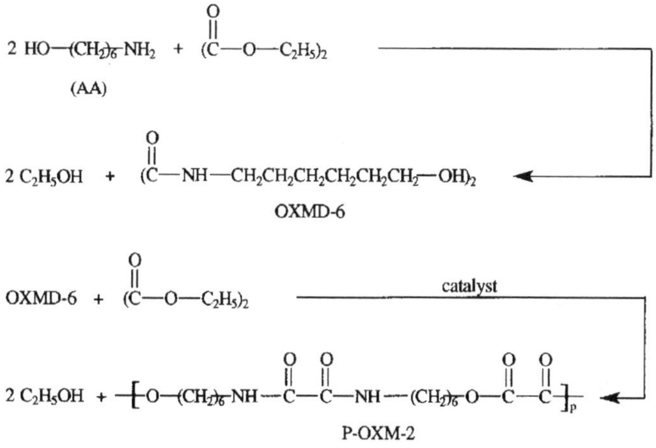

Figure 1.11   Synthesis of a typical polyoxamate (P-OXM-2).

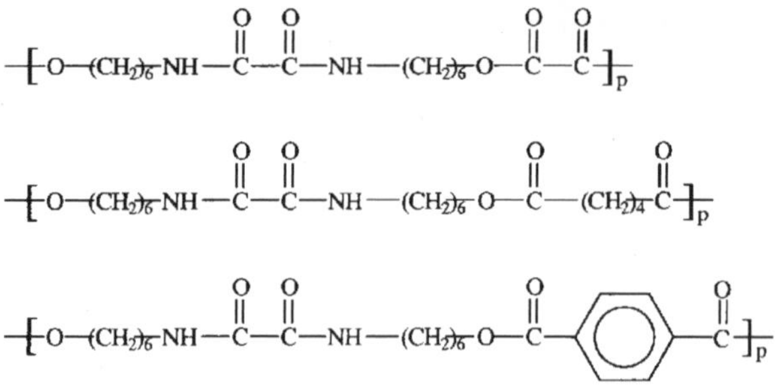

Figure 1.12   Structures of absorbable and nonabsorbable polyoxamates.

amino alcohol (AA) with diethyl oxalate is achieved in a methanol solution to produce bis-oxamidodiol (OXMD-6) in an almost quantitative yield. The diol is then condensed with diethyl oxalate, diethyl adipate, or dimethyl terephthalate in the presence of a tin catalyst (e.g., dibutyltin oxide) to produce polyesteramides P-OXM-2, P-OXM-6 and P-OXM-T, respectively (Figure 1.12). Thermal analysis data of P-OXM-2 are summarized in Table 1.7.

All three polymers could be made as high molecular weight materials and were converted to strong fibers with vastly different propensities to hydrolytic degradation. Hydrolysis or absorption of P-OXM-6 and P-OXM-T led to their classification as slightly absorbable and nonabsorbable materials, respectively. On the other hand, P-OXM-2 was considered to be absorbable. Mechanical properties of a monofilament suture made of P-OXM-2 are given in Table 1.7.

1.4   Radiostabilized Absorbable Polyesters

Sterility of biomedical devices based on synthetic polymers can usually be achieved by exposure to heat, ethylene oxide, or high-energy radiations. To a lesser extent, ultraviolet radiations and gas plasma can be used for sterilization with increased complexity of biomedical products. However, the use of gamma radiations was promoted as the most preferred method of sterilization [146]. Due to their tendency to degrade in the presence of gamma radiation during a typical sterilization cycle, a number of biomedically significant polymers could not be radiation-sterilized [2,14,16,146–149]. This led Shalaby and co-workers to explore new approaches to radiostabilize absorbable polyesters that are made by ring-opening [8,150–161] or step-growth [162] polymerization.

Primary studies on the development of radiostabilized polyesters dealt with the synthesis of partially aromatic polymeric radiostabilizers RS-1, RS-2, and RS-3 shown in Figure 1.13. Whereas RS-2 and RS-3 are a viscous liquid and amorphous solid, respectively, RS-1 is semicrystalline and can be converted to strong fibers [8,151–154]. In vivo and/or in vitro absorption studies indicate RS-1 and RS-3 are slow-absorbing materials. Copolymerization of RS-1 with glycolide resulted in crystalline,

Table 1.7   Physical Properties of P-OXM-2 Monofilaments [144,145]

• Polymer properties	
η_{inh} (HFIP)	1.51
Thermal data	
Initial heat:	$T_g = 50°C$
	$T_m = 168°C$
Reheat:	$T_g = 41°C$
	$T_m = 165°C$
• Mechanical properties of annealed fibers (7 mil. diam.)	
Straight tensile strength (kpsi)	62
Knot strength (kpsi)	44
Elongation (%)	21
Young's modulus (kpsi)	529

segmented copolymer, which could be converted to fibers with exceptional strength [8,151,152]. A representative example of these systems is a segmented copolymer based on about 10 and 90 mol% of the RS-1 and P-G components (10/90, RS-1/P-G). Suture braids made of this copolymer were shown to withstand gamma radiation during a typical sterilization cycle [8,151,152]. More importantly, their in vivo breaking strength profile (a highly sensitive property to radiation degradation) was equivalent or superior to nonsterile similarly constructed sutures (Table 1.8). Copolymers were also made of RS-3 and converted to strong braided sutures which were equally radiostable. The radiostability of both braids reflected in their in vivo strength retention is compared with that of P-G sutures in Table 1.8. Using the reduction in the braid cross-sectional area as a measure of absorption, sutures made of 10/90 RS-1/P-G and 10/90 RS-2/P-G were shown to absorb in about 120 days. Monofilaments 10/90 RS-1/P-G and RS-3/P-G exhibited similar radiostability as reflected in their in vivo strength retention data (Table 1.9).

To impart radiostability to P-PD, RS-1, RS-2, and RS-3 were either copolymerized with PD to form segmented copolymers or melt blended with suture-grade P-PD

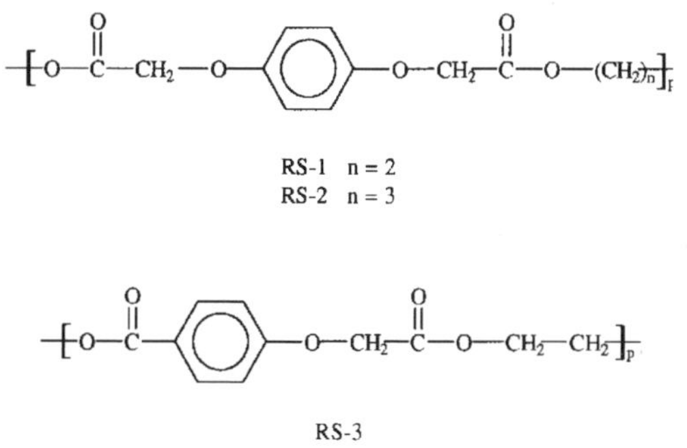

RS-1 n = 2
RS-2 n = 3

RS-3

Figure 1.13   Chemical structures of polymeric radiostabilizers [151,152].

Table 1.8 Properties of Gamma-Irradiated (2.5 Mrad) and Nonirradiated Suture Braids (Size 2–0) Made of P-G and Segmented Copolymers[a]

	P-G		10/90 RS-1/P-G		10/90 RS-3/P-G	
	Nonirrad.	Co-60	Nonirrad.	Co-60	Nonirrad.	Co-60
Fiber Properties						
Straight tensile strength (kpsi)	131	113	106	100	152	154
Knot strength (kpsi)	74	60	65	61	92	89
Elongation (%)	19	16	24	22	18	17
Biological Properties[b]						
In vivo absorption						
Percent cross-sectional area						
remaining at:						
5 days	100	100	—	100	—	100
70 days	—	60	—	92	—	92
91 days	12	20	—	49	—	49
119 days	0	3	—	1	—	1
In vivo BSR						
Percent retained at:						
1 week	112	87	122	113	—	—
2 weeks	74	38	105	88	—	73
3 weeks	51	0	97	72	—	55

[a] References [8,10,152].
[b] Postimplantation in rat subcutis.

[12,155,160,161]. For the segmented copolymers, incorporation of 5–10 mol% of a polymeric radiostabilizer not only imported radiostability but also increased the polymer compliance when the copolymers were tested as oriented monofilament sutures [155,160,161]. Due to the high reactivity of P-PD, its melt blending with RS-1 and RS-2 produced microscopically isotropic melts containing substantial fractions of segmented P-PD chains. The radiostability of both groups of the P-PD systems was reflected in their in vivo strength retention when tested as monofilament sutures (Table 1.9).

Earlier studies on alkylene oxalate polymers indicated, in certain instances, that radiation-sterilized monofilaments display rapid decrease in breaking strength when implanted in the subcutis of rats [8,137]. In a more recent study by Johnson [162], copolymers of alkylene oxalate with terephthalate or isophthalate sequences were made to determine the effect of the aromatic moieties on the chain radiostability. Preliminary data suggest that the aromatic components may impart some radiostability.

1.5   Analysis of Contemporary Issues and Emerging Trends

Following the brief review, in the previous sections, of the patent and technical literature on absorbable polyesters since their first conversion to useful devices, it

Table 1.9 In Vivo Breaking Strength Retention (BSR) Data of G- and PD-based Gamma-sterilized (2.5 Mrad) Monofilaments of Homopolymers, Segmented Copolymers, and Melt Blends[a]

Polymer system[b]	Radiostabilizer		BSR (%) at:			
	Type	Molar ratio	1 Week	2 Weeks	3 Weeks	4 Weeks
I-A	—	0	75	16	0	—
I-B	RS-1	10	—	—	62	42
I-C	RS-3	10	91	73	55	—
II-A	—	0	—	43	30	25
II-B	RS-1	12	—	62	55	40
II-C	RS-2	11	—	79	72	57
III-A	—	0	—	55	32	—
III-B	RS-1	10	—	85	51	—

[a] References [8,10,12,152,155,160,161].
[b] I-A = polyglycolide (P-G)
 I-B = copolymer, 10/90 RS-1/P-G
 I-C = copolymer, 10/90 RS-3/P-G
 II-A = poly(*p*-dioxanone)(P-PD)
 II-B = copolymer, 12/88 RS-1/P-PD
 II-C = copolymer, 11/89 RS-2/P-PD
 III-A = P-PD melted for 10 min.
 III-B = copolymer melt blended for 20 min, 10/90 RS-1/P-PD.

became apparent that over the past 5 years there has been an extraordinary growth of interest in these materials as well as new concerns about their applications. This conclusion is based on most recent information highlighted in a number of key patents, original research articles, clinical reports, and critical reviews which have been already published or expected to be published in 1993 [17,79–84,99,131,132,135,146,149, 162–196].

Discussed below are a few of the emerging trends related to the development of new materials and biomedical products with unique features, as well as new aspects concerning the biological performance and safe use of absorbable systems.

1.5.1   Segmented Copolymers as Low-Modulus Materials

The development of low-modulus fiber-forming materials is driven by the desire to develop compliant monofilament sutures with comparable pliability to braided sutures. Toward this goal the concept of chain segmentation is being explored to increase the compliance of traditional absorbable polymers such as P-PD and P-G [168,170,173]. Although this technology was driven by the suture manufacturers, it should be extendable to the development of high compliance and/or flexible conduits.

1.5.2   Segmented Copolymers as Hydrophilic Substrates

Incorporation of hydrophilic components in traditional absorbable polyesters was first pursued primarily by the suture producers who developed low molecular weight absorbable coating by the copolymerization of polyethylene glycol (P-EG) with traditional monomers, such as *dl*-lactide, glycolide, and/or *p*-dioxanone [174,175]. Interest in P-EG as a hydrophilic component in segmented or block absorbable copolyesters is likely to grow with the proliferation of data on its adequate clearance from the body [176]. A major use of this type of polymer, including its crosslinked hydrogels, appears to intensify in the area of controlled drug delivery systems, blood contacting surfaces, and tissue adhesives [177–179]. Modulation of the hydrolytic stability and, hence, absorption profile, may be controlled through manipulating the phase mixing of those polymers with P-EG in solution [180–182].

1.5.3   Extrinsic Factors Contributing to Polymer Degradation

In addition to the intrinsic factors known to affect polymer degradation (e.g., chemical structure and morphology), the role of a number of extrinsic factors is now being explored [98,183–185]. These include the effects of ionic strength in sodium chloride and magnesium chloride solutions on retarding tensile strength loss of oriented fibers [183], of enzyme and bacteria on absorption [184,185], of external mechanical stresses on strength retention [184,185], and of implant site and tissue environment on strength retention [98].

1.5.4   Reactive and Nonreactive Melt Blending

Adopting the successful practice of producing polyblends in the traditional polymer industry, biomaterials scientists are now exploring blending of absorbable polyesters to develop unique combinations of materials with new attributes using well-accepted polymers or copolymers. Thus, P-G, P-LL, and/or copolymer rich in glycolide or lactide was used successfully to produce unique multicomponent systems that were converted to absorbable surgical staples [166–168]. Melt blending of such highly reactive polyesters is expected to be associated with ester–ester interchange reactions leading to the formation of a microdispersion of one component, which can be manipulated to form a microfibril microcomposite system. Study of the reactive-blending of absorbable polymers is being explored in this laboratory, using critical data on functional interchange in chain molecules and catalyzed ring-opening polymerization [186].

Studies on properties of polymer blends made by solution mixing of several polyesters and polyanhydrides were pursued to produce polymeric matrices for drug delivery systems with controlled release profiles [187–189].

1.5.5   Polymeric Prodrugs with Biodegradable Spacers

An interesting and growing use of biodegradable systems entails the covalent binding of active drugs to water-soluble polymers via a biodegradable segment or spacer [190,193]. The viability and promising potential of this area of research was demonstrated in the design of efficient systems for the delivery of antibiotics [190] and chemotherapeutic agents [193]. In the latter case, an enzymatically degradable oligopeptide segment was used as a spacer for adriamycin of a copolymeric chain of *N*-hydroxypropyl methacrylamide. It should be noted that such a controlled delivery system was shown to be quite effective in addressing the undesirable cumulative dose-dependent cardiomyopathy of the drug, without compromising its therapeutic activity. The original article [193] should be consulted to appreciate other promising attributes of this approach to controlled drug delivery.

1.5.6   Self-Reinforced Composites

The need to develop high-modulus absorbable orthopedic implants is discussed earlier in the chapter [17,79–82,84,131,135,165]. On the other hand, two issues, namely, the formation of sterile sinuses at the surgical sites of relatively large-mass implants and the criteria for producing high-performance self-reinforced composites, need to be addressed. It is our strong belief that the data available on both topics are insufficient for drawing potentially damaging conclusions as to the clinical efficacy and safety of absorbable orthopedic devices. Related to the development of basic data pertinent to the safety issue, preliminary studies in this laboratory indicate that leachable components are present in P-PD, P-G [62,162], and a few oxalate polymers [162], as determined by monitoring in vivo and/or in vitro mass loss of molded articles. This may be relevant to the premature tensile loss reported on polyoxalate sutures [138,139]. In recent studies on the tissue response to local high concentration of absorbable polymers, such as P-LL and a polyesteramide, it was reported that subcutaneously implanted particles do not initiate necrosis [194]. It was also noted that chronic inflammatory response observed in the study was transitory. Understanding the physical and chemical events associated with the formation of self-reinforced and similar composites using simple models is highly recommended. To this end, preliminary studies on the interfacial bonding in absorbable composites have been initiated in this laboratory [165].

1.5.7   Radiation-Sterilizable Polymers

Although an adequate discussion of this topic is given in the previous section, it must be noted here that this is an issue of strategic importance to the continued growth of absorbable polymer technology. Having radiation-sterilizable polymer is likely to determine the commercial viability of future orthopedic and cardiovascular implants, as well as drug delivery systems and hybrid biomolecules [146].

1.5.8   Absorbable Substrates for Tissue Regeneration

A promising approach for tissue regeneration is illustrated in the work of Vacanti and co-workers [69,70,194–196], who used bioabsorbable, biocompatible polymers as templates onto which cells are seeded and allowed to grow in three dimensions. In a typical case, braided Vicryl sutures were seeded with chondrocytes to produce macroscopic plates of up to 100 mg of new cartilage subcutaneously in mice [194]. Other interesting examples include the regeneration of pericardial tissue on absorbable polymer patches implanted into the pericardial sac of sheep [196]. Although it is too early to assess the clinical relevance and viability of these systems they do hold a great promise for providing means to regenerate critical tissues such as bone, cartilage, and liver.

REFERENCES

1.   Vainionpää, S.; Rokkanen, P.; Törmälä P. *Prog. Polym. Sci.* **44**, 679 (1989).
2.   Shalaby, S.W. In *Encyclopedia of Pharmaceutical Technology;* Swarbrick, J.; Boylan, J.C., eds., Marcel Dekker, New York, **1**, 46 (1988).
3.   Gupta, M.C.; Desmuth, V.G. *Polymer* **24**, 827 (1983).
4.   Törmälä, P.; Vasenius, J.; Vainionpää, S.; Laiho, J.; Pohijonen, T.; Rokkanen, P. *J. Biomed. Mater. Res.* **25**, 1 (1991).
5.   Palmer, G.R.; Baier, R.E. *Am. Conf. Eng. Med.* **29**, 139 (1987).
6.   Jamiolkowski, D.D.; Shalaby, S.W. *Polym. Prepr.* **32**, 329 (1990).
7.   Chu, C.C. In *C.R.C. Critical Reviews in Biocompatibility.* Williams, D.F., Ed-in-Chief. CRC Press, Boca Raton, FL **1**, 3, 261 (1985).
8.   Shalaby, S.W.; Jamiolkowski, D.D. U.S. Patent, (to Ethicon, Inc.) 4,435,590 (1984).
9.   Jamiolkowski, D.D.; Shalaby, S.W. *Polym. Prepr.* **26**, 200 (1985).
10. Bezwada, R.S.; Shalaby, S.W.; Jamiolkowski, D.D. U.S. Patent (to Ethicon, Inc.) 4,510,295 (1985).
11. Bezwada, R.S.; Shalaby, S.W.; Jamiolkowski, D.D. U.S. Patent (to Ethicon, Inc.) 4,532,928 (1985).
12. Koelmel, D.F.; Jamiolkowski, D.D.; Shalaby, S.W. U.S. Patent (to Ethicon, Inc.) 4,559,945 (1985).
13. Kafrawy, A.; Jamiolkowski, D.D.; Shalaby, S.W. *J. Bioact. Biocomp. Polym.* **2**, 305 (1987).
14. Shalaby, S.W. *Macromol. Rev.* **14**, 406 (1979).
15. Chu, C.C. In *Concise Encyclopedia of Medical and Dental Materials* Williams, D.F., ed. Pergamon Press, London (in press).
16. Shalaby, S.W. In *High Technology Fibers—Part A,* Chap. 3. Lewin, M.; Preston, J.; eds. Marcel Dekker, New York (1985).
17. Daniels, A.U.; Chang, M.K.O.; Andriano, K.P. *J. Appl. Biomater.* **1**, 57 (1990).
18. Dittrich, W.; Schulz, R.C. *Angew. Makromol. Chem.* **15**, 109 (1971).
19. Kricheldorf, H.R. *Polym. Bull. (Berlin)* **14**, 497 (1985).

20. Gilding, D.K.; Reed, A.M. *Polymer* **20,** 1459 (1979).
21. Kricheldorf, H.R.; Dunsing, R. *Makromol. Chem.* **187,** 1611 (1986).
22. Leenslag, J.W.; Pennings, A.J. *Makromol. Chem.* **188,** 1809 (1987).
23. Schindler, A.; Hibionada, Y.M.; Pitt, C.G. *J. Polym. Sci. Chem. Ed.* **20,** 319 (1982).
24. Shalaby, S.W. Unpublished work.
25. Doddi, N.; Versfelt, C.C.; Wasserman, D. U.S. Patent (to Ethicon, Inc.) 4,052,988 (1977).
26. Albertsson, A.C.; Ljungquist, O. *J. Macromol. Sci. Chem.* A-25, 467 (1988).
27. Shalaby, S.W. U.S. Patent (to Ethicon, Inc.) 4,190,720 (1980).
28. Albertsson, A.C. *Polym. Mater. Sci. Eng.* **62,** 409 (1990).
29. Shalaby, S.W.; Koelmel, D.F.; Arnold, S. U.S. Patent (to Ethicon, Inc.) 5,082,925 (1992).
30. Boyle, W.J.; Mares, F.; Patel, K.M.; Tang, T. U.S. Patent (to Allied-Signal) 4,916,207 (1990).
31. Kotliar, A.M.; Boyle, W.J.; Tang, R.T.; Mares, F.; Patel, K.M. U.S. Patent (to Allied-Signal) 4,891, (1990).
32. Patel, W.J. U.S. Patent (to Allied-Signal) 4,916,193 (1990).
33. Shalaby, S.W.; Koelmel, D.F. U.S. Patent (to Ethicon, Inc.) 4,441,496 (1984).
34. Schmitt, E.E.; Polistina, R.A. U.S. Patent (to American Cyanamid) 3,297,033 (1967).
35. Schmitt, E.E.; Epstein, M.; Polistina, R.A. U.S. Patent (to American Cyanamid) 3,422,871 (1969).
36. Frazza, E.J.; Schmitt, E.E. *J. Biomed. Mater. Res. Symp.* **1,** 43 (1971).
37. Craig, P.H.; Williams, J.A.; Davis, K.W.; Magoun, A.D.; Levy, A.J.; Bogdansky, S.; Jones, J.P. *Surg. Gynecol. Obstet.* **141,** 1 (1975).
38. Casey, D.J.; Roby, M.S. U.S. Patent (to American Cyanamid), 4,429,080 (1984).
39. Katz, A.R.; Mukherjee, D.P.; Kaganov, A.L.; Gordon, S. *Surg. Gynecol. Obstet.* **161,** 213 (1985).
40. Chujo, K.; Kobayashi, H.; Suziki, J.; Tokuhara, S.; Tanake, M. *Makromol. Chem.* **100,** 262 (1967).
41. Chujo, K.; Kobayashi, H.; Suziki, J.; Tokuhara, S. *Makromol. Chem.* **100,** 267 (1967).
42. Reed, A.M.; Gilding, D.K. *Polymer* **22,** 494 (1981).
43. Nieuwenhuis, J. *Clin. Mater.* **10,** 59 (1992).
44. Benicewicz, B.C.; Shalaby, S.W.; Clemow, A.J.T.; Oser, Z. *Polym. Prepr.* **30,** 1499 (1989).
45. Tunc, D.C. *Trans. Soc. Biomater.* **6,** 47 (1983).
46. Tunc, D.C. *Polym. Prepr.* **27,** 431 (1986).
47. Salthouse, T.N. *J. Biomed. Mater. Res.* **10,** 197 (1976).
48. Salthouse, T.N.; Matlaga, B.F. *Surg. Gyn. Obstet.* **142,** 544 (1976).
49. Williams, D.F.; Mort, E.J. *Bioeng.* **1,** 231 (1977).
50. Williams, D.F. *J. Bioeng.* **1,** 279 (1977).
51. Williams, D.F. *J. Bioeng. Mater. Res.* **14,** 329 (1980).
52. Williams, D.F. *J. Mater. Sci.* **10,** 1233 (1982).
53. William, D.F. *Clin. Mater.* **10,** 9 (1992).
54. Ali, S.A.M.; Doherty, P.J.; Williams, D.F. *Proc. Fourth World Biomater. Congr.* 407 (1992).
55. Vert, M. *Clin. Mater.* **10,** 1 (1992).
56. Li, S.M.; Garreau, H.; Vert, M. *J. Mater. Sci. Mater. Med.* **1,** 123 (1990).
57. Li, S.M.; Garreau, H.; Vert, M. *J. Mater. Sci. Mater. Med.* **1,** 198 (1990).
58. Fredericks, R.J.; Melveger, A.J.; Dolegiewitz, L.J. *J. Polym. Sci., Polym. Phys. Ed.* **22,** 57 (1984).

59. Chu, C.C. *J. Appl. Polym. Sci.* **26,** 1727 (1981).
60. Benicewicz, B.C.; Hopper, P.K. *J. Bioact. Compat. Polym.* **5,** 453 (1990).
61. Benicewicz, B.C.; Hopper, P.K. *J. Bioact. Compat. Polym.* **6,** 64 (1991).
62. Pletzer, J.S.; Von Recum, A.F.; Shalaby, S.W. *Trans. Soc. Biomater.* **16,** 20 (1993).
63. Klinger, M.; Lambrecht, J.T. *Clin. Mater.* **10,** 81 (1992).
64. Chujo, K.; Kobayashi, H.; Suziki, J.; Tokuhara, S. *Makromol. Chem.* **100,** 267 (1967).
65. Katz, A.R.; Turner, R.J. *Surg. Gynecol. Obstet.* **131,** 701 (1970).
66. Wasserman, D.; Shalaby, S.W.; Bousma, O.J. Eur. Patent Appl. (to Ethicon, Inc.) 5,307,160 (1990).
67. Yu, T.J.; Chu, C.C. *Proc. Fourth World Biomater. Congr.* 59 (1992).
68. Greisler, H.P. In *New Biologic and Synthetic Vascular Prostheses*. CRC Press, Boca Raton, FL (1992).
69. Cima, L.G.; Ingber, D.E.; Vacanti, J.P.; Langer, R. *Biotech. Bioeng.* **38,** 145 (1991).
70. Vacanti, J.P.; Stein, J.E.; Gilbert, C.; Vacanti, C.A.; Ingber, D.E. *Polym. Prepr.* **32,** 227 (1991).
71. Ray, J.A.; Doddi, N.; Regula, D.; Williams, J.A.; Melveger, A. *Surg. Gynecol. Obstet.* **153,** 497 (1981).
72. Beroff, H.; Doddi, N.; Jewusiak, S. U.S. Patent (to Ethicon, Inc.) 4,418,694 (1984).
73. Plaga, B.R.; Caskey, P.M. *Trans. Orthop. Res. Soc.* **16,** 448, (1991).
74. Bezwada, R.S.; Shalaby, S.W.; Newman, H.D. U.S. Patent (to Ethicon, Inc.) 4,653,497 (1987).
75. Bezwada, R.S.; Shalaby, S.W.; Newman, H.D.; Kafrauy, A. U.S. Patent (to Ethicon, Inc.) 4,643,191 (1987).
76. Schneider, A.K. U.S. Patent (to DuPont) 3,636,956 (1972).
77. Hoffman, G.O. *Clin. Mater.* **10,** 75 (1992).
78. Vert, M.; Chabot, F.; Leray, J.; Christel, P. *Makromol. Chem. Suppl.* **5,** 30 (1981).
79. Illi, O.E.; Weigum, H.; Misteli, F. *Clin. Mater.* **10,** 69 (1992).
80. Amecke, B.; Bendix, D.; Entenmann, G. *Clin. Mater.* **10,** 47 (1992).
81. Gogolewski, S. *Clin. Mater.* **10,** 13 (1992).
82. Törmälä, P. *Clin. Mater.* **10,** 29 (1992).
83. Raiha, J.E. *Clin. Mater.* **10,** 41 (1992). 84. Claes, L.E. *Clin. Mater.* **10,** 41 (1992).
85. Gay, B.; Bucher, H. *Unfallchirurg.* **88,** 126 (1985).
86. Vainionpää, S.; Vihtonen, K.; Mero, M.; Pätiäla, H.; Rokkanen, P.; Kilpikari, J.; Törmälä, P. *Arch. Orthop. Traumat. Surg.* **106,** 1 (1986).
87. Matsusue, Y.; Yamamuro, T.; Oka, M.; Ikada, Y.; Hyon, S.H.; Shikinami, Y. *J. Appl. Biomater.* **2,** 1 (1991).
88. Rokkanen, P.; Böstman, O.; Vainionpää, S.; Vihtronen, K.; Törmälä, P.; Laiho, J.; Kilpikari, J.; Tamminmaki, M. *Lancet* **1,** 1422 (1985).
89. Böstman, O.; Vainionpää, S.; Hirvenalo, E.; Mäkelaä, A.; Vihtonen, K.; Törmälä, P.; Rokkanen, P. *J. Bone Jt. Surg.* **69-B,** 615 (1987).
90. Vert, M.; Christel, P.; Chabot, F.; Leray, J. In *Macromol. Biomater.* Hastings, C.W.; Ducheyne, P., eds. CRC Press, Boca Raton, FL, p. 120 (1984).
91. Leenslag, J.W.; Pennings, A.J.; Bos, R.R.M.; Rozema, F.R.; Boering, G. *Biomaterials* **8,** 70 (1987).

92. Vainionpää, J.; Kilpikali, J.; Laiho, J.; Heleverta, P.; Rokkanen, P.; Törmälä, P. *Bioma-terials* **8**, 40 (1987).
93. Törmälä, P.; Vasenius, J.; Vainionpää, S.; Laiho, J.; Pohjonen, T.; Rokkanen, P. *J. Biomed. Mater.* **25**, 1 (1991).
94. Pohjonen, T.; Heponen, V.P.; Pellinen, M.; Rokkanen, P.; Vainionpää, P. *Trans. Soc. Biomater.* **11**, 219 (1988).
95. Tunc, D.C.; Jodhav, B. In *Progress in Biomedical Polymers* Gebelein, C.C.; Dunn, R.L., eds. Plenum Press, New York, p. 239 (1990).
96. Matsusue, Y.; Yamamuro, T.; Oka, M.; Shikinami, Y.; Hyon, S.H.; Ikada, Y. *J. Biomed. Mater. Res.* **26**, 1553 (1992).
97. Vasenius, J.; Vainionpää, S.; Vitohnen, K.; Mäkelä, A.; Rokkanen, P.; Mero, M.; Törmälä, P. *Biomaterials* **11**, 501 (1990).
98. Bhatia, S.; Powers, D.L.; Ferguson, R.; Shalaby, S.W. *Proc. 12th Southern Biomed. Eng. Conf.*, New Orleans, LA, (1993).
99. Vasenius, J.; Vainionpää, S.; Vitohnen, K.; Mero, J.; Mikkola, A.; Rokkanen, P.; Törmälä, P. *Clin. Mater.* **4**, 307 (1989).
100. Mattie, F.V. U.S. Patent (to Ethicon, Inc.) 4,027,676 (1977), 4,301,216 (1980).
101. Gogolewski, S.; Pennings, A.J. *Colloid Polym. Sci.* **261**, 477 (1983).
102. Gogolewski, S.; Pennings, A.J.; Lommen, E.; Wildevuur, C.R.H.; Nieuwenhuis, P. *Makromol. Chem. Rapid Commun.* **4**, 213 (1983).
103. Van der Lei, B.; Wildevuur, Ch.R.H. In *Prosthetic Substitution of Blood Vessels.* Kogel, H.C., ed. Quintessenz-Verlag, München, Germany, p. 67 (1991).
104. Wakiyama, N.; Juni, K.; Nakano, M. *Chem. Pharm. Bull.* **30**, 3719 (1982).
105. Anderson, L.C.; Wise, D.L.; Howes, J.F. *Contraception* **13**, 375 (1976).
106. Langer, R. In *Medical Application of Controlled Release.* Langer, R.S.; Wise, D.L., eds. CRC Press, Boca Raton, FL (1984).
107. Brekke, J.A.; Olson, R.A.J.; Scully, J.R.; Osbon, D.B. *Oral Surg. Oral Med. Path.* **56**, 240 (1983).
108. Von Oepen, R.; Michaeli, W. *Clin. Mater.* **10**, 21 (1992).
109. Kulkarni, R.K.; Pani, K.C.; Neuman, C.; Leonard, F. *Arch. Surg.* **93**, 839 (1966).
110. Hyon, S.-H.; Jamshidi, K.; Ikada, Y. In *Polymers as Biomaterials* Shalaby, S.W.; Hoffman, A.S.; Ratner, B.D.; Horbett, T.A., eds. Plenum Press, New York, p. 51 (1985).
111. Jamshidi, K.; Hyon, S.H.; Nakamura, T.; Ikada, Y.; Shimizu, Y.; Teramatsu, T. In *Adv. Biomater. Biol. Biomech. Perf. Biomater.* Christel, P.; Meunier, A.; Lee, A.J.C., eds. Elsevier, Amsterdam, p. 227 (1986).
112. Kulkarni, R.K.; Moore, E.G.; Hegyelli, A.F.; Leonard, F. *J. Biomed. Mater. Res.* **5**, 169 (1971).
113. Eling, B.; Gogolewski, S.; Pennings, A.J. *Polymer* **23**, 1587 (1982).
114. Gogolewski, S.; Pennings, A.J. *J. Appl. Polym. Sci.* **28**, 1045 (1983).
115. Leenslag, J.W.; Gogolewski, S.; Pennings, A.J. *J. Appl. Polym. Sci.* **29**, 2829 (1984).
116. Leenslag, J.W.; Pennings, A.J. *Makromol. Chem.* **188**, 1809 (1987).
117. Benicewicz, B.C.; Oser, Z.; Clemow, A.J.T.; Shalaby, S.W. Eur. Patent Appl. (to J & J), 241,252 (1987).
118. Boyle, W.J.; Mares, F.; Patel, K.M.; Tang, R.T. World Patent Appl. (to Allied-Signal) 89-05-664 (1989).
119. Shieh, S.J.; Zimmerman, M.C.; Parsons, J.R. *J. Biomed. Mater. Res.* **24**, 789 (1990).
120. Tang, R.; Boyle, Jr., W.J.; Mares, F.; Chiu, T.H. *Trans. Soc. Biomater.* **13**, 191 (1990).

121. Shieh, S.J.; Law, J.K.; Poandl, T.M.; Zimmerman, M.C.; Parsons, J.R. *Trans. Soc. Biomater.* **13**, 192 (1990).
122. Tang, R.; Mares, F.; Akelman, E.; Cannistra, L.; Herndon, J.; Sidman, R.; Madison, R.; Williams, S.; Tang, M. *Proc. Third World Biomater. Congr.* (Japan), p. 378 (April, 1988).
123. Akelman, E.; Cannistra, L.M.; Tang, M.; Williams, S.; Mares, F.; Tang, R.; Herndon, J.H. *Proc. Ann. Mtg. Orthop. Res. Soc.* p. 468 (Feb., 1988).
124. Shalaby, S.W. Unpublished work.
125. Bezwada, R.S.; Hunter, A.W.; Shalaby, S.W. U.S. Patent (to Ethicon, Inc.) 5,037,950 (1991).
126. Matsusue, Y.; Yamamuro, T.; Oka, M.; Shikinami, Y.; Hyon, S.-H.; Ikada, Y. *Proc. Fourth World Biomater. Cong.* 404 (1992).
127. Jukkala, K.; Portio, E.; Hervensalo, E.; Törmälä, P.; Rokkanen, P. *Proc. Fourth World Biomater. Congr.* 613 (1992).
128. Cutright, D.E.; Hunsuck, E.E. *J. Oral. Surg.* **29**, 393 (1971).
129. Cutright, D.E.; Hunsuck, E.E. *J. Oral. Surg.* **31**, 134 (1971).
130. Cutright, D.E.; Beasley, J.D.; Perez, B. *J. Oral. Surg.* **32**, 165 (1971).
131. Böstman, O.; Hirvensalo, S.; Vainionpää, E.; Vihtonen, K.; Törmälä, P.; Rokkanen, P. *Internat. Orthopedic* (SICOT) **14**, 1 (1990).
132. Santavira, S.; Konttinen, Y.T.; Sito, J.; Gröblad, M.; Partio, E.K.; Kimppinen, P.; Rokkanen, P. *J. Bone Jt. Surg.* **72-B**, 597 (1990).
133. Claes, I. In *Clinical Implants Materials.* Heimke, G.; Soltesz, U.; Lee, A.J.C.; eds. Elsevier, Amsterdam, p. 161 (1990).
134. Claes, I. In *Biodegradable Implants in Orthopedic Surgery.* Hoffman, G.O., ed. Technik & Kommunikation, Berlin, p. 83 (1990).
135. Partio, E.K.; Böstman, O.; Vainionpää, S.; Pätiälä, H.; Hirvensalo, E.; Vihtonen, K.; Törmälä, P.; Rokkanen, P. *Acta Orthop. Scand.* **59** (Suppl. 227), 18 (1988).
136. Shalaby, S.W.; Jamiolkowski, D.D. U.S. Patent (to Ethicon, Inc.) 4,105,034 (1978).
137. Shalaby, S.W.; Jamiolkowski, D.D. U.S. Patent (to Ethicon, Inc.) 4,186,189 (1979).
138. Shalaby, S.W.; Jamiolkowski, D.D. U.S. Patent (to Ethicon, Inc.) 4,205,399 (1980).
139. Shalaby, S.W.; Jamiolkowski, D.D. U.S. Patent (to Ethicon, Inc.) 4,130,639 (1978).
140. Shalaby, S.W.; Jamiolkowski, D.D. U.S. Patent (to Ethicon, Inc.) 4,186,189 (1980).
141. Shalaby, S.W.; Jamiolkowski, D.D. U.S. Patent (to Ethicon, Inc.) 4,141,087 (1979).
142. Shalaby, S.W.; Jamiolkowski, D.D. U.S. Patent (to Ethicon, Inc.) 4,208,511 (1980).
143. Shalaby, S.W.; Jamiolkowski, D.D. *Trans. Soc. Biomater.* **8**, 212 (1985).
144. Shalaby, S.W.; Jamiolkowski, D.D. U.S. Patent (to Ethicon, Inc.) 4,209,607 (1980).
145. Shalaby, S.W.; Jamiolkowski, D.D. U.S. Patent (to Ethicon, Inc.) 4,226,243 (1980).
146. Shalaby, S.W. In *Irradiation of Polymeric Materials* Reichmanis, E.; Frank, C.W.; O'Donnell, J.H., eds,. Vol. 527, p. 315. ACS Symposium Series Am. Chem. Soc., Washington, D.C. (1993).
147. Gupta, M.C.; Desmuth, V.G. *Polymer* **24**, 827 (1983).
148. Clough, R.L.; Shalaby, S.W., eds., In *Radiation Effects on Polymers Vol.* **475**, *ACS Symposium Series,* Am. Chem. Soc., Washington, D.C. (1991).
149. Shalaby, S.W. *Polym. News* 16, 238 (1991).
150. Jamiolkowski, D.D.; Shalaby, S.W. In *Radiation Effects on Polymers Vol.* **475**, *ACS Symposium Series,* (Clough, R.L.; Shalaby, S.W., eds.) Am. Chem. Soc., Washington, D.C. p. 300 (1991).
151. Shalaby, S.W.; Jamiolkowski, D.D. U.S. Patent (to Ethicon, Inc.) 4,689,424 (1987).

152. Jamiolkowski, D.D.; Bezwada, R.S.; Shalaby, S.W. In *Irradiation of Polymeric Materials*. Reichmanis, E.; Frank, C.W., eds. Vol. 527, p. 320 *ACS Symposium Series. Am. Chem. Soc.* Washington, D.C. (1993).

153. Bezwada, R.; Shalaby, S.W.; Jamiolkowski, D.D. U.S. Patent (to Ethicon, Inc.) 4,510,295 (1985).

154. Bezwada, R.; Shalaby, S.W.; Jamiolkowski, D.D. U.S. Patent (to Ethicon, Inc.) 4,532,928 (1985).

155. Koelmel, D.F.; Jamiolkowski, D.D.; Shalaby, S.W.; Bezwada, R. U.S. Patent (to Ethicon, Inc.) 4,546,152.

156. Kafrawy, S.; Jamiolkowski, D.D.; Shalaby, S.W. *J. Bioact. Biocomp. Polym.* 2, 305 (1987).

157. Jamiolkowski, D.D.; Shalaby, S.W. *Polym. Prepr.* **31,** 327 (1990).

158. Jamiolkowski, D.D.; Shalaby, S.W. *Polym. Prepr.* **31,** 329 (1990).

159. Bezwada, R.S.; Jamiolkowski, D.D.; Shalaby, S.W. *Trans. Soc. Biomater.* **14,** 186 (1991).

160. Koelmel, D.F.; Jamiolkowski, D.D.; Shalaby, S.W.; Bezwada, R.S. U.S. Patent (to Ethicon, Inc.) 4,649,921 (1987).

161. Koelmel, D.D.; Jamiolkowski, D.D.; Shalaby, S.W.; Bezwada, R.S. *Polym. Prepr.* **32,** 235 (1991).

162. Johnson, R.A.; Drews, M.; Shalaby, S.W. *Polym. Prepr.* **33,** 450 (1992).

163. Shalaby, S.W. *J. Appl. Biomater.* **3,** 73 (1992).

164. Shalaby, S.W. *Indian J. Tech.* **31,** 464 (1993).

165. Leadbetter, K.J.; Shalaby, S.W. *Bioact. Compat. Polym.* **8,** 132 (1993).

166. Smith, C.; Gaterud, M.T.; Jamiolkowski, D.D.; Shalaby, S.W.; Newman, Jr. H.D. U.S. Patent (to Ethicon, Inc.) 4,741,337 (1988).

167. Jamiolkowski, D.D.; Gaterud, M.T.; Newman, Jr., H.D.; Shalaby, S.W. U.S. Patent (to Ethicon, Inc.) 4,889,119 (1989).

168. Hirashima, T.; Eto, T.; BenBesten, L. *Am. J. Surg.* **150,** 381 (1985).

169. Bezwada, R.S.; Shalaby, S.W.; Newman, H.D. *Agriculture and Synthetic Polymers*. Glass, J.E.; Swift, G., eds. **Vol. 433,** *ACS Symposium Series,* Am. Chem. Soc., Washington, D.C. (1990).

170. Bezwada, R.S.; Shalaby, S.W.; Newman, H.D., Jr.; Kafrawy, A. *Trans. Soc. Biomater.* **13,** 194 (1990).

171. Jamiolkowski, D.D.; Shalaby, S.W.; Bezwada, R.S.; Newman, H.D., Jr. *Trans. Soc. Biomater.* **13,** 193 (1990).

172. Benicewicz, B.C.; Shalaby, S.W.; Clemow, A.J.; Oser, Z. *In-vitro and In-vivo, Agriculture and Synthetic Polymers*. Glass, J.E.; Swift, G., eds. **Vol. 433,** *ACS Symposium Series,* Am. Chem. Soc., Washington, D.C. (1990).

173. Bezwada, R.S.; Shalaby, S.W.; Erneta, M. U.S. Patent (to Ethicon, Inc.) 5,047,048 (1991).

174. Bezwada, R.S.; Shalaby, S.W. U.S. Patent (to Ethicon, Inc.) 5,019,094 (1991).

175. Casey, D.J.; Jarrett, P.K.; Rosati, L. U.S. Patent (to American Cyananid) 4,716,203 (1987).

176. Yamaoka, T.; Ikada, T. *Trans. Soc. Biomater.* **15,** 79 (1992).

177. Cohen, D.; Younes, H. *J. Biomed. Mater. Res.* **22,** 993 (1988).

178. Sawhney, A.S.; Hubbel, J.A. *J. Biomed. Mater. Res.* **24,** 1397 (1990).

179. Sawhney, S.C.; Pathak, C.P.; Hubbel, A. *Macromolecules* **26,** 581 (1993).

180. Shas, S.S.; Cha, Y.; Pitt, C.G. *J. Controlled Release* **18,** 261 (1992).

181. Desai, N.P.; Hubbel, J.A. *Biomaterials* **12,** 144 (1991).

182. Shah, S.S.; Zhu, K.J.; Pitt, C.G. *J. Biomater. Sci. Polym. Ed.* (in press).

183. Pratt, L.; Chu, C.; Auer, J.; Chu, A.; Kim, J.; Zollweg, J.A.; Chu, C.C. *J. Polym. Sci. Part A. Polym. Chem.* **31**, 1759 (1993).
184. Smith, R.; Oliver, C.; Williams, D.F. *J. Biomed. Mater. Res.* **21**, 699 (1988).
185. Chu, C.C.; Browning, A. *J. Biomed. Mater. Res.* **22**, 699 (1988).
186. Kim, S.H.; Han, Y.K.; Kim, Y.A.; Hong, S.I. *Makromol. Chem.* **193**, 1623 (1992).
187. Holland, S.J.; Yasin, M.; Tighe, B. *J. Biomaterials* **11**, 206 (1990).
188. Yasin, M.; Tighe, B. *J. Biomaterials* **13**, 9 (1992).
189. Domb, A.J. *J. Polym. Sci., Polym. Chem. Sci.* (in press).
190. Cassidy, J.; Duncan, R.; Morrison, G.J.; Strohalm, D.P.; Kopecek, J.; Kaye, S.B. *Biochem. Pharmacol.* **38**, 875 (1989).
191. Duncan R. *Anti-Cancer Drugs* **3**, 175 (1992).
192. Subr, V.; Strohalm, J.; Ulbrick, K.; Duncan, R.; Hume, I.C. *J. Controlled Rel.* **18**, 123 (1992).
193. Krinick, N.L.; Sun, Y.; Joyner, D.; Spikes, J.D.; Straight, R.C.; Kopecek, J. *J. Biomat. Sci. Polym. Ed. (in press).*
194. *Gibbons, D.; Barrow, T.; Truong, M. Proc. Fourth World Biomater. Cong.* 408 (1992).
195. Vacanti, C.A.; Langer, R.; Schloo, B.; Vacanti, J.P. *Plast. Reconstr. Surg.* **88**, 753 (1991).
196. Malm, T.; Bowald, S.; Bylock, A.; Saldeen, T.; Busch, C. *Scand. J. Thorac. Cardiov. Surg.* **26**, 15 (1992).

CHAPTER 2

Poly(ortho esters)

J. Heller and A.U. Daniels

2.1   Introduction

The first example of a totally synthetic polymer specifically designed for biomedical applications was poly(glycolic acid) [1]. Development of this polymer was followed by the development of poly(lactic acid) and copolymers of lactic and glycolic acids [2].

Although these materials were originally developed as suture materials, they have been and continue to be used extensively as bioerodible matrices for the controlled release of drugs and as fracture fixation devices. Despite the fact that for many applications these materials do not have optimal properties, they degrade to the natural metabolites glycolic and lactic acid, and their extensive use is thus largely driven by their demonstrated benign toxicology.

Commencing in 1970, an effort was launched first at the Alza Corporation and then at SRI International centered on the development of another synthetic polymer that would be specifically designed as a bioerodible matrix for the controlled release of drugs physically incorporated into the polymer. In designing this polymer, the following desirable properties were considered: (1) the polymer should degrade by a well-defined reaction to small, water-soluble molecules that must be toxicologically benign; (2) hydrolysis rate of the polymer should be adjustable within wide limits by simple manipulations of polymer structure or use of excipients; (3) the polymer should be capable of undergoing surface erosion; and (4) mechanical properties should be variable by simple changes in polymer structure.

J. Heller, Controlled Release and Biomedical Polymers Department, SRI International, Menlo Park, California 94025, U.S.A., and A.U. Daniels, Department of Orthopedic Surgery, University of Utah School of Medicine, Salt Lake City, Utah 84132, U.S.A.

In a polymer that is capable of undergoing surface erosion, polymer hydrolysis in the surface layers must occur at significantly higher rates than that in the bulk material. Thus, the polymer must not only be highly hydrophobic to limit water penetration into the bulk material, but it must also contain linkages that are capable of undergoing rapid hydrolysis. Two such linkages are anhydrides and ortho esters.

Because ortho ester linkages are acid sensitive and are also stable in base, erosion rates of polymers based on ortho ester bonds can be accelerated by the addition of acidic excipients or stabilized by the addition of basic excipients. This simple means of controlling hydrolysis rate was considered an attractive feature and poly(ortho esters) were thus selected for development. At this time four families of such polymers have been described. These will be designated as poly(ortho ester) I, poly(ortho ester) II, poly(ortho ester) III, and poly(ortho ester) IV.

2.2   Development of Poly(ortho esters)

2.2.1   Poly(ortho ester) I

2.2.1.1   Chemical Aspects

Because alkoxy groups of an ortho ester linkage readily transesterify, poly(ortho esters) can be prepared by the reaction shown in Scheme 1.

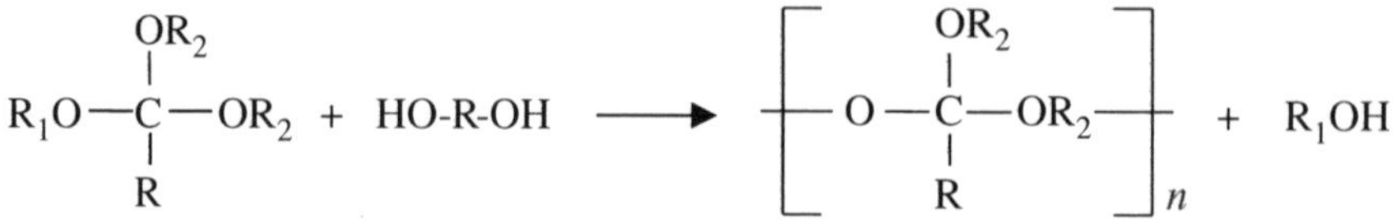

Scheme 1

However, linear polymers can be prepared only if the reactivity of the alkoxy group OR_2 is significantly reduced relative to that of OR_1. Because alkoxy groups in an ortho ester linkage rapidly equilibrate, such a decreased reactivity can be realized only by using a cyclic structure such as diethoxytetrahydrofuran shown in Scheme 2 [3–6].

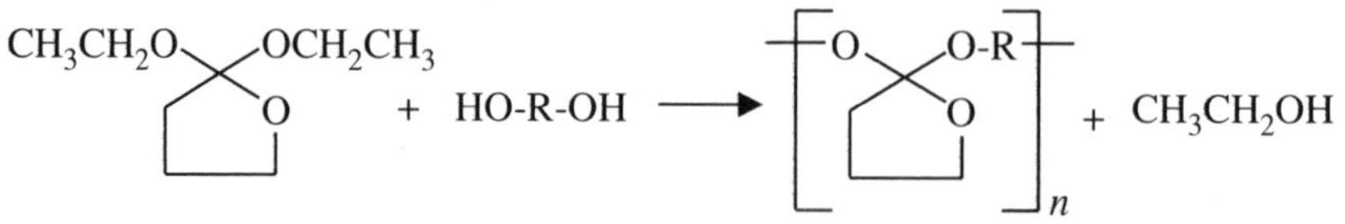

Scheme 2

When poly(ortho esters) prepared by this procedure are placed in an aqueous environment, an initial hydrolysis to a diol and γ-butyrolactone takes place. The γ-butyrolactone then rapidly hydrolyzes to γ-hydroxybutyric acid. This reaction path is shown in Scheme 3.

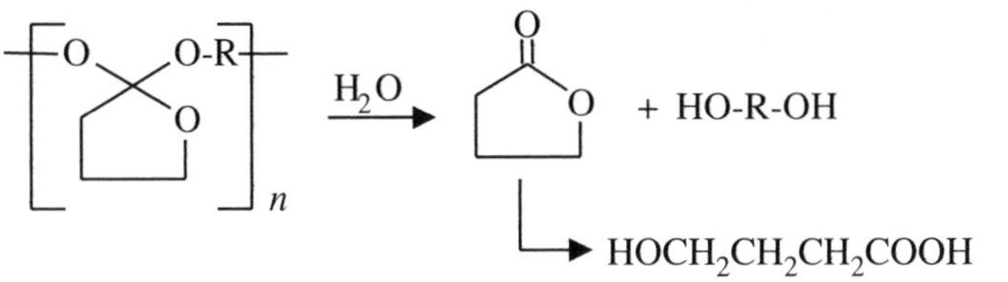

Scheme 3

Thus, hydrolysis is an autocatalytic process because the γ-hydroxybutyric acid hydrolysis product accelerates hydrolysis of the acid-sensitive ortho ester linkages. Therefore, to prevent autoacceleration, a base must be incorporated into the polymer to neutralize this acidic hydrolysis product.

2.2.1.2   Physical Properties

One potential problem with poly(ortho esters) is that the carbon–oxygen linkage in the polymer allows free rotation about the C–O bond, so that polymer chains are highly flexible and the polymers will have a relatively low glass transition temperature unless a highly rigid component is incorporated into the polymer chain. Because the tetrahydrofuran component of the polymers shown in Scheme 2 does not provide the necessary rigidity, not even when a rigid diol such as *trans*-cyclohexanedimethanol shown in Scheme 4 is used, the polymer has a glass transition temperature of about 40°C which is very close to the body temperature of 37°C.

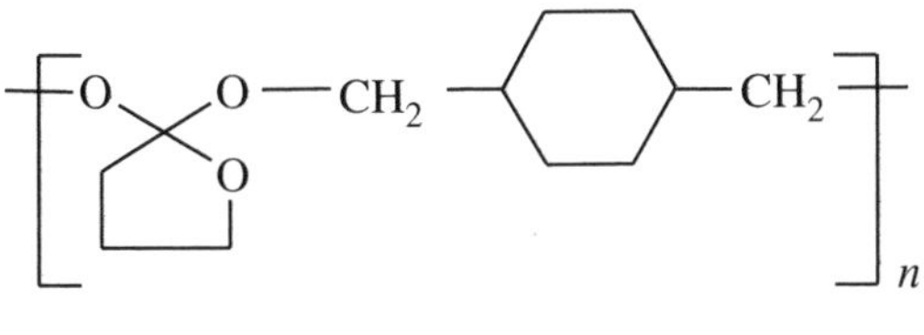

Scheme 4

When a flexible diol such as 1,6-hexanediol is used, the polymer has a glass transition temperature below room temperature and the polymer is a viscous, ointment-like material. The structure of this polymer is shown in Scheme 5.

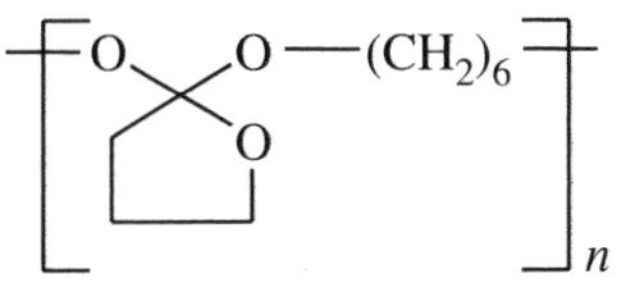

Scheme 5

Because solid implants designed for drug delivery should undergo minimal deformation and should thus have glass transition temperatures well above body temperature, the utility of this polymer as a solid implant appears to be somewhat limited.

However, the ointment-like material has useful properties that can be exploited in specific applications.

2.2.1.3   Applications

Although these polymers were first prepared about 20 years ago, there is very little published information that provides details of polymer structure, or indeed detailed experimental procedures. The polymers are usually designated by trade names such as C111 and C101ct. However, it can be inferred that C101ct has the structure shown in Scheme 4 where the diol is *cis/trans*-cyclohexanedimethanol of undefined *cis/trans* ratio and C111 refers to a polymer having the structure shown in Scheme 5.

The C101ct polymer has been used in the development of bioerodible contraceptive devices [7]. Initial toxicological evaluations of devices with incorporated norethindrone and Na_2CO_3 in dogs, rats, and primates have been carried out but were somewhat equivocal in that there was evidence of swelling in dogs after a latent period of about 2 weeks. There was no swelling in baboons, but some swelling did occur in rhesus monkeys. Following these preliminary studies, a small local irritation study in human volunteers was carried out. None of the volunteers experienced itching with the placebo devices, but itching did occur in some volunteers in whom active devices had been implanted [8]. Norethindrone blood plasma level of the human volunteers is shown in Figure 2.1 [7].

Because the only adverse effects were itching and some redness which rapidly resolved when the devices were removed, another human study using levonorgestrel was carried out [9]. During that study, itching and redness were again observed and further work was discontinued. The reasons for the irritation were never elucidated, but it is unlikely that it is caused by the polymer or its degradation products.

The C101ct polymer has also been investigated as a bioerodible naltrexone delivery system [10]. In this study 20 wt% naltrexone was dispersed into the polymer, pressed

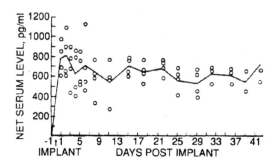

Figure 2.1   Serum concentration of norethindrone in four women receiving two poly(ortho ester) norethindrone devices each. Solid line represents the mean. (Reproduced with permission from ref. 7. © 1983, Raven Press.)

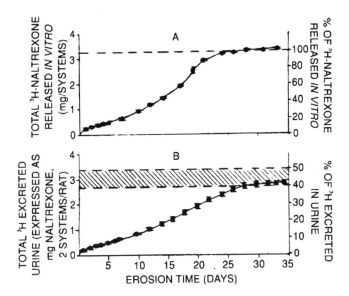

Figure 2.2   Cumulative release of [³H]naltrexone in vivo (A) and urinary excretion of ³H in rats (B). Polymer/[³H]naltrexone (20 w/w) system, rods 1 ×1 ×14 mm. (A) Each value is the mean ± standard deviation for five systems eroded in vitro. The dashed line represents the total naltrexone content; (B) each value is the mean ± standard deviation for five rats, two systems in the subcutaneous tissue of each rat. The dashed line represents the range of expected total ³H excretion in urine. (Reproduced with permission from ref. 10. © 1978, Gordon and Breach Publishers.)

into films, and 1 ×1 ×14 mm devices punched from the films using a heated punch. Although not stated, the devices very likely also contained 10 wt% Na_2CO_3. Sterilization was achieved by ⁶⁰Co 2.5 Mrad irradiation.

Figure 2.2 shows in vitro release of [³H]naltrexone and urinary ³H release in rats. The pattern of urinary excretion follows closely the in vitro data for the first 10 days. Then, despite the accelerated in vitro release, urinary excretion remains relatively constant until day 25, when it drops rapidly. Only 2% of the initial ³H was found in devices retrieved from the rats at the termination of the experiment at 34 days. However, no erosion data were presented so that it is not clear how much naltrexone is released by diffusion and how much by erosion of the matrix. Recovery of ³H in the urine amounted to 43% and it was presumed that the remainder was excreted in the feces.

One of the more interesting studies is the use of C111 in the treatment of burns [11]. In this study, the C111 polymer has been investigated as a bioerodible ointment for the controlled release of 4-homosulfanilamide in the management of *Pseudomonas aeruginosa* burn wound sepsis. The 4-homosulfanilamide was mixed into the C111 polymer and the mixture impregnated into a polyester double knit material. The impregnated fabric was then placed on rat burns infected with a strain of *Pseudomonas aeruginosa* and changed daily. This treatment was compared to that with a commercially available

Sulfamylon hydrophilic cream and controls consisting of C111 polymer alone and polyester fabric alone. Results of these studies are shown in Figure 2.3. Clearly, survival time with the C111 polymer containing 4-homosulfanilamide is superior to that achieved with the commercial preparation. However, no in vitro kinetics of drug release studies have been presented and no further work with this system has been reported.

2.2.2   Poly(ortho ester) II

2.2.2.1   Chemical Aspects

An alternate reaction that can be used for the preparation of ortho esters is the reaction between a ketene acetal and an alcohol as shown in Scheme 6.

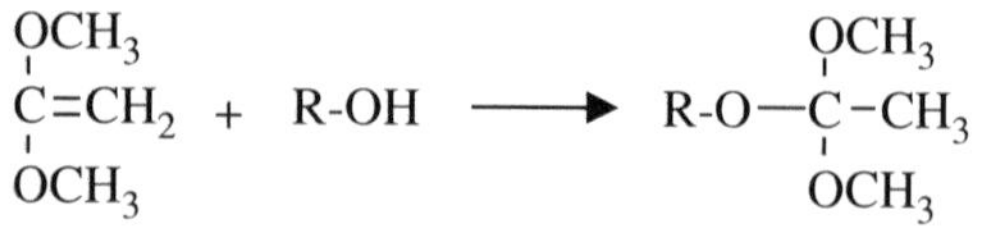

$$
\begin{array}{c}
OCH_3 \\
| \\
C=CH_2 \;+\; R\text{-}OH \longrightarrow \\
| \\
OCH_3
\end{array}
\qquad
\begin{array}{c}
OCH_3 \\
| \\
R\text{-}O\text{--}C\text{-}CH_3 \\
| \\
OCH_3
\end{array}
$$

Scheme 6

The only diketene acetal described in the literature is 1,1,4,4-tetramethoxy-1,3-butadiene which has been prepared, in very poor overall yield, by the multistep synthesis shown in Scheme 7 [12].

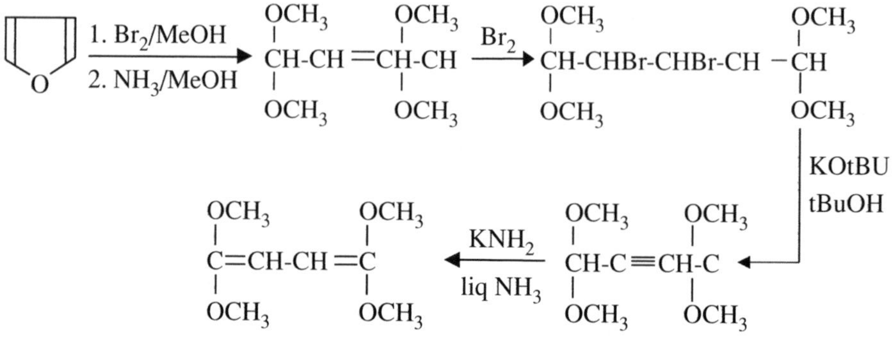

Scheme 7

It was expected that when this diketene acetal is reacted with a diol, the polymer shown in Scheme 8 would form.

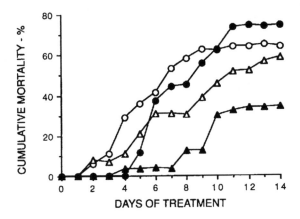

Figure 2.3 Cumulative mortality of rats (○), C111 polymers, (●), polyester fabric; (Δ), sulfamylon cream containing homosulfanilamide acetate; (▲), C111 polymer with homosulfanilamide free base impregnated in polyester fabric. (Reproduced with permission from ref. 11. © 1976, Mosby-Yearbook Publishing Co.)

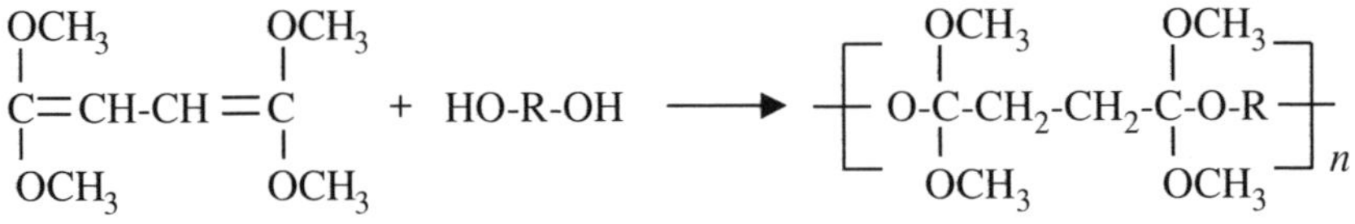

$$
\begin{array}{c}
\underset{\displaystyle \underset{OCH_3}{|}}{\overset{\displaystyle \overset{OCH_3}{|}}{C}}=CH\text{-}CH=\underset{\displaystyle \underset{OCH_3}{|}}{\overset{\displaystyle \overset{OCH_3}{|}}{C}} \quad + \quad HO\text{-}R\text{-}OH \quad \longrightarrow \quad \left[\, O\text{-}\underset{\displaystyle \underset{OCH_3}{|}}{\overset{\displaystyle \overset{OCH_3}{|}}{C}}\text{-}CH_2\text{-}CH_2\text{-}\underset{\displaystyle \underset{OCH_3}{|}}{\overset{\displaystyle \overset{OCH_3}{|}}{C}}\text{-}O\text{-}R \,\right]_n
\end{array}
$$

Scheme 8

Unfortunately, due to the extreme ease with which alkoxy groups of ortho ester linkages transesterify, only the crosslinked material shown in Scheme 9 could be isolated.

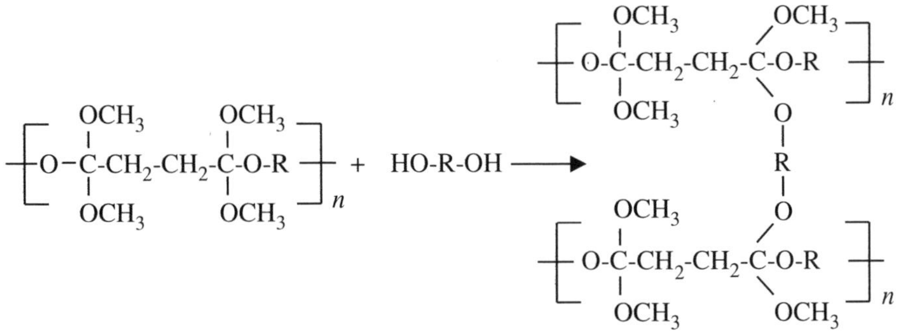

$$
\left[\, O\text{-}\underset{OCH_3}{\overset{OCH_3}{C}}\text{-}CH_2\text{-}CH_2\text{-}\underset{OCH_3}{\overset{OCH_3}{C}}\text{-}O\text{-}R \,\right]_n \; + \; HO\text{-}R\text{-}OH \longrightarrow
$$

Scheme 9

To prevent this transesterification, the pendant alkoxy groups in the final polymer were made part of a cyclic structure. The synthesis of such a polymer necessitated the preparation of the cyclic diketene acetal, 3,9-bis(methylene) 2,4,8, 10-tetraoxaspiro[5.5]undecane, as shown in Scheme 10 [13].

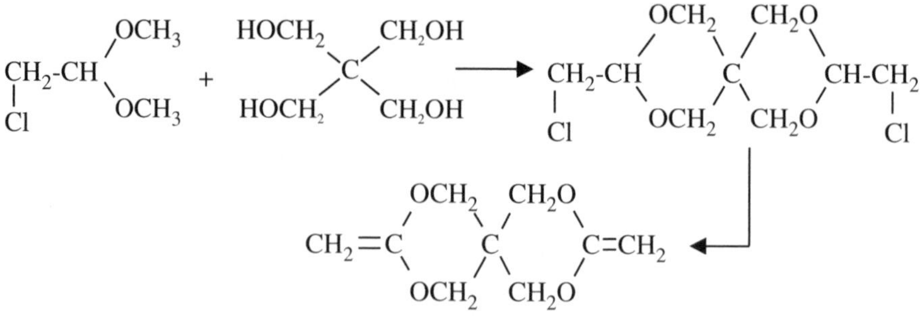

$$CH_2\text{-}CH\begin{smallmatrix}OCH_3\\OCH_3\end{smallmatrix}\;\;+\;\;HOCH_2\;\;C\;\;CH_2OH\;\;\longrightarrow\;\;CH_2\text{-}CH\begin{smallmatrix}OCH_2\;\;CH_2O\\C\\OCH_2\;\;CH_2O\end{smallmatrix}CH\text{-}CH_2$$

with Cl groups on the CH positions

$$CH_2{=}C\begin{smallmatrix}OCH_2\;\;CH_2O\\C\\OCH_2\;\;CH_2O\end{smallmatrix}C{=}CH_2$$

Scheme 10

However, this diketene acetal has two electron donor alkoxy groups on each double bond which makes it highly susceptible to a cationic polymerization which is initiated by even trace amounts of extremely weak acids. This is a serious complication because the addition of an alcohol to a ketene acetal is an acid-catalyzed reaction, usually carried out with acids such as p-toluenesulfonic acid. This complication has been overcome by replacing p-toluenesulfonic acid by iodine in pyridine which is able to catalyze the addition of alcohols to ketene acetals without the concomitant catalysis of a cationic polymerization [14]. However, despite the fact that linear polymers could be prepared, the extreme sensitivity of this diketene acetal toward acids makes its synthesis, purification, and handling extremely difficult and precludes scale-up necessary for eventual commercialization.

The structure of a polymer prepared using 3,9-bis(methylene) 2,4,8,10-tetraoxaspiro[5.5]undecane and 1,6-hexanediol is shown in Scheme 11. The polymer has been characterized by ^{13}C-NMR spectroscopy as shown in Figure 2.4 [14].

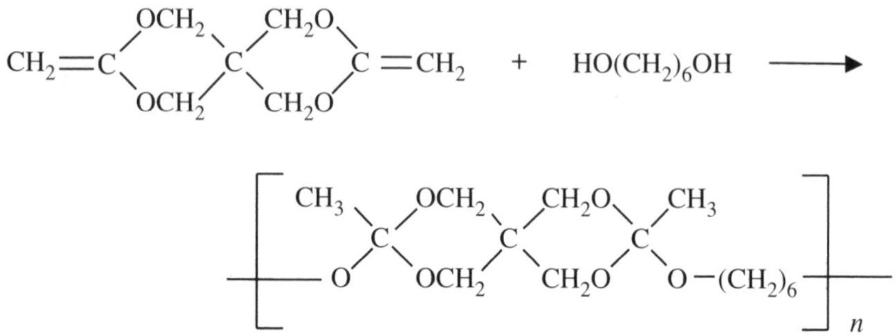

$$CH_2{=}C\begin{smallmatrix}OCH_2\;\;CH_2O\\C\\OCH_2\;\;CH_2O\end{smallmatrix}C{=}CH_2\;\;+\;\;HO(CH_2)_6OH\;\;\longrightarrow$$

$$\left[\!\!\begin{array}{c}CH_3\;\;\;OCH_2\;\;CH_2O\;\;CH_3\\C\;\;\;\;\;C\;\;\;\;\;C\\O\;\;\;\;OCH_2\;\;CH_2O\;\;\;O{-}(CH_2)_6\end{array}\!\!\right]_n$$

Scheme 11

To convert this monomer to a more useful one, it was necessary to block the facile cationic polymerization which was achieved by the introduction of steric hinderance about the double bond by replacing a hydrogen by a methyl group. The structure of the new monomer as compared to the previous one is shown in Scheme 12.

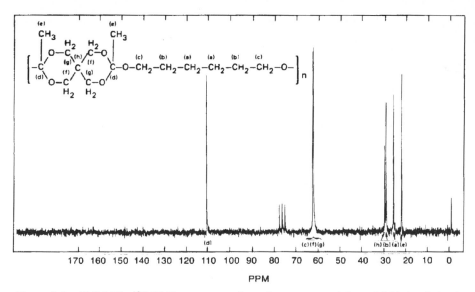

Figure 2.4   25.2 MHz ¹³C-NMR spectrum of a polymer prepared from 3,9-bis(methylene)-2,4,8,10-tetraoxaspiro[5.5]undecane and 1,6-hexanediol in CDCl₃ at room temperature. (Reproduced with permission from ref. 14. © 1980, John Wiley & Sons.)

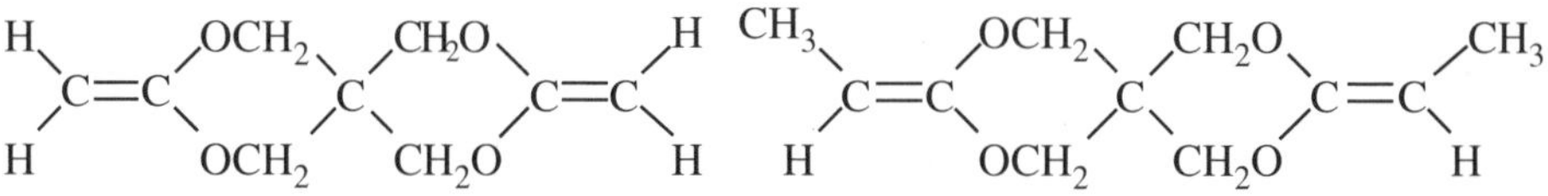

Scheme 12

This new monomer, 3,9-bis(ethylidene) 2,4,8,10-tetraoxaspiro[5.5]undecane, is prepared as shown in Scheme 13 [15].

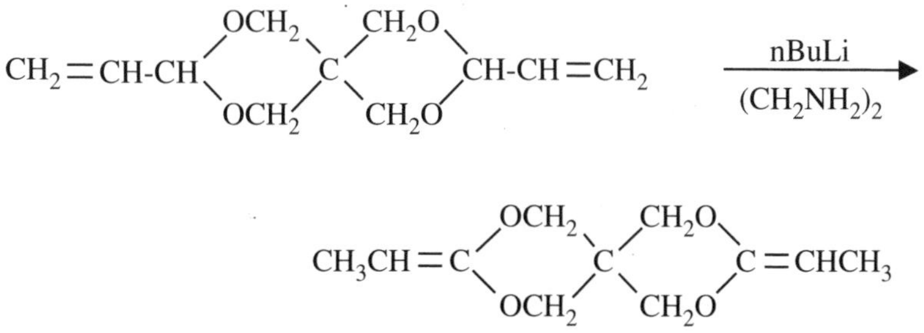

Scheme 13

The introduction of the methyl group was sufficient to prevent the facile cationic polymerization and linear polymers could be readily prepared using acid catalysts such as *p*-toluenesulfonic acid. Further, this diketene acetal could be easily handled, purified, and the synthesis readily scaled up.

Formation of poly(ortho esters) using the diketene acetal 3,9-bis(ethylidene) 2,4,8, 10-tetraoxaspiro[5.5]undecane is shown in Scheme 14.

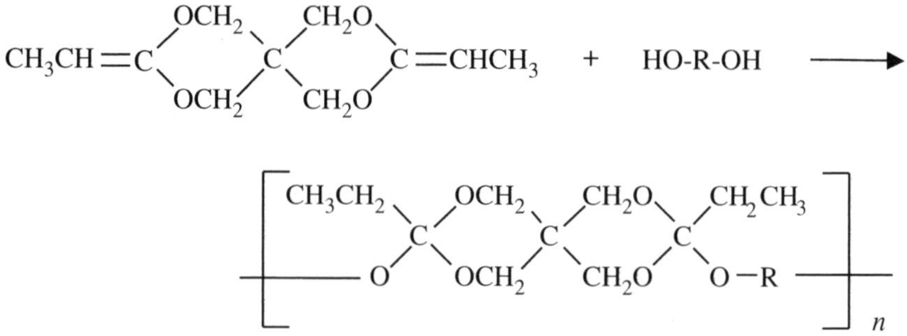

$$CH_3CH=C \underset{OCH_2}{\overset{OCH_2}{<}} C \underset{CH_2O}{\overset{CH_2O}{>}} C=CHCH_3 \quad + \quad HO-R-OH \longrightarrow$$

$$\left[O \underset{OCH_2}{\overset{CH_3CH_2}{\underset{C}{<}}} C \underset{CH_2O}{\overset{OCH_2}{<}} C \underset{O-R}{\overset{CH_2CH_3}{>}} \right]_n$$

Scheme 14

The reaction proceeds readily at room temperature and to prepare polymers it is necessary merely to dissolve the monomers in a polar solvent such as tetrahydrofuran and to add a trace of an acid catalyst. Polymerization is exothermic and high molecular weight polymers are formed virtually instantaneously [15].

Because the condensation between a diketene acetal and a diol, just like that between a diisocyanate and a diol, proceeds without the evolution of volatile byproducts, dense, crosslinked materials can be produced by using reagents having a functionality greater than 2.

To prepare crosslinked materials, a molar excess of the diketene acetal is used and the resulting prepolymer with ketene acetal end groups is reacted with a triol or a mixture of diols and triols [16]. This synthesis is shown in Scheme 15.

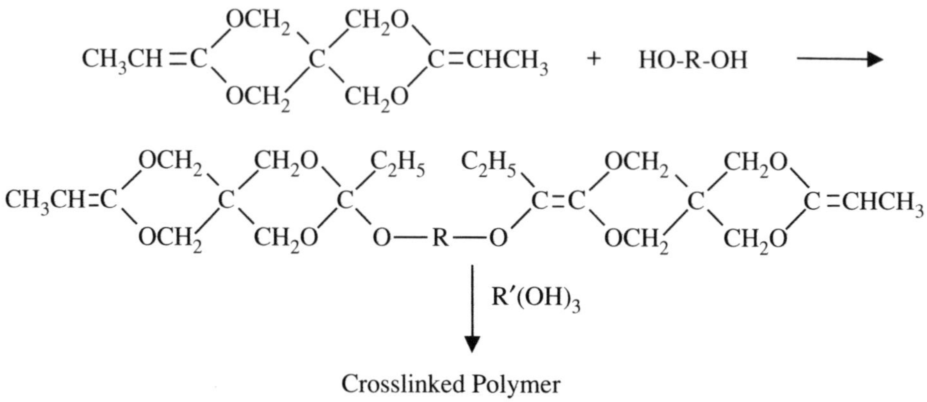

$$CH_3CH=C \underset{OCH_2}{\overset{OCH_2}{<}} C \underset{CH_2O}{\overset{CH_2O}{>}} C=CHCH_3 \quad + \quad HO-R-OH \longrightarrow$$

$$CH_3CH=C \underset{OCH_2}{\overset{OCH_2}{<}} C \underset{CH_2O}{\overset{CH_2O}{<}} C \underset{O-R-O}{\overset{C_2H_5}{<}} \overset{C_2H_5}{C=C} \underset{OCH_2}{\overset{OCH_2}{<}} C \underset{CH_2O}{\overset{CH_2O}{>}} C=CHCH_3$$

$$\downarrow R'(OH)_3$$

Crosslinked Polymer

Scheme 15

When these linear or crosslinked poly(ortho esters) are placed in an aqueous environment they hydrolyze as illustrated in Scheme 16 for the linear polymer [17].

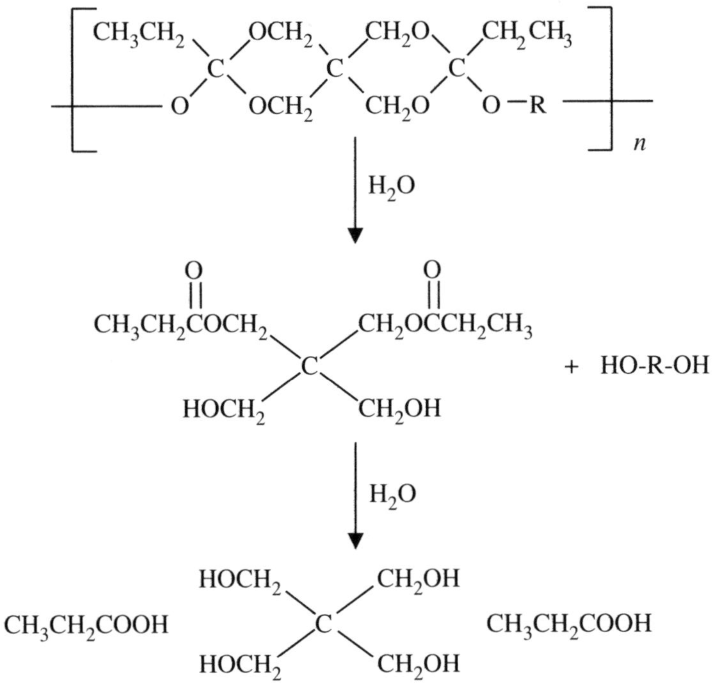

Scheme 16

Even though the hydrolysis eventually produces an acid, polymer erosion rate is controlled by hydrolysis of the ortho ester bonds. The subsequent hydrolysis of the ester bonds takes place at a much slower rate so that the neutral, low molecular weight reaction products can diffuse away from the implant before hydrolysis to an acid takes place. Thus, unlike the poly(ortho ester) system I, no autocatalysis is observed and it is not necessary to use basic excipients to neutralize the acidic hydrolysis products.

2.2.2.2   Physical Properties

Mechanical properties of the polymer can be readily controlled by an appropriate choice of the diols that are used in the condensation reaction. Use of the rigid diol *trans*-cyclohexane dimethanol produces a rigid polymer having a glass transition temperature of 110°C whereas use of the flexible diol 1,6-hexanediol produces a soft material having a glass transition temperature of 20°C. Mixtures of the two diols produces polymers that have intermediate glass transition temperatures [17]. Variation of the glass transition temperature with composition of the diol mixture is shown in Figure 2.5.

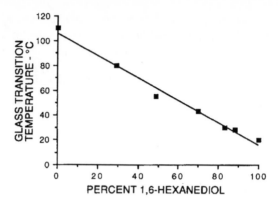

Figure 2.5 Glass transition temperature of 3,9-bis(ethylidene)-2,4,8,10-tetraoxaspiro[5.5]
undecane/*trans*-cyclohexanedimethanol/1,6-hexanediol polymer as a function of mol% 1,6-
hexanediol. (Reproduced with permission from ref. 17. © 1983, MTP Press Ltd.)

As in all condensation polymerizations, molecular weight of the polymer is a
function of stoichiometry and can be readily varied by appropriate skewing. This
is important because an exact equivalence of functional groups can lead to polymers
having molecular weights in excess of 100,000, which for some fabrication techniques
is too high due to an undesirably high melt viscosity. Figure 2.6 shows a relationship

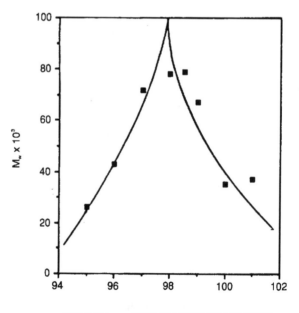

Figure 2.6   Effect of stoichiometry on weight average molecular weight of a polymer prepared
from 3,9-bis(ethylidene)2,4,8,10-tetraoxaspiro[5.5]undecane and *trans*-cyclohexanedimethanol.
(Reproduced with permission from ref. 18. © 1983 Plenum Publishing Co.)

between polymer molecular weight and stoichiometry [18]. In carrying out these experiments it was assumed that both the diketene acetal and the diol were 100% pure even though gas chromatographic analysis of the monomers indicated that whereas the diol was 100% pure, the diketene acetal was only 98.1% pure. The fact that the molecular weight peaks at about 98 mol% of *trans*-cyclohexane dimethanol verifies that the gas chromatographic analysis of the diketene acetal is accurate.

Figure 2.7 shows a relationship between log [melt viscosity] and log [weight average molecular weight] as determined by light scattering. As shown, melt viscosity can be readily adjusted by controlling polymer molecular weight [18].

Even though poly(ortho esters) contain hydrolytically labile linkages, they are highly hydrophobic materials and for this reason are very stable and can be stored without careful exclusion of moisture. However, the ortho ester linkage in the polymer is inherently thermally unstable and at elevated temperatures is believed to dissociate into an alcohol and a ketene acetal. A possible mechanism for the thermal degradation

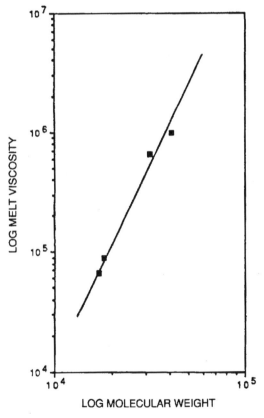

Figure 2.7   Melt viscosity versus molecular weight of a polymer prepared from 3,9-bis(ethyli-dene)-2,4,8,10-tetraoxaspiro[5.5]undecane and a 70/30 mole ratio of *trans*-cyclohexanedimeth-anol and 1,6-hexanediol measured at 150°C. (Reproduced with permission from ref. 18. © 1983 Plenum Publishing Co.)

is shown in Scheme 17 [18]. This thermal degradation is similar to that observed with polyurethanes [19].

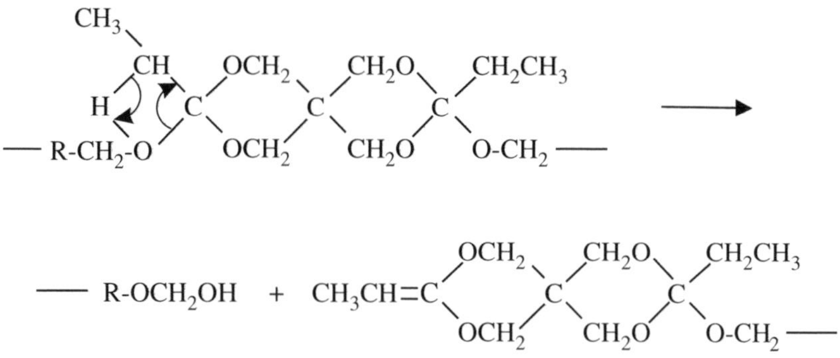

Scheme 17

When acidic or latent acidic excipients (anhydrides) are incorporated into the polymer to control erosion rate, the polymers become quite sensitive to moisture and heat and must be processed in a dry environment. A rigorous exclusion of moisture is particularly important with materials that contain a high (5 wt%) loading of an acidic catalyst [20].

Thus, catalyzed poly(ortho esters) must be processed and then packaged in a dry environment. Processing in an environment that has a relative humidity (RH) of < 20% at 70°F is desirable, with lower RH values leading to improved device properties. Prior to fabrication, the polymer, drug, and excipient must be dried and the final device packaged in a dry environment using a high moisture barrier aluminum laminated strip packaging material [20].

2.2.2.3   Applications

Control of Polymer Erosion Rate   The sorption of water by poly(ortho esters) has been found to be relatively small, about 0.30–0.75% with a diffusion coefficient ranging from a high of 4.07×10^{-8} cm^2 s^{-1} for a polymer based on 1,6-hexanediol (T_g 22°C) to a low of 2.11×10^{-8} cm^2 s^{-1} for a polymer based on *trans*-cyclohexanedimethanol (T_g, 122°C) [21]. Because these poly(ortho esters) are extremely hydrophobic, ortho-ester linkages that are quite labile in solution become very unreactive when they are part of the polymer matrix. This remarkable lack of reactivity is shown in Figure 2.8, which shows the very slow weight loss of a polymer disk in a pH 7.4 buffer at 37°C [22].

However, polymer hydrolysis can be accelerated by the addition of acidic excipients, by increasing the hydrophilicity of the polymer matrix, or both of these. It is also possible to retard polymer hydrolysis by using basic excipients which stabilize ortho ester linkages [23].

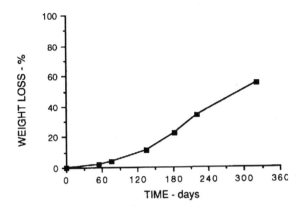

Figure 2.8   Weight loss of polymer disks prepared from 3,9-bis(ethylidene) 2,4,8,10-tetraoxa-spiro[5.5]undecane and 1,6-hexanediol at pH 7.4 and 37°C.

When a hydrophobic polymer with physically dispersed maleic anhydride excipient is placed into an aqueous environment, water will diffuse into the polymer, dissolve the acidic excipient, and accelerate polymer hydrolysis [24]. This process is shown in Figure 2.9, where rate of polymer erosion is measured as the rate of release of timolol maleate which was physically incorporated into the polymer. Clearly, use of an acidic excipient allows excellent control over rate of matrix hydrolysis.

The hydrolysis proceeds by diffusion of water into the surface layers of the polymer, where the lowered pH will accelerate hydrolysis of the ortho ester bonds [25]. The

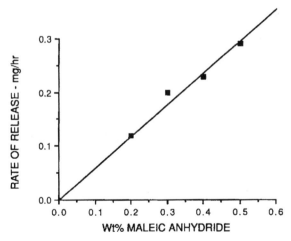

Figure 2.9   Release rate of timolol maleate from a 7:3 blend of polymers prepared from 3,9-bis(ethylidene)-2,4,8,10-tetraoxaspiro[5.5]undecane and 1,6-hexanediol and 3,9-bis(ethylidene)-2,4,8,10-tetraoxaspiro[5.5]undecane and *trans*-cyclohexanedimethanol at pH 7.4 and 37°C. Drug loading 2 wt%. (Reproduced with permission from ref. 24. © 1984, Butterworth-Heine-mann Ltd.)

process is shown schematically in Figure 2.10 where it has been analyzed in terms of the movement of two fronts, V_1, the movement of a hydrating front, and V_2, the movement of an erosion front. The ultimate behavior of a device will then be determined by the relative movement of these two fronts. If $V_1 > V_2$, the thickness of the reaction zone will gradually increase and at some time, the matrix will be completely permeated by water. At that point, all ortho ester linkages will hydrolyze at comparable rates and bulk hydrolysis will take place. However, if $V_1 = V_2$, then hydrolysis is confined to the surface layers and only surface hydrolysis will take place. In this latter case, rate of polymer erosion will be completely determined by the rate at which water intrudes into the polymer.

Bioerodible Drug Delivery Applications   Acidic excipients are useful in the development of drug delivery devices that have lifetimes of 2–4 weeks. The development of devices that have significantly longer lifetimes requires the use of basic excipients. The use of acidic excipients is illustrated with devices that deliver 5-fluorouracil and naltrexone, whereas the use of basic excipients is illustrated with a device that delivers levonorgestrel.

5-Fluorouracil is a well known antineoplastic agent that finds applications in cancer chemotherapy [26] and in the prevention of fibroblast proliferation following glaucoma filtration surgery [27]. Devices were fabricated by dispersing 10 wt% 5-fluorouracil in a polymer prepared from 3,9-bis(ethylidene) 2,4,8,10-tetraoxaspiro[5,5]undecane and 1,6-hexanediol followed by compression molding to form thin disks [28]. Figure 2.11 shows results of a study where both 5-fluorouracil release and weight loss of the disks were determined. The data show that with this particular system, concomitant drug release and polymer erosion has been achieved. Further, because the molecular weight of the residual polymer remains unchanged, the hydrolysis process is confined to the outer surface of the device and surface erosion has been achieved.

Naltrexone is a narcotic antagonist that occupies the same receptors as morphine but produces no euphoric effects. Therefore, a patient on naltrexone therapy experiences no euphoria on intake of heroin which is rapidly metabolized in the body to morphine. Naltrexone therapy is currently a method of choice in the rehabilitation of opiate-dependent individuals because it provides an enforced opiate-free life [29]. However,

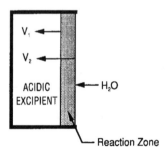

Figure 2.10   Schematic representation of water intrusion and erosion for one side of a bioerodible device containing dispersed acidic excipient. (Reproduced with permission from ref. 25. © 1985, Elsevier Science Publishers.)

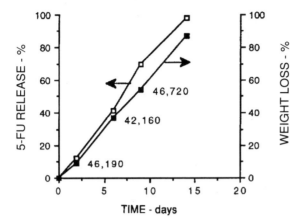

Figure 2.11 Cumulative weight loss (●) and cumulative release of 5-fluorouracil (5FU) (□) from polymer disks prepared from 3,9-bis)ethylidene)-2,4,8,10-tetraoxaspiro[5.5]undecane and 1,6-hexanediol at pH 7.4 and 37°C. Devices contain 10 wt% 5FU and 0.15 wt% suberic acid. Numbers indicate weight average molecular weight of residual polymer. (Reproduced with permission from ref. 28. © 1990, Elsevier Science Publishers.)

continuing a maintenance of oral naltrexone therapy requires very strong motivation because discontinuing therapy produces no withdrawal effects. For this reason, it is desirable to develop a naltrexone-releasing implant that can be implanted by a physician so that removal by the patient would be impossible, or at least very difficult.

Because naltrexone is a basic drug and because poly(ortho esters) are stable in base, it cannot be used as such and for this reason the slightly acidic salt naltrexone pamoate was used. Figure 2.12 shows release of naltrexone pamoate as a function of amount of incorporated suberic acid [30]. The data show that naltrexone pamoate can catalyze polymer erosion and that erosion rate can be further controlled by the addition of small amounts of suberic acid.

Levonorgestrel is a steroid that finds application as a contraceptive for humans. Because devices with a lifetime of 1 year were desired, development of such devices requires the use of a base that will stabilize the interior of the device and allow erosion to take place only in the outer layers from which the base has been depleted by diffusion.

In the development of these devices a crosslinked polymer was used. Devices were fabricated by first preparing a ketene acetal terminated prepolymer derived from two equivalents of the diketene acetal 3,9-bis(ethylidene) 2,4,8,10 tetraoxaspiro[5.5]undecane and one equivalent of the diol 3-methyl-1,5-pentanediol. Then, 30 wt% levonorgestrel, 7 wt% $Mg(OH)_2$, and a 30 mol% excess of 1,2,6-hexanetriol were mixed into the prepolymer and this mixture extruded into 2.4 ID diameter Teflon tubing. After curing at 60°C overnight, the rods were removed from the tubing and cut into 2-cm lengths. Erosion and drug release from these devices was studied by implanting the devices subcutaneously into rabbits, explanting at various time intervals, and measuring weight loss and residual drug [16].

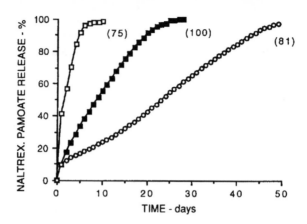

Figure 2.12   Cumulative release of naltrexone pamoate from polymer slabs (25 ×4 ×25 mm) prepared from 3,9-bis(ethylidene)-2,4,8,10-tetraoxaspiro[5.5]undecane and 1,6-hexanediol at pH 7.4 and 37°C. Numbers in parentheses indicate percent weight loss. Devices contain 50 wt% drug and varying amounts of suberic acid (SA). (□), 3 wt% SA; (■), 1 wt% SA; (○), no SA. (Reproduced with permission from ref. 30. © 1990, Elsevier Science Publishers.)

Levonorgestrel blood plasma levels determined by radioimmunoassay are shown in Figure 2.13 [23]. The steady-state plasma shown is below the therapeutically effective level so that a more rapidly eroding polymer was needed. This was achieved by using a material containing 7 wt% $Mg(OH)_2$ and 1 mol% copolymerized 9,10-dihydroxystearic

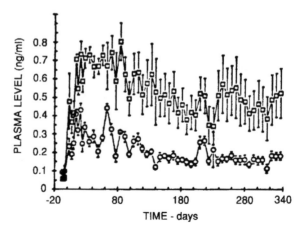

Figure 2.13   Daily rabbit blood plasma levels of levonorgestrel from a crosslinked polymer prepared from 3,9-bis(ethylidene)-2,4,8,10-tetraoxaspiro[5.5]undecane/3-methyl-1,5-pentanediol prepolymer crosslinked with 1,2,6-hexanetriol. Polymer rods 2.4 ×20 mm containing 30 wt% levonorgestrel and 7.1 mol% $Mg(OH)_2$. Devices implanted subcutaneously in rabbits (○), one device/rabbit; (□), two devices/rabbit. (Reproduced with permission from ref. 23. © 1986, Plenum Press.)

acid. When these devices were implanted into rabbits and levonorgestrel blood plasma level determined, higher levels as shown in Figure 2.14 were obtained [20].

Devices explanted from rabbits were examined by scanning electron microscopy [25]. As shown in Figure 2.15, the devices clearly indicate surface erosion as development of voids around the periphery of the rod-shaped device and a progressive diminution of a central uneroded zone. The presence of voids suggests that once erosion starts, generation of hydrophilic degradation products at that location accelerates further polymer hydrolysis.

Hard and Soft Tissue Fixation Applications   Metallic implant devices have been used for bone fracture fixation for many years. However, load sharing between bone and the device is in proportion to bone and device structural stiffness and complete healing of the bone is prevented as long as the device is present and bears part of the load normally seen by the bone [31]. For this reason, replacing metallic fracture fixation devices with a bioabsorbable polymer that has an appropriate combination of initial strength and stiffness is of considerable interest because the load will be gradually transferred to the healing bones and, because the device is completely absorbed, surgical removal is not necessary.

The first polymer used in the construction of a resorbable fracture fixation device was poly(DL-lactic acid) and since that report, many other resorbable polymers have been investigated. A detailed review of this field has recently been published [32]. The most recent addition to polymers useful in the fabrication of fracture fixation devices are poly(ortho esters) [33,34]. Their usefulness in this application is derived from

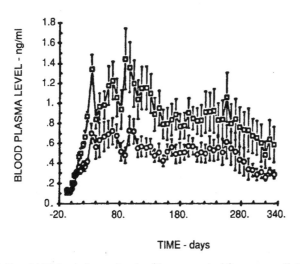

Figure 2.14   Daily rabbit blood plasma levels of levonorgestrel from a crosslinked polymer prepared from 3,9-bis(ethylidene)-2,4,8,10-tetraoxaspiro[5.5]undecane/3-methyl-1,5-pentanediol prepolymer crosslinked with 1,2,6-hexanetriol. Prepolymer contains 1 mol% copolymerized 9,10-dihydroxysteric acid. Polymer rods 2.4 ×20 mm containing 30 wt% levonorgestrel and 7.1 mol% $Mg(OH)_2$. Devices implanted subcutaneously in rabbits. (○), one device/rabbit; (□), two devices/rabbit. (Reproduced with permission from ref. 20. © 1990, Marcel Dekker.)

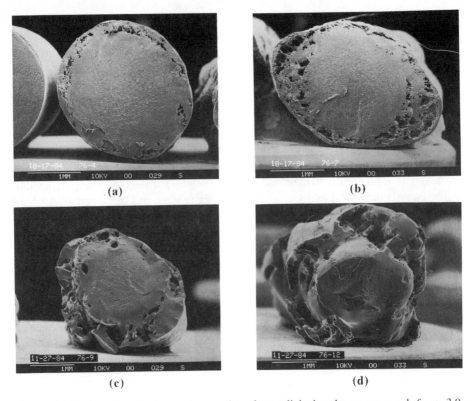

Figure 2.15 Scanning electron micrographs of crosslinked polymer prepared from 3,9-bis(ethylidene)-2,4,8,10-tetraoxaspiro[5.5]undecane/3-methyl-1,5-pentanenediol   prepolymer crosslinked with 1,2,6-hexanetriol. Prepolymer contains 1 mol% copolymerized 9,10-dihydroxy-stearic acid. Polymer rods 2.4 ×20 mm containing 30 wt% levonorgestrel and 7.1 mol% of $Mg(OH)_2$. Devices implanted subcutaneously in rabbits. (a) after 6 weeks, 30×; (b) after 9 weeks, 30 ×; (c) after 12 weeks, 25×; (d) after 16 weeks, 25 ×. (Reproduced with permission from ref. 25. © 1985, Elsevier Science Publishers.)

their extreme hydrophobicity, the ability to vary polymer mechanical properties by a simple adjustment in the ratios of the rigid and flexible diols used in the synthesis, and from the fact that the initial degradation products are neutral which may eliminate problems associated with the development of excessive acidity recently reported for devices prepared from poly(DL- or L-lactic acid).

Figures 2.16 and 2.17 show mean flexural modulus and mean flexural strength of compression-molded specimens that were placed in tris-buffered saline maintained at 37°C for various lengths of time after which the specimens were removed and their mechanical properties determined. As is clear from the figures, both mean flexural modulus and mean flexural strength show only a slight decrease during the 12-week duration of the experiment. This lack of significant change was observed for both the physiological pH of 7.4 and the acidic pH of 5.0.

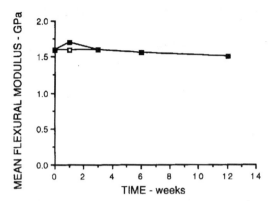

Figure 2.16   Change in mean flexular modulus of a polymer prepared from 3,9-bis(ethylidene)-2,4,8,10-tetraoxaspiro[5.5]undecane and a 60/40 mole ratio of *trans*-cyclohexanedimethanol and 1,6-hexanediol measured in tris-buffered saline at 37°C. (■), pH 7.4; (□), pH 5.0.

The insensitivity of the polymer to an external acidic environment is due to the extreme hydrophobicity of the polymer which makes the acid-sensitive ortho ester linkages virtually inaccessible to the external acid. The lack of change of mechanical properties is supported by weight loss data shown in Figure 2.18 which shows virtually no weight loss at pH 7.4 and only minimal weight loss at pH 5.0.

However, data shown in Figures 2.16 and 2.17 were obtained using specimens that were immersed in the tris-buffered saline without loading. As this is not representative of actual use in fracture fixation where a device is exposed to either a static load, and, more likely, to a dynamic load, experiments were conducted where the compression-molded specimens were subjected to no load, a static load, and a dynamic load. Results of these experiments are shown in Figures 2.19 and 2.20. The data show that devices subjected to dynamic loading undergo a significant decrease in

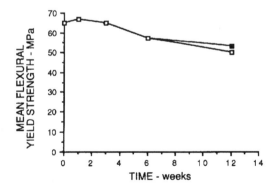

Figure 2.17   Change in mean flexular yield strength of a polymer prepared from 3,9-bis(ethylidene)-2,4,8,10-tetraoxaspiro[5.5]undecane and a 60/40 mole ratio of *trans*-cyclohexanedimethanol and 1,6-hexanediol measured in tris-buffered saline at 37°C. (■), pH 7.4; (□), pH 5.0.

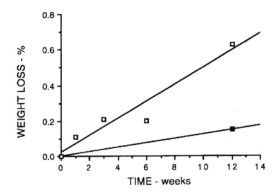

Figure 2.18   Weight loss of a polymer prepared from 3,9-bis(ethylidene)-2,4,8,10-tetraoxaspiro [5.5]undecane and a 60/40 mole ratio of *trans*-cyclohexanedimethanol and 1,6-hexanediol measured in tris-buffered saline at 37°C. (■), pH 7.4; (□), pH 5.0.

mechanical properties relative to devices subjected to no load conditions or to static load conditions.

The decrease in mechanical properties with time is very likely due to the generation of microscopic cracks which form when the device undergoes flexing. These microscopic cracks then provide a path for water to penetrate into the device which allows polymer hydrolysis to take place with consequent deterioration of mechanical properties.

It is well known that mechanical properties of polymers can be greatly enhanced by the incorporation of reinforcing fibers into the polymer. The mechanical properties of these composites depend on the fiber's chemical composition, on its volume percent, on its orientation, and on its dimensions, particularly its aspect ratio. For this

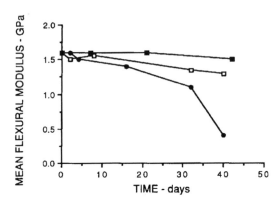

Figure 2.19   Change in mean flexular modulus of a polymer prepared from 3,9-b is(ethylidene)-2,4,8,10-tetraoxaspiro[5.5]undecane and a 60/40 mole ratio of *trans*-cyclohexanedimethanol and 1,6-hexanediol measured in tris-buffered saline at 37°C and pH 7.4. (■), no load; (□), static load; (●), dynamic load.

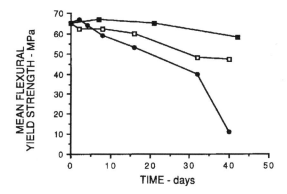

Figure 2.20   Change in mean flexular yield strength of a polymer prepared from 3,9-bis(ethyl-idene)-2,4,8,10-tetraoxaspiro[5.5]undecane and a 60/40 mole ratio of *trans*-cyclohexanedimeth-anol and 1,6-hexanediol measured in tris-buffered saline at 37°C and pH 7.4. (■), no load, (□), static load; (●) dynamic load.

reason, we have investigated mechanical properties and stability of poly(ortho esters) reinforced with randomly oriented calcium sodium metaphosphate (CSM) fibers. These fibers are biodegradable and solubilize in an in vivo environment by exchange of calcium for sodium. In the incorporation of fibers into the polymer, a diamine coupling agent was used in order to provide a bond between the fiber and polymer matrix. Fiber loading studies have established that mechanical properties peak at about 30 vol% so that loading was selected for further studies.

Figures 2.21 and 2.22 compare mechanical properties of CSM composites to those of polymers without CSM fibers. As shown, reinforced polymer shows a very significant increase in both mean flexural yield strength and mean flexural modulus, but on

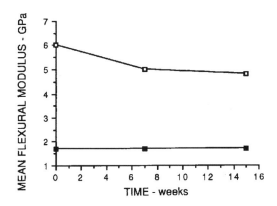

Figure 2.21   Effect of 30 vol% incorporation of calcium sodium metaphosphate crystalline microfibers on mean flexular modulus of a polymer prepared from 3,9-bis(ethylidene)-2,4,8,10-tetraoxaspiro[5.5]undecane and a 60/40 mole ratio of *trans*-cyclohexane dimethanol and 1,6-hexanediol measured in tris-buffered saline at 37°C and pH 7.4 (■), unloaded; (□), loaded.

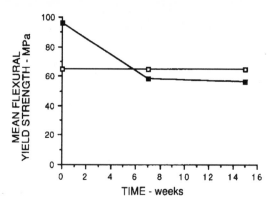

Figure 2.22   Effect of 30 vol% incorporation of calcium sodium metaphosphate crystalline microfibers on mean flexular yield strength of a polymer prepared from 3,9-bis(ethylidene)-2,4,8,10-tetraoxaspiro[5.5]undecane and a 60/40 mole ratio of *trans*-cyclohexane dimethanol and 1,6-hexanediol measured in tris-buffered saline at 37°C and pH 7.4. (□), unloaded; (■), loaded.

immersion in pH 7.4 tris buffer at 37°C, mean flexural modulus and mean flexural yield strength dramatically decrease after only a few weeks and the reinforcing effect is completely lost. This loss in properties is caused by water penetration into the device which affects the bonding between the CSM fiber and the polymer.

Water penetration can be delayed by coating of the composite with a thin layer of unfilled poly(ortho ester). Preliminary results of retention of properties of the coated specimens are shown in Figures 2.23 and 2.24.

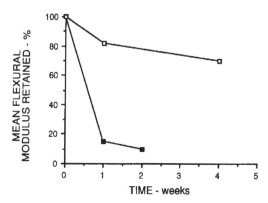

Figure 2.23   Effect of coating on retention of mean flexular modulus of a polymer filled with 30 vol% calcium sodium metaphosphate crystalline microfibers in tris-buffered saline at 37°C and pH 7.4. Polymer prepared from 3,9-bis(ethylidene)-2,4,8,10-tetraoxaspiro[5.5]undecane and a 60/40 mole ratio of *trans*-cyclohexane dimethanol and 1,6-hexanediol. (■), uncoated; (□), coated.

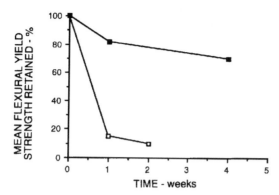

Figure 2.24   Effect of coating on retention of mean flexural yield strength of a polymer filled with 30 vol% calcium sodium metaphosphate crystalline microfibers in tris-buffered saline at 37°C and pH 7.4. Polymer prepared from 3,9-bis(ethylidene)-2,4,8,10-tetraoxaspiro[5.5]undecane and a 60/40 mole ratio of *trans*-cyclohexane dimethanol and 1,6-hexanediol. (□), uncoated; (■), coated.

2.2.3   Poly(ortho ester) III

2.2.3.1   Chemical Aspects

Another means of preparing poly(ortho esters) is shown in Scheme 18 [35].

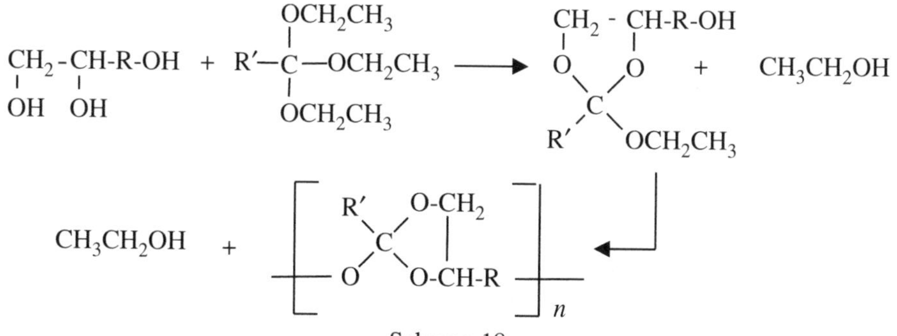

$$
\begin{array}{c}
\underset{\substack{|\;\;\;| \\ \text{OH}\;\;\text{OH}}}{\text{CH}_2\text{-CH-R-OH}} \;+\; \underset{\substack{| \\ \text{OCH}_2\text{CH}_3}}{\overset{\text{OCH}_2\text{CH}_3}{\text{R}'\text{-C-OCH}_2\text{CH}_3}} \longrightarrow
\end{array}
$$

Scheme 18

The intermediate does not have to be isolated and continuous reaction produces a polymer.

When a polymer prepared from 1,2,6-hexanetriol is placed in an aqueous environment, hydrolysis as shown in Scheme 17 takes place [36]. Initial hydrolysis occurs at the labile ortho ester bonds to generate one or more isomeric monoesters of the triol. This initial hydrolysis is followed by a much slower hydrolysis of the monoesters to produce a carboxylic acid and a triol. Because the neutral monoesters diffuse away from the implant site before hydrolysis to a triol and a carboxylic acid takes place, as with poly(ortho ester) II, no autocatalysis takes place.

2.2.3.2   Physical Properties

When flexible triols such as 1,2,6-hexanetriol are used in the synthesis shown in Scheme 16, highly flexible polymers that have ointment-like properties even at relatively high molecular weights are obtained.

Preliminary determination of polymer viscosity using a Rheometric RDS-II rheometer is shown in Figure 2.25. These data indicate that the polymer, even at relatively low molecular weights, is very viscous.

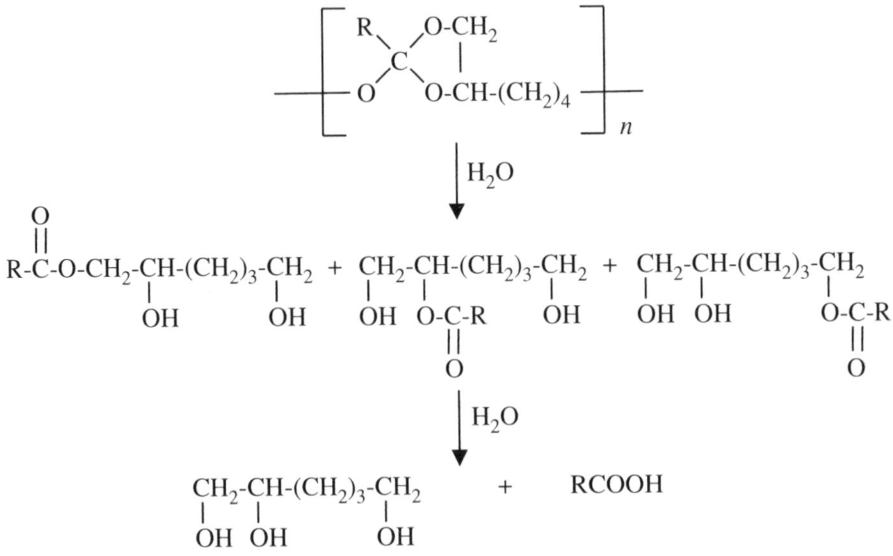

$$\left[\begin{array}{c} \mathrm{R} \quad\quad \mathrm{O\text{-}CH_2} \\ \diagdown \mathrm{C} \diagup | \\ \mathrm{O}\mathrm{O\text{-}CH\text{-}(CH_2)_4} \end{array} \right]_n$$

$$\Big\downarrow \mathrm{H_2O}$$

$$\begin{array}{l} \overset{\displaystyle O}{\overset{\displaystyle \|}{\mathrm{R\text{-}C}}}\text{-}\mathrm{O\text{-}CH_2\text{-}CH\text{-}(CH_2)_3\text{-}CH_2} \; + \; \mathrm{CH_2\text{-}CH\text{-}(CH_2)_3\text{-}CH_2} \; + \; \mathrm{CH_2\text{-}CH\text{-}(CH_2)_3\text{-}CH_2} \end{array}$$

(with substituents: OH, OH; OH, O-C-R (‖O); OH, OH; OH, O-C-R (‖O))

$$\Big\downarrow \mathrm{H_2O}$$

$$\mathrm{CH_2\text{-}CH\text{-}(CH_2)_3\text{-}CH_2} \quad + \quad \mathrm{RCOOH}$$

(OH, OH, OH)

Scheme 19

2.2.3.3   Applications

Ointment-like materials allow the incorporation of therapeutic agents by a simple mixing procedure without the need to use solvents or elevated temperatures. Thus, these materials are of interest for the delivery of sensitive materials such as proteins that can undergo loss of tertiary structure when elevated temperatures or solvents are used. Drug delivery from these materials can encompass both topical use and systemic applications.

Topical Applications   Because the polymer is highly hydrophobic, topical application to a moist surface is not possible. However, the polymer–drug mixture can be impregnated into a double-knit Dacron cloth and the cloth placed on the moist surface.

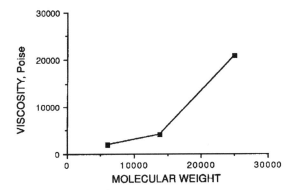

Figure 2.25 Viscosity of acetate polymer determined on a Rheometrics RDS-II rheometer at shear rate of 3.162 1/sec using cone and plate at room temperature.

One interesting application is the controlled delivery of 4-homosulfanilamide for the management of *Pseudomonas aeruginosa* burn wound sepsis. This compound is available as both the acidic hydrochloride and as the basic free drug. As expected, and as shown in Figure 2.26, when 10 wt% of the hydrochloride is incorporated into the polymer, rapid release occurs whereas virtually no release takes place when the free base is used. These data suggest that a blend of the two forms of 4-homosulfanilamide should produce a formulation where the drug is released at an intermediate rate. As shown in Figure 2.26 mixture of the acidic and basic forms produces intermediate release rates in accordance with the composition of the mixture.

Systemic Applications   The mild conditions under which therapeutic agents can be incorporated into the ointment-like polymer makes this system ideally suited for the delivery of proteins. Initial work designed to establish the feasibility of releasing

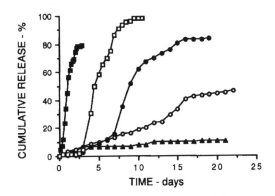

Figure 2.26   Release of 4-homosulfanilamide free base and hydrochloride from a 24,400 molecular weight polymer prepared from 1,2,6-hexanetriol and triethylorthoacetate at pH 7.4 and room temperature. Polymer contains 10 wt% drug. (■), 4-homosulfanilamide hydrochloride; (□), 90/10 mixture of hydrochloride and free base; (●), 75/25 mixture of hydrochloride and free base; (○), 50/50 mixture of hydrochloride and free base; (Δ), 4-homosulfamlamide.

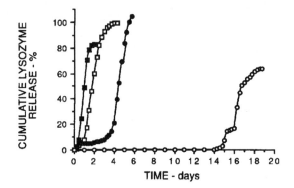

Figure 2.27   Release of lysozyme from a polymer prepared from 1,2,6-hexanetriol and triethyl orthoacetate at pH 7.4 and room temperature. Molecular weights are as indicated. (■), 5350; (□), 6800; (●), 12,000; (○), 24,400. Lysozyme loading 5 wt%.

proteins without loss of activity was carried out with the enzyme lysozyme because any loss of activity can be readily detected by noting changes in the rate of lysis of the enzyme substrate *Micrococcus lysodeikticus* [36].

Figure 2.27 shows release of the enzyme from an acetate polymer of four different molecular weights. These data show that the enzyme is released from the matrix in about 24 h but that release occurs after an induction period that varies from about 6 h for the lowest molecular weight polymer to about 14 days for the highest molecular weight polymer. As shown in Figure 2.28, by increasing the size of the alkyl group R′ the induction period for a low molecular weight polymer can be increased to 1 month.

Stability studies of lysozyme immobilized in a polymer where R′ is methyl are shown in Figure 2.29. The activity of the released lysozyme was assayed by noting rate of lysis of *Micrococcus lysodeikticus*. The control was a solution made up at the start of the experiment and stored at room temperature. This solution was sampled at

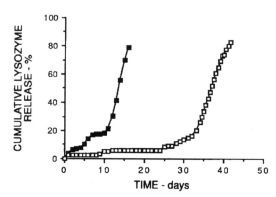

Figure 2.28   Release of lysozyme from a polymer prepared from 1,2,6-hexanetriol and triethyl orthovalerate at pH 7.4 and room temperature. Molecular weights are as indicated. (■), 5500; (□), 9300. Lysozyme loading 5 wt%.

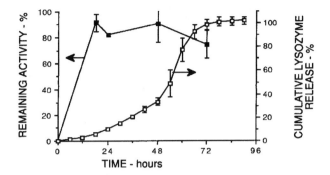

Figure 2.29   Assay of lysozyme activity following release from a polymer prepared from 1,2,6-hexanetriol and triethyl orthoacetate at pH 7.4 and room temperature. Activity assayed by noting change in rate of lysis of *Micrococcus lysodeikticus*. Lysozyme loading 10 wt% (Reproduced with permission from ref. 36. © 1992, Elsevier Science Publishers.)

each time point of the release experiment. As shown, virtually complete retention of enzyme activity was achieved indicating that the mixing procedure and release from the polymer does not lead to enzyme deactivation. However, whether more sensitive proteins can also be incorporated without loss of activity still needs to be determined.

2.2.4   Poly(ortho ester) IV

2.2.4.1   Chemical Aspects

The general synthetic procedure described for poly(ortho ester) III can also be used for the preparation of solid polymers. To do so, it is necessary only to replace the flexible triol with a rigid one, such as 1,1,4-cyclohexanetrimethanol, as shown in Scheme 20 [37]. As before, the intermediate does not have to be isolated and continuing reaction produces a high molecular weight polymer.

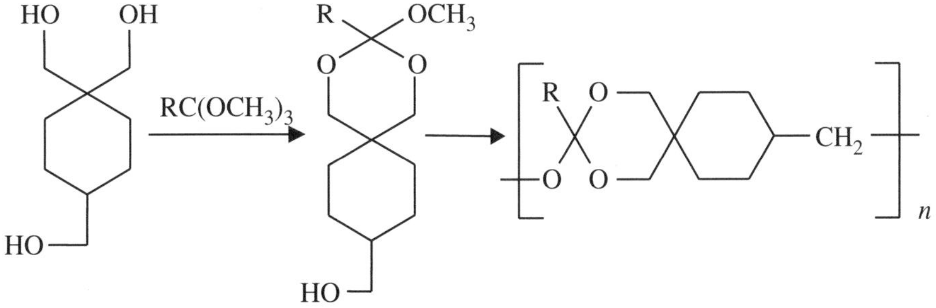

Scheme 20

The triol, 1,1,4-cyclohexanetrimethanol, is prepared as shown in Scheme 21.

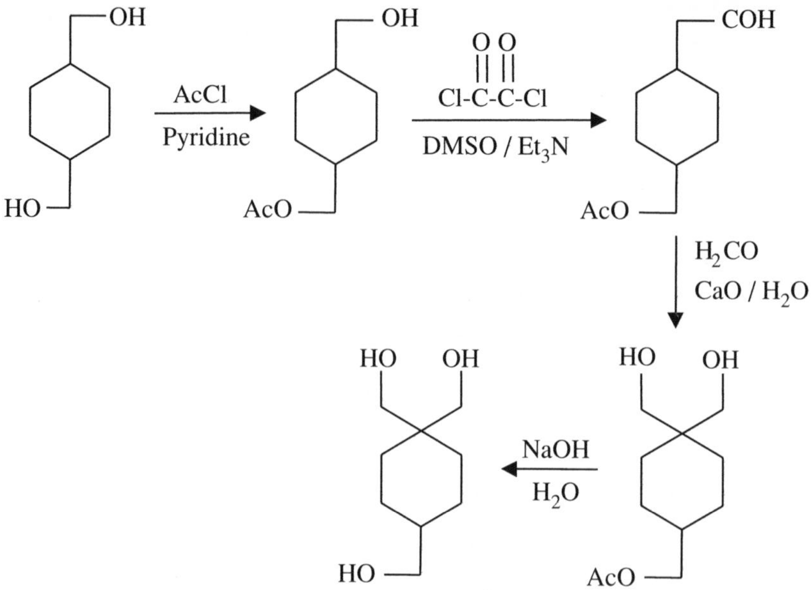

Scheme 21

When these poly(ortho esters) are placed in an aqueous environment, they hydrolyze as shown in Scheme 22.

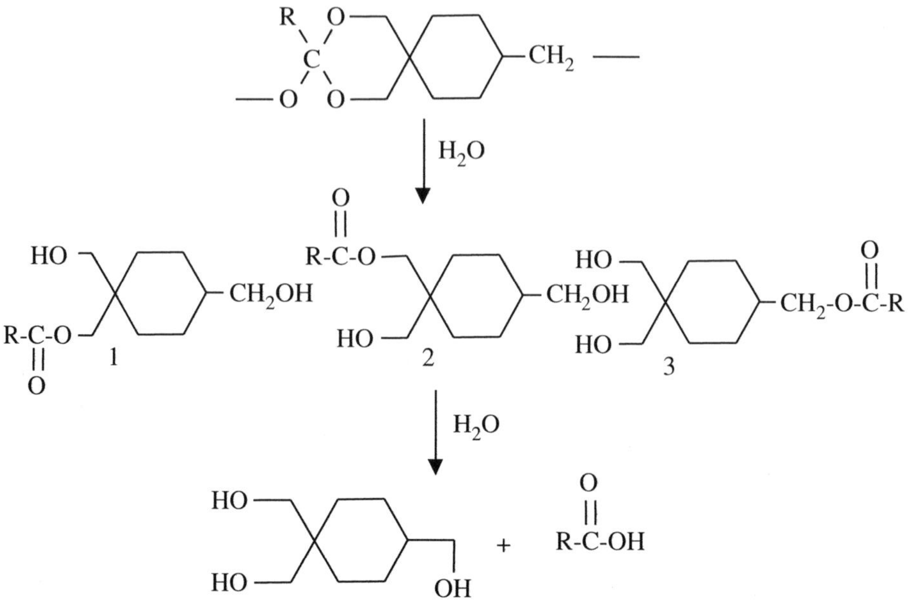

Scheme 22

As with the other poly(ortho ester) systems, initial cleavage occurs exclusively at the ortho ester bonds and the subsequent final hydrolysis takes place later. Thus, hydrolysis is not autocatalytic because the neutral hydrolysis products can diffuse away from the polymer before the carboxylic acid is formed.

When R = CH_3, the polymer is crystalline and precipitates out of the reaction mixture. The precipitated polymer is completely insoluble in the usual organic solvents such as methylene chloride, chloroform, ether, tetrahydrofuran, ethyl acetate, acetone, dimethylformamide, and dimethyl sulfoxide. When R = CH_3CH_2, the polymer remains in solution and as it is isolated, is amorphous and readily soluble in the usual organic solvents. The same observation holds for polymers with higher alkyl groups. Figures 2.30a and 2.30b show powder X-ray diffraction patterns of the methyl and ethyl

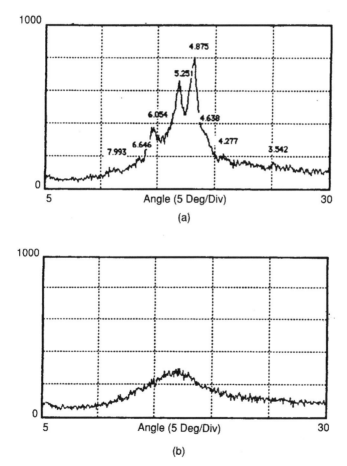

Figure 2.30   X-ray diffractometer traces of (a) acetate polymer (R–CH_3) and (b) propionate polymer (R–CH_3CH_2). (Reproduced with permission from ref. [37]. © 1992, American Chemical Society.)

polymer. Attempts to anneal the ethyl polymer were not successful, and the polymer remained amorphous.

REFERENCES

1. Frazza, E.J.; Schmitt, E.E. *J. Biomed. Mater. Res.*, **1**, 43 (1971).
2. Lewis, D.H. In *Biodegradable Polymers as Drug Delivery Systems.* Chasin, M.; Langer, R., eds. Marcel Dekker, New York, p. 1, (1990).
3. Choi, N.S.; Heller, J. U.S. Patent, 1978, 4,093,709.
4. Choi, N.S.; Heller, J. U.S. Patent, 1978, 4,131,648.
5. Choi, N.S.; Heller, J. U.S. Patent, 1979, 4,138,344.
6. Choi, N.S.; Heller, J. U.S. Patent, 1979, 4,180,646.
7. Gabelnick, H.L. In *Long-Acting Contraception.* Mishell, D.R., Jr., ed. Raven Press, New York, p. 149 (1983).
8. 10th Annual World Health Organization Report, p 62 (1981).
9. 11th Annual World Health Organization Report, p 61 (1982).
10. Capozza, R.C.; Sendelbeck, S.L.; Balkenhol, W.J. In *Polymeric Delivery Systems,* Kostelnik, R.J., ed. Gordon and Breach Publishers, New York, p. 59 (1978).
11. Vistness, L.M.; Schmitt, E.E.; Ksander, G.A.; Rose, E.H.; Balkenhol, W.J.; Coleman, C.L. *Surgery,* **79**, 690 (1976).
12. Scheeren, J.W.; Aben, R.W. *Tetrahed. Lett.* **12**, 1019 (1974).
13. Yasnitskii, B.G.; Sarkisyants, S.A.; Ivanyua E.G. *Zhurnal Obshchei Khimii* **34**, 1940 (1946).
14. Heller, J.; Penhale, D.W.H.; Helwing, R.F. *J. Polymer Sci., Polym. Lett.* **18**, 82 (1980).
15. Ng, S.Y.; Penhale, D.W.H.; Heller, J. *Macromol. Synth.* **11**, 23 (1992).
16. Heller, J.; Fritzinger, B.K.; Penhale, D.W.H.; Ng, S.Y.; Helwing, R.F.J. *Controlled Release* **1**, 233 (1985).
17. Heller, J.; Penhale, D.W.H.; Fritzinger, B.K.; Rose, J.E.; Helwing, R.F. *Contracept. Deliv. Syst.* **4**, 43 (1983).
18. Heller, J.; Penhale, D.W.H.; Fritzinger, B.K.; Rose, J.E. In *Polymers in Medicine* Chielini, E.; Giusti, P., Eds.; Plenum Publishing Co., New York, p. 169 (1983).
19. Fabris, H.J. In *Advances in Urethane Science and Technology.* Technomic Publishing Co., Westport, CT, **Vol 4** (1976).
20. Heller, J.; Sparer, R.V.; Zentner, G.M. In *Biodegradable Polymers as Drug Delivery Systems* Chasin, M.; Langer, R. eds. Marcel Dekker, New York, p. 121 (1990).
21. Nguyen, T.H.; Himmelstein, K.J.; Higuchi, T. *Int. J. Pharmaceutics* **25**, 1 (1985).
22. Heller, J.; Penhale, D.W.H.; Helwing, R.F.; Fritzinger, B.K. *Polymer Eng. Sci.* **21**, 727 (1981).
23. Heller, J. In *Polymers in Medicine II.* Chielini, E.; Giusti, P.; Migliaresi, C.; Nicholais, L.L., eds. Plenum Press, New York, p. 357 (1986).
24. Shih, C.; Himmelstein, K.J. *Biomaterials* **5**, 237 (1984).
25. Heller, J.J. *Controlled Release* **2**, 167 (1985).
26. Soloway, M.S. *Cancer Res.* **37**, 2918 (1977).
27. Gressel, M.G.; Parrish, R.K.; Folberg, R. *Ophthalmology* **91**, 378 (1984).
28. Maa, Y.F.; Heller, J.J. *Controlled Release* **13**, 11 (1990).
29. Martin, W.R.; Jasinski, D.R.; Mansky, P.A. *Arch. Gen. Psychiatry* **28**, 784 (1973).
30. Maa, Y.F.; Heller, J. *J. Controlled Release* **14**, 21 (1990).

31. Bradley, G.W.; McKenna, G.B.; Dunn, H.K.; Daniels, A.U.; Statton, W.O. *J. Bone Jt. Surg.* **61-A,** 866 (1979).

32. Daniels, A.U.; Chang, M.K.O.; Andriano, K.P.; Heller, J. *J. Appl. Biomat.* **1,** 57 (1990).

33. Andriano, K.P.; Daniels, A.U.; Heller, J. *J. Appl. Biomat.* **3,** 206 (1992).

34. Daniels, A.U.; Andriano, K.P.; Heller, J. *J. Appl. Biomat.* **3,** xx (1992).

35. Heller, J.; Ng, S.Y.; Fritzinger, B.K.; Roskos, K.V. *Biomaterials* **11,** 235 (1990).

36. Wuthrich, P.; Ng, S.Y.; Fritzinger, B.K.; Roskos, K.V.; Heller, J. *J. Controlled Release* **21,** 191 (1992).

37. Heller, J.; Ng, S.Y.; Fritzinger, B.K. *Macromolecules* **25,** 3362 (1992).

CHAPTER 3

Polyanhydrides as Carriers of Drugs

Abraham J. Domb, Shimon Amselem, Robert Langer, and Manoj Maniar

3.1   Introduction

The development of biodegradable polymers capable of releasing physically entrapped drugs by well-defined kinetics is a major area of research and has been receiving increasing attention. In the past, research focussed primarily on the polymers based on lactic and glycolic acids because of the innocuous nature of the degradation products.

Abraham J. Domb, Hebrew University of Jerusalem, School of Pharmacy, Faculty of Medicine, Jerusalem 91120, Israel, and Drug Delivery Laboratories, Nova Pharmaceutical Corporation, Baltimore, Maryland 21224, U.S.A.; Shimon Amselem, Manoj Maniar, Drug Delivery Laboratories, Nova Pharmaceutical Corporation, Baltimore, Maryland 21224, U.S.A.; and Robert Langer, Massachusetts Institute of Technology, Department of Chemical Engineering, Cambridge, Massachusetts 02139, U.S.A.

However, the release kinetics from these polymers are not very well defined because the polymers undergo bulk erosion. Also, it is very difficult to formulate a delivery system, based on these polymers, for water-labile low molecular weight drugs and high molecular weight peptides and proteins because of the penetration of water into the bulk of the matrix. The predominant mode of release from these polymers is by diffusion and there is a very weak relationship between the rate of polymer hydrolysis and the rate of drug release.

The drawbacks of the bulk eroding polymers led researchers to develop novel polymers in which the release rate of the drug would coincide with the degradation rate of the polymer. At present, the increasing trend is in the development of genetically engineered drugs that are essentially polypeptides. Thus, the development of surface eroding polymers is becoming crucial due to the recognition that high molecular weight polypeptides do not diffuse from dense, hydrophobic polymers at useful rates.

About 10 years ago, it was recognized that polyanhydrides could be developed as a truly surface eroding biodegradable drug delivery system. Since then, tremendous effort has been focussed on developing new polyanhydride-based polymers. Because of the continued research, the chemistry of polyanhydrides has been considerably developed and a large selection of polymers are available for a variety of biomedical applications. Also, significant advances have been achieved in the understanding of the release mechanism of the drug and degradation of the polymer.

Numerous toxicity studies have been conducted with two classes of polyanhydrides and extensive biocompatibility studies have been done on many polyanhydrides. The two polyanhydrides poly(carboxyphenoxy propane-sebacic acid) and poly(fatty acid dimer-sebacic acid) are at advanced stages of clinical development, and the drug delivery systems based on these polymers have been scaled up and methods have been developed to produce them economically.

This chapter reviews the recent developments in the synthesis of polyanhydrides, the variety of clinically suitable applications, and the biocompatibility of the polymers.

3.2   Polymer Synthesis

Polyanhydrides are the polymeric products of the dehydration of diacid molecules. Methods used for the synthesis of polyanhydrides are listed in Table 3.1. The most effective and widely used method for the synthesis of polyanhydrides is melt poly-condensation [1–3]. The polymerization is carried out in two steps. In the first step a purified diacid monomer is reacted with excess acetic anhydride to form an acetic acid mixed anhydride oligomer that is then polymerized under vacuum at 180°C to yield a high molecular weight polymer [3]. Simple diacids of the structure $HOOC-(CH_2)_x-COOH$, such as sebacic acid (SA), form polymers of average molecular weight of over 100,000. Glutaric and succinic acids do not form polymers but rather cyclic monomers. Adipic acid forms a polymer of molecular weight 14,000 but the polymer contains an appreciable amount of the cyclic adipic anhydride. The polymerization reaction can

Table 3.1   Methods for the Synthesis of Polyanhydrides

A. $HOOC-R-COOH + (CH_3-CO)_2O \longrightarrow CH_3-CO-(O-CO-R-CO-)_mO-CO-CH_3$

   (I) $\xrightarrow{\text{180°C/>1 mm Hg}} CH_3-CO-(O-CO-R-CO-)_nO-CO-CH_3$

   $m = 1\text{--}20; \quad n = 100\text{--}1000$

B. $x \text{ HOOC-R-COOH} + y \text{ Cl-CO-R}'-\text{CO-Cl} \xrightarrow{\text{Et}_3\text{N/0°C}}$

   $[(OC-R-COO)_x(OC-R-COO)_y]n + \text{Et}_3\text{N:HCl}$

C. $n \text{ Cl-CO-R}'-\text{CO-Cl} + n/2 \text{ H}_2\text{O} \longrightarrow [CO-R'-COO]n + n \text{ HCl}$

D. $HOOC-R-COOH \xrightarrow{\text{dehydrat. agent}} [CO-R'-COO]n$

E. $HOOC-R-COOH + \text{Cl-CO-Cl} + \text{Et}_3\text{N} \xrightarrow{\text{0°C}} [CO-R'-COO]n + \text{Et}_3\text{N:HCl} + CO_2$

   R = aliphatic, aromatic, and a heterocyclic organic residue

be catalyzed to produce higher molecular weight polymers in the range of 300,000 within 30 min of polymerization at 180°C [3]. Cadmium acetate, $\text{ZnEt}_2\text{-H}_2\text{O}$ 1:1, and calcium carbonate were the most effective. The acid catalyst *p*-toluenesulfonic acid did not show any effect whereas the basic catalyst 4-dimethylaminopyridine caused a decrease in molecular weight. The reaction between diacid monomers with stoichiometric amounts of diacid chlorides in the presence of acid acceptors yielded polymers of moderate molecular weights [4]. The polymerization of dicarboxylic acid monomers with phosgene or diphosgene and crosslinked poly(4-vinylpyridine), as acid acceptor, at 0°C yielded pure polymers with a molecular weight up to 13,950 in one step. Similarly, the use of an appropriate solvent where the polymer is exclusively soluble but the corresponding byproduct (e.g., $\text{Et}_3\text{N:HCl}$) is insoluble with phosgene as coupling agent yielded pure polymers in one step. Under these conditions polymerization of SA gave the best results in *N, N*-dimethylformamide and in toluene. This method is useful for the synthesis of heat-sensitive monomers that may be destroyed in the melt condensation process.

3.3   Polymer Structures

Hundreds of polymers have been reported since the discovery of polyanhydrides [1,2]. The newly synthesized polymers are listed in Table 3.2. These polymers were designed for use as carriers of drugs and were synthesized from nontoxic monomers. Polymers were also synthesized from unsaturated monomers such as fumaric acid (FA) and acetylenedicarboxylic acid. The unsaturated homopolymers were crystalline and insoluble in common organic solvents. However, the copolymers with aliphatic diacids were less crystalline and were soluble in chlorinated hydrocarbons. The copolymers of

Table 3.2   Representative New Polyanhydrides

Structure	References
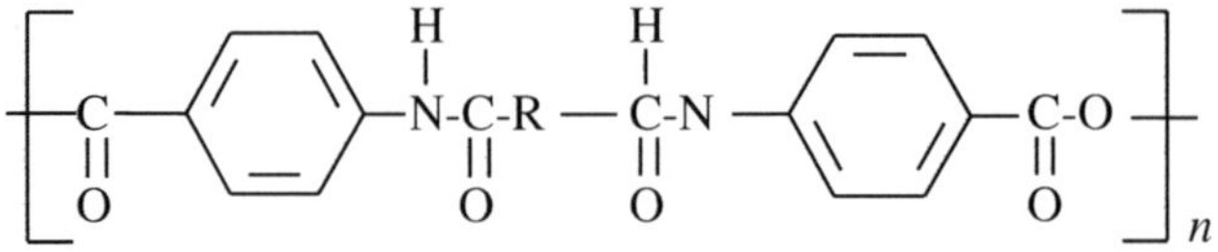	[5]
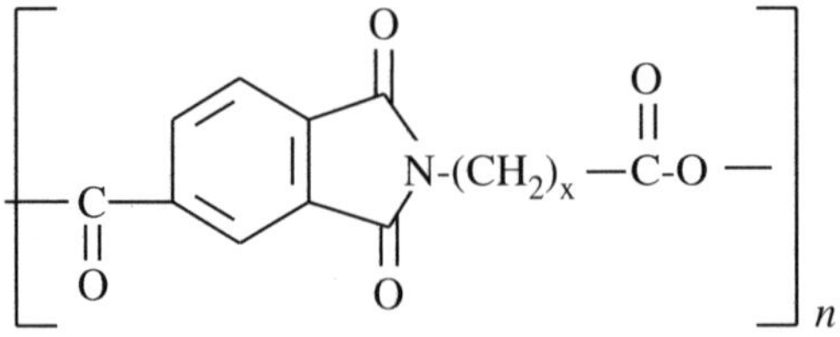	[6]
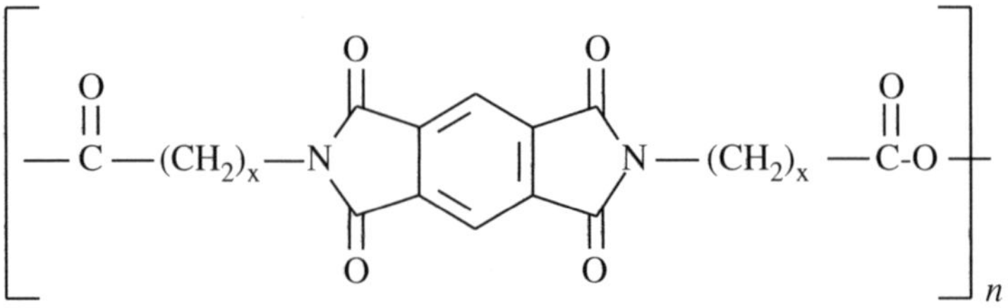	[7]
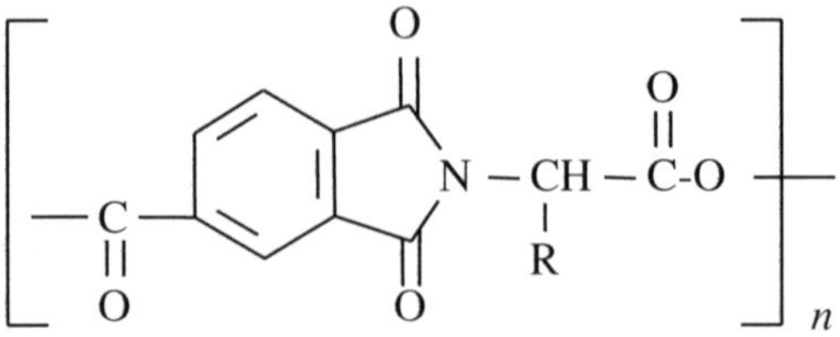	[7–9]
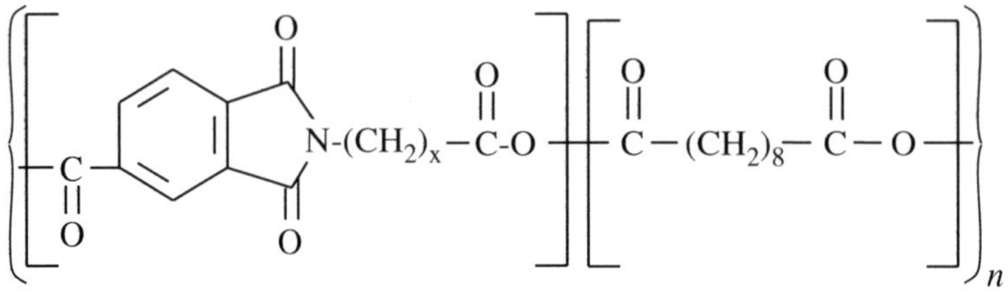	[7–9]
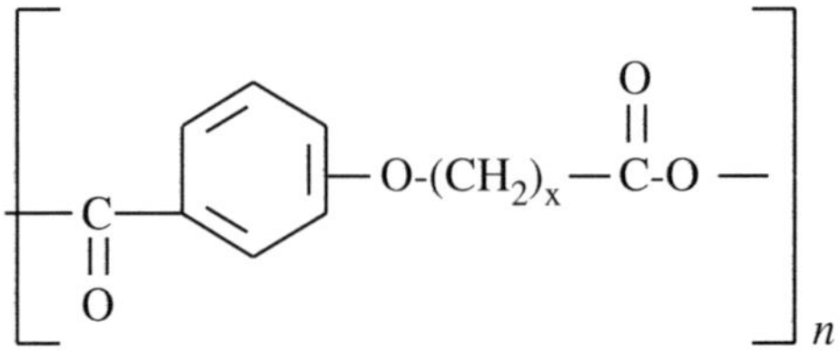	[7–9]

Continued

Table 3.2   Continued

Structure	References

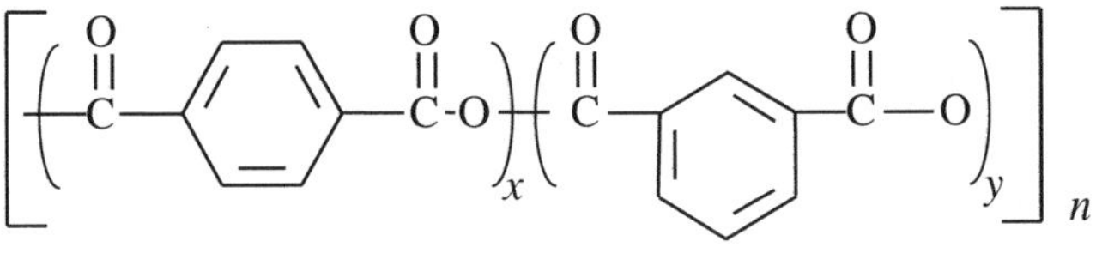

TA   IPA   [11]

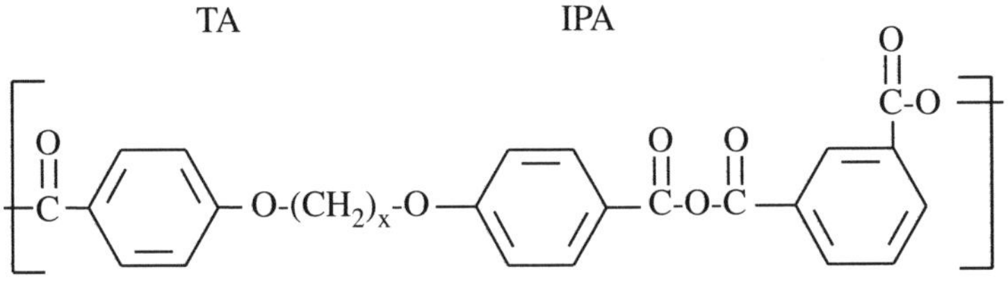

[11]

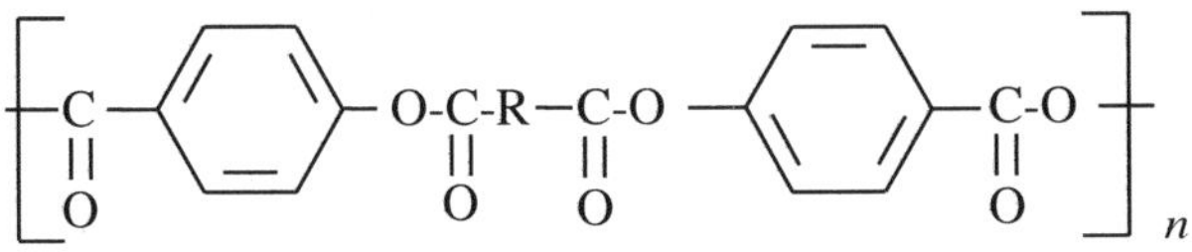

[12]

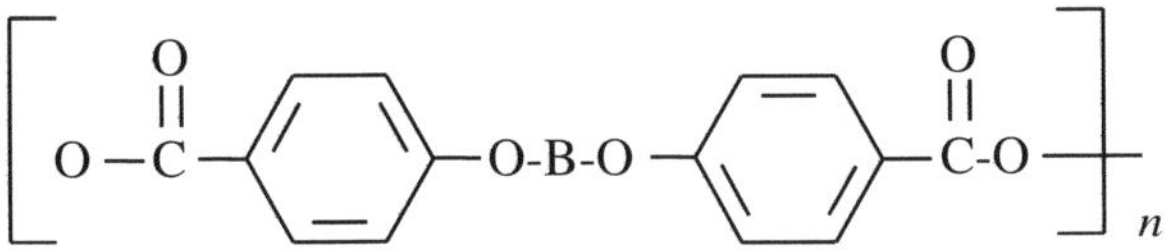

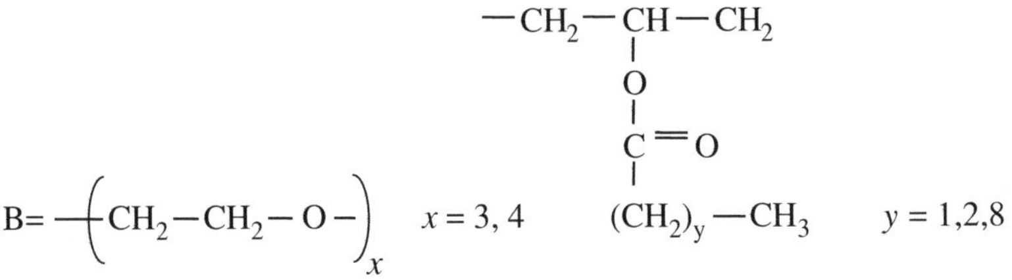

[13]

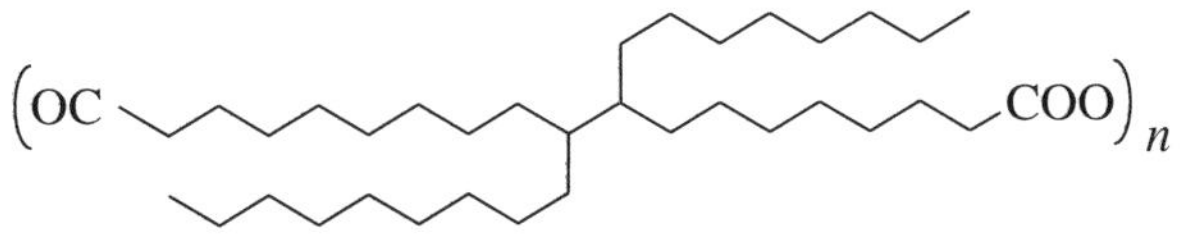

$$-CH_2-CH-CH_2$$
$$\overset{|}{O}$$
$$\overset{|}{C}=O$$
$$\overset{|}{(CH_2)_y}-CH_3 \quad y = 1,2,8$$

$$B = -\left(CH_2-CH_2-O-\right)_x \quad x = 3,4$$

[14]

FA and SA degraded by surface erosion within 2–15 days. The degradation products were the corresponding SA and FA, indicating that the monomers were not altered during the polymerization process.

Several polymers derived from amino acids were synthesized [15]. The amino acids were converted into diacid monomers by reacting the amino groups with an acid-containing moiety. Useful copolymers of common aromatic diacids such as terephthalic acid and isophthalic acid were synthesized by melt condensation [11]. Nonlinear aliphatic dimer and trimer fatty acids were used for the synthesis of polyanhydrides.

Homopolymers of dimer fatty acids are liquids; copolymerization with aliphatic diacids such as SA forms pliable and film forming polyanhydrides, melting within a range of 25–75°C. Films and beads made of these polymer have been used as implantable drug carriers [14,16].

3.4   Polymer Characterization and Properties

Homopolyanhydrides of aliphatic or aromatic diacid monomers possess some degree of crystallinity [17,18]. Copolymers of aromatic and aliphatic monomers show a decrease in crystallinity that is dependent on the copolymer composition. A quantitative analysis of the effect of polymer composition on crystallinity was performed [17,18]. Polymers based on SA, p-carboxyphenoxypropane (CPP), p-carboxyphenoxyhexane (CPH), and FA were investigated. The crystallinity was determined by X-ray diffraction, a combination of X-ray and DSC, and data generated from ^{1}H nuclear magnetic resonance (NMR) spectroscopy using the Flory equilibrium theory [17]. Homopolyan-hydrides of aromatic and aliphatic diacids were crystalline (> 50% crystallinity). Copolymers possess a high degree of crystallinity at high molar ratios of either aliphatic or aromatic diacids. The glass transition, T_g; the melting point, T_m; and the heat of fusion were determined by DSC. The crystallinity, X_c, was calculated from the DSC and X-ray powder diffraction. Heat of fusion values for the polymers demonstrated a sharp decrease as CPP is added to SA or vice versa. The trend of decreasing crystallinity, as one monomer is added, appeared using the X-ray or DSC methods. The decrease in crystallinity is a direct result of the random presence of other units in the polymer chain. A detailed analysis of the copolymers of SA with the aromatic and unsaturated monomers, CPP, CPH, FA, and trimellitic-amino acid derivative was reported [8]. Copolymers with high ratios of SA and CPP, TMA-Gly, or CPH were crystalline whereas copolymers with equal ratios of SA and CPP or CPH were amorphous. The poly(FA–SA) series displayed high crystallinity regardless of comonomer ratio.

The melting points of a large number of polyanhydrides have been reported [1,2]. Aliphatic polymers with a straight chain melt at temperatures below 100°C. The replacement of methylene groups with sulfur atoms yielded a small decrease in the melting point [1]. The introduction of sulfon groups increases the melting point consid-erably. Aromatic polymers melt at a much higher temperature, in general between 150 and 400°C (Table 3.1). With the introduction of aliphatic substitutes in the aromatic monomers a decrease in the melting point was found. As a result, a large number of polyanhydrides with aromatic rings were synthesized [1,2]. The melting point (MP) of copolymers of aromatic and aliphatic diacids is characterized by a minimum at a certain composition that is lower than that of the respective homopolymers. Similar results were obtained in aromatic copolymers as seen in Figure 3.1. The melting point of polyterephthalic acid was reduced from 400°C to about 120°C by copolymerization with 20% isophthalic acid or FA [11].

Aliphatic polyanhydrides are soluble in chlorinated hydrocarbons and in tetrahydro-furan (THF). Aromatic homopolymers are insoluble in common organic solvents; they

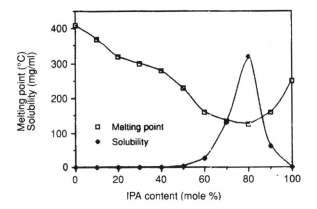

Figure 3.1   Solubility in dichloromethane and melting point of carboxyphenoxypropane (CPP) and isophthallic acid (IPA) copolymers.

are soluble in the more polar solvents such as *m*-cresol and *N,N*-dimethylformamide. Copolymers of aromatic with aliphatic or other aromatic monomer are soluble in chloroform or methylene chloride [3].

The molecular weights of polyanhydrides were determined by viscosity measurements and gel permeation chromatography (GPC) [17]. Attempts to determine the molecular weight using vapor pressure osmometry (VPO) were not successful because of a change in the polymer molecular weight during the experiment via a depolymerization process. The weight average (Mw) molecular weight of polyanhydrides ranges from 5000 to 300,000 with a polydispersity of 2–15 which increases with the increase in Mw. The intrinsic viscosity [η] increases with the increase in Mw. Aliphatic diacid monomers produce high molecular weight polymers of up to Mw = 300,000 by the melt condensation method. Aromatic monomers produce, in general, lower molecular weight polymers, in the range of 5000–40,000. Copolymers of aliphatic and aromatic monomers produce polymers with intermediate molecular weights, where the molecular weight decreases with the increase in aromatic content.

The Mark–Houwink relationship for poly(CPP–SA) was calculated from the viscosity data and the Mw values as determined by universal calibration of the GPC data using polystyrene standards [17].

$$[\eta]_{CHCl_3}^{23°C} = 3.88 \times 10^{-7} \times Mw^{0.658}$$

Anhydrides present characteristic peaks in the infrared (IR) and Raman spectra. In general, aliphatic polymers absorb at 1740 and 1810 cm^{-1} and aromatic polymers at 1720 and 1780 cm^{-1}. A typical IR spectrum of aliphatic and aromatic copolymers that contain aliphatic and aromatic anhydride bonds may present three distinct peaks where the aromatic peak is shown at 1780 cm^{-1} and the peaks at 1720–1740 cm^{-1} in general overlap. The presence of carboxylic acid groups in the polymer can be determined from the presence of a peak at 1700 cm^{-1}. The degradation of polyanhydrides can be followed by IR from the ratio between the anhydride peak at 1810 and the acid peak at 1700 cm^{-1}. The Raman spectra for a variety of polyanhydrides were studied by Davies et al. [19,20]. The Raman spectra of polyanhydrides were similar to the IR spectra for the same compounds. Polyanhydrides show two distinctive carbonyl Raman bands corresponding to the symmetric and asymmetric vibrations of the carbonyl groups with the separation of the pair between 50 and 70 cm^{-1}.

Fourier-transform Raman spectroscopy (FTRS) was used to characterize an homologous series of aliphatic polyanhydrides, polycarboxyphenoxyalkanes, and copolymers of CPP and SA. All anhydrides show two diagnostic carbonyl bands. The aliphatic polymers have the carbonyl pairing at 1803/1739 cm^{-1}, and the aromatic polymers have the band pair at 1764 and 1712 cm^{-1}. All the homo- and copolymers show methylene bands due to deformation, stretching, rocking, and twisting, and the spectra for the aromatic polyanhydrides such as PCPP also show diagnostic benzene para-substitution bands. It is possible to differentiate between aromatic and aliphatic anhydrides bonding and in conjunction with other diagnostic bands to monitor the change in individual monomer composition within a copolymer mixture.

FTRS was used to study the hydrolytic degradation of polyanhydrides [19]. Poly (sebacic acid) PSA rods exposed to water for 15 days were analyzed daily by FTRS. The carbonyl anhydride band pair (1803/1739 cm^{-1}) diminishes in intensity from days zero to 15 with the emergence of the complementary acid carbonyl band (1640 cm^{-1}) which increases in intensity over the same period. Similarly, the increase in the intensity of the C–C deformation at 907 cm^{-1} with hydrolysis reflects the increased freedom of the methylene chain in the low molecular weight oligomers.

The morphology of polyanhydrides was studied by scanning electron microscopy (SEM) to elucidate the mechanism of polymer degradation and drug release from polyanhydrides [21]. Microspheres prepared by three different techniques—solvent removal, solvent evaporation, and melt encapsulation—were analyzed by SEM. The degradation process in buffer was followed by SEM. Microspheres showed distinctive morphological characteristics induced by the fabrication method [21].

The surface chemical structure of aliphatic polyanhydride films has been examined using time-of-flight secondary ion mass spectroscopy (ToF–SIMS) and X-ray photoelectron spectroscopy (XPS) [22,23]. The XPS data confirmed the purity of the surface, and the experimental surface elemental ratios were in good agreement with the known stoichiometry of the polyanhydrides. The ToF–SIMS spectra of the polyanhydrides are shown to reflect the polymer structure. The SIMS data conform to a systematic fragmentation, in both negative- and positive-ion SIMS spectra, occurring throughout the entire series of the polyanhydrides examined. Radical cations were observed in the positive-ion spectra. The lower mass ranges of the negative-ion SIMS spectra contain ions at m/z 12, 13, 16, 17, 24, 25, 41, 43, and 45 that may be assigned to C^-, CH^-, O^-, OH^-, C_2^-, C_2H^-, C_2HO^-, $C_2H_3O^-$, and CHO_2^-. The ion at m/z 71 may arise from the fragmentation of the anhydride unit, CH_2–CHCOO– and it was seen for all polyanhydrides. At higher mass a general fragmentation pattern was observed, and the major ions were MH^+, M^+, $M\text{-}CO_2^{\pm}$, and $M\text{-}OH^+/MH\text{-}H_2O^{\pm}$.

The combined use of ToF–SIMS and XPS is shown to provide a detailed insight into the interfacial chemical structure of polyanhydrides. Little data were reported on the mechanical properties of polyanhydrides [1–3]. Fibers of poly(bis-carboxyphenoxy ethane) showed a tensile strength of 4000 kg/cm^2 with an elongation of 17.2% and a Young modulus of 50,600 kg/cm^2 [24]. The effects of copolymer composition and molecular weight on the mechanical strength of copolymers of SA and CPP were studied [3]. Increasing either the CPP content in the copolymer or the molecular weight increases the tensile strength. The elongation at break of films made of these polymers ranged from 17 to 23%.

Poly(imide–anhydrides) possess good mechanical properties as seen in Table 3.3 [9]. The copolymer of SA with trimellitic imide diacids of the general structure showed a tensile strength of up to 3531 kg/cm^2 with elongation at break of 339% for a 50:50 copolymer. The imide–anhydride homopolymer had poor mechanical properties. These polymers degrade rapidly within 2 weeks to the respective diacids.

3.5   Applications

The release of drug from a matrix type polymeric carrier system is primarily controlled by the diffusion of drug through the polymer and the dissolution rate of the drug. An additional factor that governs the release rate of the drug, in the case of a biodegradable polymeric system, is the erosion rate of the polymer. In case of biodegradable polyanhydrides it was hypothesized that the extreme hydrophobicity of the polymer would prevent the penetration of water into the bulk of the matrix and thus the release of the entrapped drug would primarily be due to the degradation of the polymer. To test this hypothesis, polyanhydride matrices were prepared by compression molding containing p-nitroaniline, a highly water-soluble marker. The polymer used in formulating the matrix was poly(carboxyphenoxyacetic acid), P(CPA). The cumulative release of p-nitroaniline and degradation of P(CPA) were measured by absorbance at 380 and 235 nm, respectively. The release and degradation profile is shown in Figure 3.2, and it clearly demonstrates that the release of drug and the degradation of the polymer occurs simultaneously [25]. The surface eroding property was established for a variety of polyanhydrides and was found to be a function of monomer composition. Depending on the monomers the biodegradation period of the polymers ranged from days to years.

Table 3.3   Mechanical and Thermal Properties of Poly(imide-coanhydride) Fibers Containing Imide Units of Various Spacer Lengths (*n*=1,2,3,4,5,10,11) Copolymerized Either with Sebacic Acid (SA) or Carboxyphenoxyhexane (CPH)[a]

TMA-n		SA		Elongation	Tensile strength	T_g
n	%	(%)	mol wt	at break (%)	(kg/cm^2)	(° C)
1	30	70	<5,000	13	98	12
1	39	61	<5,000	143	693	13.9
1	50	50	<5,000	44	205	29.3
2	50	50	21,390	298	1872	26
3	50	50	18,060	339	3531	29.9
4	100	0	14,600	54	1202	81
5	100	0	12,350	21	1890	63
10	53	47	17,760	37	1069	9.3
10	100	0	10,680	7	96	21.4
11	100	0	19,250	23	71	25
2	33	67% CPH	17,850	143	1731	38.8
2	51	49% CPH	20,210	99	659	36
2	69	31% CPH	14,850	226	897	32.4

[a] Data taken from ref. [9].

The features of the polyanhydride system that distinguish it from other biodegradable polymeric systems are surface erodibility; protection of unstable drug from the surrounding environment; and the availability in a variety of forms such as sheets, rods, microspheres, or wafers. Thus, this has led many investigators to evaluate the carrier system for delivering a wide array of drugs. Representative examples of several of these are described below.

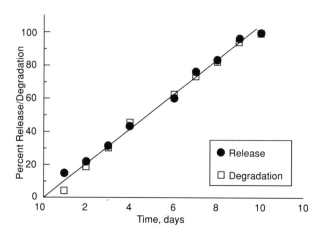

Figure 3.2   Release of *p*-nitroaniline and the degradation of poly(carboxyphenoxyacetic acid) versus time.

3.5.1   Gliadel (BCNU Brain Implant)

The chemotherapeutic treatment of brain tumors has not achieved much success. The primary reason is the presence of the blood–brain barrier, restricting the access of the systemically administered drugs to the diseased site. Local administration of the drugs, directly at the tumor site, would bypass the blood–brain barrier and thus has the potential to increase the efficacy and to decrease the systemic toxicity of these agents. A novel method of delivering cancer chemotherapeutic agents to the brain was developed by incorporating carmustine (BCNU) in P(CPP–SA); 20:80 and compressing the solid solution into wafers. Each wafer contained 3.8% carmustine and measured 1.4 cm in diameter with a thickness of 1 mm. These wafers are then implanted at the site of tumor inside the brain. This dosage form is particularly useful for the treatment of glioblastoma multiforme, a universally fatal form of brain cancer. The objective is to deliver very high local concentrations of a toxic chemotherapeutic agent directly to the surface of the tumor remaining after surgical debulking has taken place. Systemic concentrations would be expected to remain very low because of the short half-life of carmustine in vivo. The formulation, production, sterilization, and stability of the wafers has been described by Chasin et al. [26].

A quantitative autoradiography study demonstrated that delivering carmustine (BCNU) using Gliadel resulted in local concentrations of BCNU above 6 mM, about 60,000 times higher than is necessary to kill susceptible tumor lines in vitro. These high levels of BCNU were sustained for 3 weeks in the brain, surrounding the implant site. An efficacy study in rats showed that the systemic administration of BCNU extended the lives of rats in which brain tumors had been transplanted for 4 days prior to treatment from 11 to 27 days. In the same model, the Biodel delivery system extended the lives of rats to over 62 days. In addition, several animals treated with the delivery system survived over 120 days and at necropsy were found to be free of tumor [27]. The safety and efficacy of these wafers in the preclinical studies was outstanding and has led to the evaluation of the wafers in humans.

In a pursuit of developing next generation product for treating brain tumors, carboplatin-containing devices were prepared using a new family of polyanhydride, P(FAD–SA). This polyanhydride is a copolyanhydride of fatty acid dimer and SA. In vitro the devices containing 5% carboplatin showed release of drug up to 5 weeks. In vivo the efficacy of the devices was evaluated against F-98 rat glioma. The rats bearing tumors and receiving no therapy died within 16 days of implantation; the average life span of the animals receiving the polymer containing carboplatin was 56 days [28].

3.5.2   Insulin

A delivery system in the form of microspheres of P(CPP–SA); 50:50 containing insulin was developed to evaluate the in vitro release of insulin and its in vivo efficacy in rats. The loading of insulin was 15% in these microspheres and its size varied from 850 to

1000 μm in diameter. The in vitro release profile of insulin is shown in Figure 3.3. Most of the insulin is released in the first 2 days but a significant amount continues to be released over the next 4–5 days [29,30].

In vivo efficacy was evaluated by inducing diabetes in two groups of rats. They were administered 65 mg/kg of streptozotocin in 0.1 *M* citrate buffer, pH 4.5. Within several days, the animals had become diabetic, as evidenced by blood glucose levels of 400 mg/dl. One group of rats was then injected subcutaneously with 40–50 mg of 15% insulin-loaded microspheres. A second group of diabetic rats did not receive any treatment and a third group of normal rats served as the control. The results of the treatment are shown in Figure 3.4. The injection of insulin-containing microspheres decreased the blood glucose level equivalent to the levels of the normal rats for up to 5 days [31].

3.5.3   Bovine Somatotropin

Bovine somatotropin is a highly water-soluble and unstable protein with a tendency for self-aggregation and inactivation. The protein was successfully incorporated into the polyanhydride matrix of P(CPH). Bovine somatotropin was colyophilized with sucrose, dry mixed with finely powdered P(CPH), and compression molded into thin wafers. The in vitro release of bovine somatotropin is shown in Figure 3.5. It is evident that the release of bovine somatotropin was well controlled over a period of 1 month. The released protein was analyzed by high-performance liquid chromatography assay which is specific for the monomeric species and the activity of the released protein was confirmed by radioimmunoassay [32].

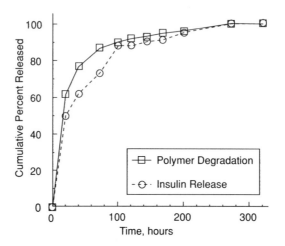

Figure 3.3   In vitro release of insulin from microspheres of poly(carboxyphenoxypropane–sebacic acid); 50:50.

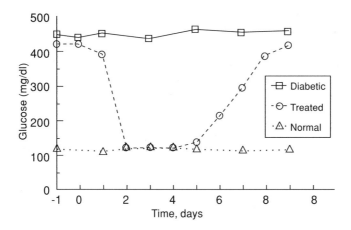

Figure 3.4   Effect of insulin released from microspheres of poly(carboxyphenoxypropane–sebacic acid); 50:50 injected subcutaneously in streptozotocin-diabetic rats.

3.5.4   Angiogenesis Inhibitors

Several drugs are known to cause inhibition of angiogenesis. The inhibitors of angio-genesis, heparin and cortisone, were incorporated into a series of polyanhydrides. The cortisone acetate was microencapsulated using poly(terephthalic acid), poly(ter-ephthalic–SA), 50:50, and poly(carboxyphenoxypropane–SA): 50:50 polymers. The

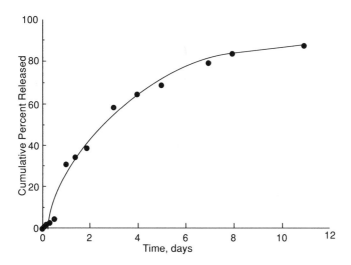

Figure 3.5   Release of bovine growth hormone from compression molded discs of poly(car-boxyphenoxyhexane).

microcapsules were prepared by an interfacial condensation of diacyl chloride in methylene chloride with the appropriate dicarboxylic acid in water. The release of cortisone acetate from the microcapsules is shown in Figure 3.6. The difference in release rates from the various polymers is due to the difference in their degradation rates. The formulation containing heparin and cortisone delivered in vivo using the polyanhydride was shown to prevent new blood vessel growth for over 3 weeks (Figure 3.7), following the implantation of the VX2 carcinoma into rabbit cornea [33].

3.5.5   Local Immunosuppression

The most common drugs used to prevent the rejection of the transplanted organs such as heart, liver, and kidney are dexamethasone, cyclosporine, and azathioprine. Systemic administration of such immunosuppressants frequently results in severe side effects. Also, the phenomenon of cardiac rejection has been recognized as an individual cellular event with extreme variability noted in rejection and myocyte destruction between adjacent cells. Hence, it would be beneficial to have a delivery system that would locally release the drug at the site of the transplanted organ. To evaluate this approach, polyanhydride films containing dexamethasone were formulated and implanted locally around the transplanted heart at the time of rat heterotopic heart transplants. Rejection was determined by the lack of transplanted heart contractions. Local administration of dexamethasone significantly delayed time to allograft rejection. The mean survival time for 0.2% loaded dexamethasone/polyanhydride films was 10.2 days when compared to 6.9 days in control animals [34]. Figure 3.8 shows an in vitro release profile of dexamethasone from the matrix of P(CPP:SA), indicating a variety of formulations are possible permitting an array of steroid drug dosages and release rates for later in vivo studies.

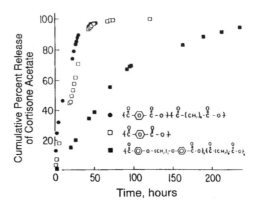

Figure 3.6   Release of cortisone acetate from 10% loaded microspheres of various polyanhydrides.

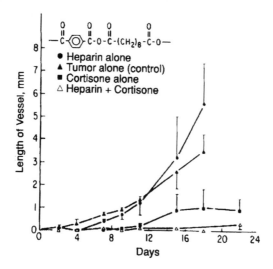

Figure 3.7   Effect of angiogenesis inhibitors on blood vessel growth.

3.5.6   Craniofacial Growth

Polyanhydride microspheres containing L-glutamate were used to study the effects of long-term glutamic acid stimulation of trigeminal motoneurons. The study was undertaken to determine the role of glutamate in possible growth disorders of the craniofacial skeleton. Rats implanted with glutamate-containing microspheres showed pronounced skeletal changes in the snout region, showing that sustained release of glutamic acid in vivo can significantly affect the developing skeleton in growing rats [35].

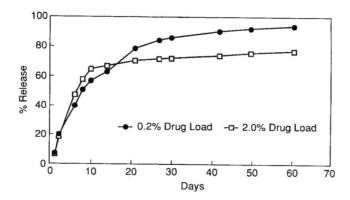

Figure 3.8   In vitro release of dexamethasone from poly(carboxyphenoxypropane–sebacic acid); 40:60.

3.5.7   Osteomyelitis

A polyanhydride-based drug delivery system, the Septacin implant, is under development for the treatment of chronic osteomyelitis. The current treatment for chronic osteomyelitis is surgical removal of dead and infected tissue followed by high-dose, intravenous antibiotic therapy for several weeks, and oral antibiotic therapy for up to 2 years. A limitation of the current therapy is that it is difficult to produce sufficiently high local concentrations of antibiotics at the site of the infection to effectively treat the disease [36,37].

Septacin implant is manufactured as a flexible chain of beads, consists of a copolymer of fatty acid dimer (FAD) and SA, and loaded with gentamicin. The implants were tested in vivo in a rabbit osteomyelitis model. A total of 75 rabbits were divided into several groups as follows:

1. Debridement and treatment with beads containing 20% gentamicin sulfate
2. Debridement and treatment with beads containing 10% gentamicin sulfate
3. IM injection of gentamicin sulfate
4. Debridement of infected bone, no treatment
5. Control

The results of the study are shown in Figure 3.9. As evident from the figure, the 10% and 20% gentamicin-loaded beads in conjunction with debridement produced a significantly higher rate of success in treating surgically induced osteomyelitis [38]. Thus, the beads combined with adequate debridement, appear to be a clinically useful delivery system that may be used in the treatment of osteomyelitis and other soft tissue infection.

3.6   Biocompatibility and Toxicity

3.6.1   General

The term "biocompatibility" has been defined as the effect that a synthetic material has on the biological environment into which it is introduced, together with any synergistic effect that environment has on the polymeric material. Both will affect the efficiency of performance of the material in its selected function, in our case to operate as a drug carrier system [39].

When foreign materials or macromolecules come in contact with a living medium, many negative biological responses such as toxicity, immunogenicity, carcinogenicity, mutagenicity, and teratogenicity can result. Biocompatibility assessment of a polymeric material includes adequate testings for these undesired responses. Since

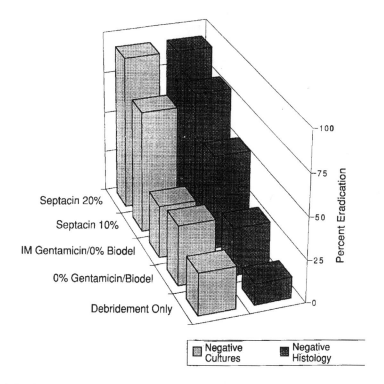

Figure 3.9   Comparison of debridement and Septacin beads loaded with 20% gentamicin and 10% gentamicin on treatment of osteomyelitis in rabbits.

biocompatibility usually results from a number of compromises, bioacceptability can also be used as an alternative term for the combination of biological responses to be evaluated for the eventual application of new polymeric materials [40].

Biocompatibility describes the surface interactions between polymer and biologic tissue, and thus is considered a surface phenomenon, and it is affected by the physicochemical properties of the polymer, design features of the medical device, and the physiologic environment [41].

To evaluate the biocompatibility and toxicology of any polymeric material, four general categories of examinations should be performed: (1) systemic toxic responses; (2) local tissue responses; (3) carcinogenic, teratogenic, and mutagenic responses; and (4) allergic responses [42]. Table 3.4 summarizes the specific tests included in these four categories that should be carried out before a polymeric material can be considered biocompatible or bioacceptable and safe for use in biomedical applications in the form of implants, devices, or delivery systems. More details about the different assays, methodologies, and biological responses mentioned in Table 3.4 can be found in the references cited.

Table 3.4   Biocompatibility and Toxicology Tests for Evaluation of Polymeric Biomaterials

Test	Methodology	Responses	References
Acute toxicity	High dose administration LD_{50} (median lethal dose) estimation	Percent survival as function of time. Toxicity rating from super toxic to practically nontoxic, depending on dose used/animal body weight	[43–45]
Tissue culture	Agar overlay	Loss of color or stained cells, and extent of lysis within the decolorized zone. Indexes scored on a scale of 0 to 5. Response index = zone index/lysis index. Lysis index related to cytotoxicity. Zone index related to leachability-diffusibility	[46–49]
Cell growth inhibition test	Incubation of polymeric extracts with mouse fibroblast cell tissue cultures	Percent inhibition of cell growth (ICG)	[46, 48, 49]
Hemolysis test	Incubation of polymeric extracts with whole oxalated rabbit blood	Percent hemolysis	[46, 49]
Intramuscular rabbit implant test	IM injection into paravertebral muscles	Inflammatory characteristics: local reaction, histopathology evaluation rating. Macroscopic reactions scored from 0 to 5	[46, 48–50]
Rabbit intracutaneous test	Dorsal surface injection	Irritant responses of polymeric extracts by observation of injection sites. Reaction scored from 0 (negative) to 5 (positive)	[46, 48, 50]
Mutagenic potential	Ames mutagenicity test. Quantitative forward mutation assay	Mutagens detection Resistance to 8-azaguanine (genetic marker)	[51] [52]
Teratogenic potential	Incubation of polymeric extracts with ascitic mouse ovarian tumor cells	Ability to inhibit attachment of ascites tumor cells to plastic surfaces coated with Concanavalin A	[53]
Carcinogenic potential	Subcutaneous implantation of polymeric material	Macroscopic observation at autopsy, and histologic examinations of organs	[49, 54–56]

Continued

Table 3.4   Continued

Test	Methodology	Responses	References
Allergenic potential	Intradermal injections in guinea pigs of polymeric extracts with and without Freund's adjuvant, followed by topical challenge applications (polymer extract in cotton patch).	Allergic reaction, irritation, redness. In the basis of percent of animals sensitized, 1 of 5 grades of allergenic potency are assigned ranging from weak (I) to extreme (V).	[49, 57]

3.6.2   Polyanhydrides

In this section we summarize all the biocompatibility and toxicology data obtained so far with polyanhydride materials based on the evaluation tests listed in Table 3.4.

3.6.2.1   Acute Toxicity

An acute toxicity test is generally conducted to determine the adverse effects of single, relatively large doses of a test material in animals. Usually mice or rats are treated one time with a range of dose levels, and over toxic manifestations such as abnormal behavior, change in body weight gains, and deaths are monitored at specific intervals for the next few days or weeks [43]. If deaths occur at more than one dose level, a median lethal dose (LD_{50}) can be estimated. Often, a formulation does not cause deaths at the maximum feasible dosage, and an LD_{50} cannot be determined. In these cases, signs of over-toxicity or exaggerated pharmacologic activity instead of lethality may be used to compare the relative toxicity and bioavailability of the formulation.

The route of administration can greatly affect the toxicity of test materials, and obviously the largest effects and the most rapid responses occur following intravenous injection. However, the route of administration used in acute toxicity studies usually is the same as the intended eventual clinical route for the material to be tested.

A toxicity study to evaluate the potential use of bioerodible polyanhydrides as implants for drug delivery was performed in rats [58]. The polyanhydride copolymer of bis(p-carboxyphenoxy)propane (CPP) and sebacic acid (SA) in a 20:80 molar ratio was chosen for this study, and its systemic and local effects were examined. Polymer implants in the form of discs (1.4 cm diameter, 1 mm thick) were administered subcutaneously at high doses, 800 and 2400 mg/kg rat, for a period of 8 weeks. The doses administered were 40 and 120 times the anticipated human dose, respectively, on the basis of polymer weight per kilogram body weight. Biocompatibility was assessed by a number of means. Through the monitoring of blood clinical chemistry and hematological parameters and postmortem tissue histology, systemic toxicological influences and effects on individual organs were evaluated. In addition, through local implant site histological examinations, the degree to which the degraded polyanhydride was compatible in vivo with surrounding tissue was determined [58].

Prenecropsy examination of all rats in the mentioned study revealed no clinical evidence of induced changes in physical appearance or activity due to implantation of the polyanhydride matrices [58]. Gross examination of the body cavities and tissues at the time of necropsy showed no evidence of polymer implantation associated changes. Microscopic examination of all organ tissues revealed no histomorphological evidence of a systemic induced or related effect of the polyanhydride copolymer implantation [58]. Palpation of the implant sites prior to necropsy showed only occasionally thickened areas in the subcutaneous tissue. Grossly, most subcutaneous implant sites of animals implanted with one and three matrices, respectively, appeared identical to controls, with the exception of occasional irregular beige foci at some implant sites. The beige foci were thought to be hemosiderin pigment; it was usually found within mononuclear macrophages around the peripheral margins of the sites [58].

The primary cellular response to the implanted test substance was an accumulation of mononuclear macrophages with minimal fibrosis. They were either clustered at the periphery of residual polyanhydride, or in those instances where no polymeric material remained, they were scattered throughout the implant site. Their cytoplasm contained fine particulate matter which appeared to be phagocytized polymeric material. There was no indication that the material was cytotoxic. The macrophages appeared morphologically normal. Minimal foreign body giant cells were noted and nonprominent fibrosis and fibroplasia were present in a few implanted rats. Heterophils, lymphocytes, and plasma cells were scarcely encountered and were not a part of the cellular response at the 8-week termination point of the study [58].

Analysis of clinical chemistry and hematology parameters representing tests of various organ systems such as liver and kidney revealed that the implantation of the polymer resulted in no significant differences in the systems under study. The neutrophil and lymphocyte levels measured for all the groups were in the normal range of Sprague–Dawley rats of the age under study. It was concluded that the polyanhydride studied, poly(CPP–SA) 20:80, had relatively minimal tissue irritation properties with no evidence of local or systemic toxicity [58].

3.6.2.2   Tissue Cultures

Response of cellular constituents to polyanhydride materials by observing in vitro growth of mammalian cells on the polymer surface was reported [59]. Endothelial cells and smooth muscle cells were grown on the surface of poly(CPP–SA); 45:55, poly(terephthalic anhydride) (PTA), and copolymer of PTA with sebacic acid P(TA–SA); 50; 50. Endothelial cells were isolated from bovine aortas by collagenase digestion and then cultured using a described procedure [60]. Smooth muscle cells were obtained by explanation of bovine aortic medial tissue [61]. The cells were plated directly onto circular pieces of polymer (1.5 cm^2 × 1 mm) and allowed to attach for 1 h in adequate growth medium. The culture medium was changed daily to avoid accumulation of the acidic polymer degradation products. After 2 weeks in culture the polymers and adherent cells were rinsed and fixed for histological examinations. The bovine aortic endothelial and smooth muscle cells grown in Petri dishes containing the polymer and degradation products displayed normal endothelial morphology of polygonal cells in a monolayer conformation. No evidence of toxic effects of the polymers on cells were found as measured either by cell morphology or ability to proliferate. Deleterious effects such as enlarging of cells, vacuole formation, and loss of contact inhibition growth patterns (for endothelial cells) were not evident [59]. Growth rates (doubling times) of the two cell types were essentially the same on polymer substrates as on the control substrate of polystyrene tissue culture plastic. Histologic examinations, by hematoxylin and eosin staining of cross sections of fixed polymer–cell complexes, revealed that the endothelial cells grew as a flattened monolayer similar in character to their appearance in blood vessels in vivo [59].

3.6.2.3   Local Irritation Tests

In addition to evaluations for untoward systemic effects, polymeric controlled release systems and implants should be evaluated for the potential to produce tissue irritation at the site of administration. Polymeric materials intended for intramuscular or dermal applications are usually tested for adverse local irritation and inflammation by muscle and skin tests (Table 3.4).

The rabbit skin is generally considered at least as sensitive as human skin, and has become the standard animal model for skin irritation tests. The reaction of the skin to the test material is evaluated by periodic gross examination of the treatment site during the testing period, and then by histologic examination of the treated skin at the end of the exposure period [43].

The local muscle irritation potential of parenteral formulations is commonly evaluated in rabbits as well. A dose of the formulation is usually injected into a sacrospinalis muscle, a large muscle located on both sides of the spinal column of rabbits. Animals are sacrificed at the end of the treatment and compared macroscopically and microscopically with controls. For gross observations, macroscopic reactions are scored from 0 to 5 in comparison to controls. Histopathologic evaluations based on inflammatory characteristics are rated from nontoxic to marked toxic reaction [42,43,46].

Another very sensitive area to evaluate inflammatory responses is the cornea. The rabbit cornea is usually used to test the biocompatibility and toxicity of biomaterials because of the easy accessibility for frequent observations, and the transparency and avascularity which allows to differentiate inflammatory reactions such as edema, neovascularization, and cellular infiltration [42].

The local tissue polymer responses of PCPP and PTA were studied through subcutaneous implantation in rats and through corneal implantation in rabbits [59]. No evidence of inflammatory cells in the tissues adjacent to the implant was observed after the subcutaneous implantation in rats over a 6-month period, and only slight tissue encapsulation by layers of fibroblastic cells was noted. Gross postmortem analysis also did not reveal any abnormalities. The polymers did not provoke inflammatory responses in the corneas over a 6-week implantation period, as confirmed by histologic examinations. Clarity of the corneas was maintained, and proliferation of new blood vessels was absent [59].

As previously mentioned, subcutaneous implantation in rats of high doses of a 20:80 mole ratio of CPP–SA anhydride copolymer for up to 8 weeks caused relatively minimal local tissue irritation, with no evidence of local or systemic toxicity. The cellular response at the implant site was chiefly mononuclear cells and minimal fibrosis [58].

Recently, new classes of polyanhydrides have been synthesized and are undergoing extensive preclinical testing, including a wide range of toxicity and biocompatibility studies. Examples of these new polyanhydride materials are polymers of sebacic acid (SA), and 1:1 copolymers of SA with fatty acid dimer [poly(FAD–SA)], fumaric acid [poly(FA–SA)], and 20:80 copolymer of SA with isophthalic acid[poly(IPA–SA)]. These new polyanhydrides were implanted intramuscularly, subcutaneously, and in the cornea of rabbits, and ocular and muscle irritation studies were performed [62]. The results were compared to those obtained with standard materials used in surgery, which have been extensively studied. These materials were Gelfoam (absorbable gelatin sponge), Surgicel (oxidized cellulose absorbable hemostat commonly used in brain surgery), and Vicryl (a synthetic absorbable suture).

The polymers were implanted in rabbits into the paravertebral muscle on one side of the spine and into the subcutaneous tissue on the other side of the spine using a sterile surgical technique. The test pieces measured $10 \times 1 \times 1$ mm. Detailed observations of over-toxicity and signs of bleeding, swelling, or infection of the incision site were conducted daily. At 1, 2, 3, and 4 weeks following implantation, animals were sacrificed and a gross necropsy examination of the tissues surrounding the implant site was conducted using a scoring scale for reaction from 0 (no reaction) to 5 (extreme reaction). All implant sites were fixed in formalin and histopathological examination was performed using the following criteria: inflammatory cell infiltrates, muscle fiber degeneration/necrosis, and fibrosis/encapsulation scored on the same scale [62]. The ocular studies were conducted by inserting a polymer disc of 1 mm diameter into the bottom of a 2-mm pouch created within the rabbit corneal stroma. The corneas were examined using a slit stereomicroscope on weeks 1, 2, and 3 postimplantation. Evidence of inflammatory response in the form of neovascularization, edema, or opacification of the cornea was recorded, and photographs of each cornea exam were taken [62].

No significant clinical signs or abnormalities of the incision sites were observed during the study period (4 weeks). No meaningful changes or differences with respect to grossly observable reactions could be seen at the implantation sites between the various polymer implants tested and the control materials [62]. Several abscesses were observed; however there was no consistent pattern of association with any particular test article.

In the cornea studies, no evidence of inflammatory response was observed with any of the implants at any time. On an average, the bulk of the polymers disappeared completely between 7 and 14 days after the implantation [62].

3.6.2.4   Carcinogenic, Mutagenic, and Teratogenic Potentials

Some polymeric materials and their chemical constituents have been known to be associated with cancer [42,54–56]. The etiology of polymer-induced tumor formation may be related to a chemical carcinogen, a solid-state surface phenomenon, or both [42,55,56]. Studies designed to assess the carcinogenic capability of a test article are generally conducted in mice and rats, usually through subcutaneous implantation. A minimal size of 0.5 cm for discs or squares of solid materials has been reported for the production of solid-state tumors from subcutaneous implantation [63]. Special attention is paid to the detection of masses or suspected tumors during the life of the animals as well as during the subsequent postmortem histopathologic examinations.

Subcutaneous implantation of PCPP in rats over a 6-month period showed no evidence of tumor formation, as evaluated from histologic examinations of host response to the polymer implanted during the study and after necropsy [59].

The mutagenicity and cytotoxicity of P(CPP–SA) and its degradation products were determined using the forward mutation assay in *Salmonella typhimurium* using 8-azaguanine resistance as a genetic marker (Table 3.4) [52]. The results showed no mutagenicity of the polymer or its degradation products either with or without the addition of a mammalian xenobiotic metabolizing system (typically used in mutation assays) [59]. In both cases, the induced mutant fraction was essentially zero, being indistinguishable from the spontaneous background. The positive controls used, benzo[a]pyrene for assays with metabolizing enzymes and 4-nitroquinoline-N-oxide for assays without the metabolizing enzymes, were in both cases 20 times higher in mutant counts. No toxicity was measured in samples without the addition of metabolic enzymes and slight, but not significant, toxicity was measured in samples with the metabolizing system [59].

The teratogenic potential of P(CPP–SA) was assessed by a recently developed in vitro assay in which the attachment efficiency of ascitic mouse ovarian tumor cells to plastic surfaces coated with concavalin A was measured. In general, nonteratogenic materials do not inhibit attachment. The teratogenic tests indicated an average decline in attachment efficiency of 35% ± 3%. The criterion for potential teratogenicity is an inhibition of attachment efficiency > 50%. The degradation products of P(CPP–SA) were therefore considered to have a low teratogenic potential [59].

3.6.2.5   Allergenic Potential

Sensitization studies are conducted to determine whether a material will produce allergic reactions. These are adverse reactions where preexposure to the agent is required before the toxic effect will occur. Initial contact with an allergy-producing material will induce the body to produce antibodies, and subsequent exposure to the same agent will result in antigen–antibody reaction or allergic response [43].

The guinea pig is the animal commonly used for this type of test. The initial phase of the test (induction phase) consists of repeatedly applying the test material to the skin of the animals. Usually a 2-week period follows in which the animals are monitored but not treated. This intermediate phase allows the animals to develop antibodies to the test material. Finally, a challenge dose of the material is applied to the animals, and signs of allergic response are monitored. The allergic response can range from swelling and inflammation to an anaphylactic shock which can lead to animal death [43]. The drawback to this test is that the guinea pig is less sensitizable than humans and therefore cannot be relied on fully for the identification of contact allergens in humans. However, there seem to be no instances in which a substance found to sensitize the guinea pig fails to do so in humans [57].

No data regarding the allergenic potential of polyanhydrides have been published yet.

3.6.2.6   Brain Biocompatibility

The brain has a natural "protection" system, the blood–brain barrier, that prevents unwanted substances or toxins from entering the brain parenchyma. Unfortunately, this barrier also limits the entry of therapeutic agents which would benefit the brain in treating cerebral disorders.

Various polyanhydrides were designed and developed to be used clinically to deliver sustained high doses of anticancer drugs directly into specific areas of the brain for the treatment of brain neoplasms. In vivo safety evaluations and brain compatibility of polyanhydride copolymers were assessed in rats [63], rabbits [64], and monkeys [65,66].

In the rat brain study, the tissue reaction of the polyanhydride P(CPP–SA); 20:80 was compared to the reaction observed with the two standard materials previously mentioned that are routinely used in surgery, Gelfoam (absorbable gelatin sponge), and Surgicel (oxidized cellulose absorbable hemostat commonly used in brain surgery). Histological evaluation of the tissue demonstrated a small rim of necrosis around the implant, and a mild to marked cellular inflammatory reaction limited to the area immediately adjacent to the implantation site, slightly more pronounced than Surgicel at the earlier time points, but noticeably less marked than Surgicel at the later times [63]. Like Surgicel the CPP–SA polymer elicited a transient, well-demarcated tissue reaction that subsided as the foreign substance was degraded. None of the rats tested showed any evidence of systemic or neurologic toxicity and all lived to the scheduled date of sacrifice. The reaction to Gelfoam was essentially equivalent to that observed in control rats.

In the rabbit brain safety study using P(CPP–SA); 50:50, even less of an inflammatory reaction was observed, and the polymer was essentially equivalent to Gelfoam [64]. In a similar brain biocompatibility study conducted in monkeys with P(CPP–SA); 20:80, no abnormalities were noted in the computer tomography scans and magnetic resonance images, nor in the blood chemistry or hematology evaluations [65]. This study was aimed to measure the direct effects of the polymer, polymer degradation products, and polymer containing the antineoplastic drug BCNU implanted in the monkey brain. No systemic effects of the implants were observed on histological examinations of any of the tissues tested [66]. No unexpected or untoward reactions to the treatments were observed.

To optimize the delivery of water-soluble drugs, the new biodegradable polyanhydride copolymer poly(FAD–SA) was recently developed and shown to be capable of controlled release of water-soluble chemotherapeutic drugs such as methotrexate, carboplatin, and 4-hydroxyperoxycyclophosphamide (4-HC) to the rat brain [67]. In vitro release kinetics studies conducted on water-soluble drugs have shown sustained release of the drugs over a period of 3–5 weeks, paralleling the degradation time of the polymer. This is particularly important for delivery to the brain in that the blood–brain barrier effectively blocks most water-soluble drugs from entering the brain substance.

The brain biocompatibility and safety of poly(FAD–SA) were studied in the rat brain and compared with that of Surgicel and Gelfoam [68]. No neurological deficits or behavioral changes suggestive of either systemic or localized toxicity were observed in the animals implanted with the new polyanhydride. A well-demarcated acute inflammatory response was seen at days 3 and 6 for poly(FAD–SA) and Surgicel implants. The inflammatory response remained well localized and resolved with total degradation of the polymer by day 36 [68]. The localized reaction evoked by poly(FAD–SA) was comparable to the one generated by the previously described polyanhydride poly(CPP–SA), and to the response to the commonly used surgical hemostatic material, Surgicel, but more pronounced than the reaction evoked by Gelfoam.

3.6.2.7   Clinical Use

With these preclinical toxicology and biocompatibility studies carried out in animals having demonstrated both the efficacy and safety of the polyanhydrides, studies involving these materials moved toward human clinical use. In 1985, a collaborative study aimed to explore the possibility of implanting polyanhydride discs containing nitrosoureas for treating brain cancer after surgery was started, and preliminary results were published in several articles [26,69–72].

In 1987, the Food and Drug Administration approved experimental use of these polyanhydrides in humans, under an Investigational New Drug clinical trial application. The first clinical study (Phase I/II clinical trial), designed to measure the safety of polyanhydride copolymer devices, was conducted at five major medical centers across the United States and involving 21 patients [72]. In these clinical trials, a polyanhydride dosage form (Gliadel) consisting of wafer polymer implants of poly(CPP–SA) 20:80 and containing the chemotherapeutic agent carmustine (BCNU) were used for the treatment of glioblastoma multiforme, a universally fatal form of brain cancer. In these studies, up to eight of these wafer implants (1.4 cm in diameter, 1 mm thick) were placed to line the surgical cavity created during the surgical removal of the bulk of the brain tumor in patients undergoing a second operation for surgical debulking of either a Grade III or IV anaplastic astrocytoma. Following surgery the BCNU is released directly into adjoining tissues that may contain cancer cells not removed during surgery. The results of this study (an open-label ascending dose tolerance) showed safety and no toxicity of these polymers either clinically or pathologically, and extending patient lifetime beyond conventional drug treatment [26,72].

The safety of these polyanhydride copolymer wafer implants in humans has been demonstrated [26,69–72]. No central or systemic side effects of doses of BCNU which would produce marked effects on the hemopoietic system when injected intravenously were observed. No adverse reactions to the Gliadel wafer implants were found in over 100 patients treated with the polyanhydride–BCNU combination. Based on these results, further studies designed to measure the efficacy of this approach to the treatment of brain cancer, in a Phase III study in 32 United States and Canadian hospitals, has been conducted, with the results not yet published [70,71].

REFERENCES

1. Cottler, R.J.; Matzner, M. *Chemisch Weekblad* **11**, 133 (1967).
2. Polyanhydrides In *Encyclopedia of Polymer Science Technology* John Wiley, New York, **10**, 630 (1969).
3. Domb, A.J.; Langer, R. *J. Polym. Sci. Polym Chem* **25**, 3373 (1987).
4. Domb, A.J.; Ron, E.; Langer, R. *Macromolecules* **21**, 1926 (1988).
5. Hartmann, M.; Schultz, V. *Makromol. Chem.* **190**, 2133 (1989).
6. Gonzalez, J.I.; de Abajo, J.; Gonzalez-Babe, S.; Fontan, J. *Agnew Makromol Chem.* **55**, 85 (1976).
7. de Abajo, J.; Gonzalez-Babe, S.; Fontan, J. *Angew Makromol. Chem.* **19**, 1259 (1972).
8. Staubli, A.; Mathiowitz, E.; Langer, R. *Macromolecules* **24**, 2291 (1991).
9. Staubli, A.; Mathiowitz, E.; Lucarelli, M.; Langer, R. *Macromolecules* **24**, 2283 (1991).
10. Domb, A.J.; Gallardo, C.F.; Langer, R. *Macromolecules* **22**, 3200 (1989).
11. Domb, A.J. *Macromolecules* **25**, 12 (1992).
12. McIntyre, J.E. (1964) British Patent 978,660
13. Ziegast, G. (1988) U.S. Patent 4,792,598
14. Domb, A.J.; Maniar, M. *J. Polym. Sci. Part A: Polym. Chem.* **31**, 1275–1285 (1993).
15. Domb, A.J.; Mathiowitz, E.; Ron, E.; Giannos, S.; Langer, R. *J. Polym. Sci. Part A: Polymer Chemistry* **29**, 571 (1991).
16. Xie, X.; Adam, M.; Maniar, M.; Domb, A.J. 1991 AAPS meeting, Washington, D.C.
17. Ron, E.; Mathiowitz, E.; Mathiowitz, G.; Domb, A.J.; Langer, R. *Macromolecules* **24**, 2278 (1991).
18. Laurencin, C.T.; Domb, A.J.; Morris, C.; Harris, M.; Lopez, L.; Langer, R. *Proceed. Intern. Symposium on Controlled Rel. of Bioactive Materials* (1990).
19. Davies, M.C.; Tudor, A.M.; Hendra, P.J.; Domb, A.J.; Langer, R. *Proc. Intern. Symp. Controlled Rel. Bioact. Mater.* **17**, 236 (1990).
20. Tudor, A.M.; Melia, C.D.; Davies, M.C.; Hendra, P.J.; Church, S.; Domb, A.; Langer, R. *Spectrochimica Acta.* **9/10**, 1335 (1991).
21. Mathiowitz, E.; Kline, D.; Langer, R. *Scanning Microscopy* **4**, 329 (1990).
22. Davies, M.C.; Domb, A.; Lynn, R.A.P.; Khan, M.A.; Paul, A.; Langer, R. (May 1991) *ISPAC, 4th International Symposium* Baltimore, MD.
23. Davies, M.C.; Khan, M.A.; Domb, A.; Langer, R.; Watts, J.F.; Paul, A. *J. Appl. Polym. Sci.* **42**, 1597 (1991).

24. Conix, A. *Makromol. Chem.* **24,** 76 (1957).
25. Leong, K.W.; Brott, B.C.; Langer, R. *J. Biomed. Mater. Res.* **19,** 941–955 (1985).
26. Chasin, M.; Hollenbeck, G.; Brem, H.; Grossman, S.; Colvin, M.; Langer, R. *Drug Dev. Ind. Pharm.* **16,** 2579–2594 (1990).
27. Tamargo, R.J.; Epstein, J.I.; Yang, M.B.; Pinn, M.L.; Chasin, M.; Brem, H. *J. Neurosurg.* **70,** 311A (1989).
28. Olivi, A.; Domb, A.; Pinn, M.L.; Ewend, M.G.; Goodman, J.H.; Brem, H. Controlled International Symposium on Advances in Neuro-Oncology. SanRemo, Italy, (1990).
29. Mathiowitz, E.; Langer, R. *J. Control. Rel.* **5,** 13–22 (1987).
30. Mathiowitz, E.; Saltzman, M.; Domb, A.; Dor, Ph.; Langer, R. *J. Appl. Polym. Sci.* **35,** 755–774 (1988).
31. Mathiowitz, E.; Leong, K.; Langer, R. *12th Int. Symp. Control. Rel. Bioact. Mater.* 183–184 (1985).
32. Ron, E.; Turek, T.; Mathiowitz, E.; Chasin, M.; Langer, R. *Proc. Contr. Rel. Soc.* **338** (1989).
33. Leong, K.W.; Kost, J.; Mathiowitz, E.; Langer, R. *Biomaterials* **7,** 364–371 (1986).
34. Bolling, S.F.; Lin, H.; Ning, X.; Levy, R.J. *Polymers for Advanced Technologies,* **3,** (1992).
35. Byrd, K.E.; Domb, A.; Sokoloff, A.; Byrd, E.L. *Polymers for Advanced Technologies,* **3,** (1992).
36. Mordeni, J. *J. Pharm. Sci.* **75,** 1028 (1986).
37. Wahlig, H. *J. Antimicrob. Chemother.* **10,** 463 (1982).
38. Nelson, C.L.; Hickmon, S.G.; Skinner, R.A. *38th Annual Meeting, Orthopedic Research Society,* pp. 431 (1992).
39. Williams, D.F., ed., *Fundamental Aspects of Biocompatibility.* CRC Press, Boca Raton, FL, 1981.
40. Vert, M. *Crit. Rev. Therap. Drug Carrier Syst.* **2,** 291–327 (1985).
41. Peppas, N.A. Biocompatibility of polymers. Controlled release technology: polymeric delivery systems for drugs, pesticides, and foods. Cambridge, MA, MIT, 1985.
42. Laurencin, C.; Pierre-Jackes, H.M.; Langer, R. *Clin. Lab. Med.* **10,** 549 (1990).
43. Majors, K.R.; Friedman, M.B. In *Polymers for Controlled Drug Delivery.* Tarcha, P.J., ed., CRC Press, Boca Raton, FL, pp. 231–239 (1991).
44. Lichfield, J.T.; Wilcoxon, F.C. *J. Pharmacol. Exp. Ther.* **96,** 99 (1949).
45. Gleason, M.; Groselin, R.; Hodges, H.; Smith, R. *Clinical Toxicology of Commercial Products,* 3rd ed. Williams & Wilkins, Baltimore (1976).
46. Autian, J. *Artif. Organs* **1,** 53–60 (1977).
47. Guess, W.L.; Rosenbluth, S.A.; Schmidt, B. *J. Pharm. Sci.* **54,** 156 (1965).
48. Langer, R. *Science* **349,** 1527–1533 (1990).
49. Lawrence, W.H. In *Controlled Drug Delivery, vols. I and II.* Bruck, S., ed. CRC Press, Boca Raton, FL 1983.
50. The United States Pharmacopeia XIX. Biological test: Plastic containers, Mack Publishing Company, Easton, PA, pp. 644–647 (1975).
51. Ames, B.N.; McCann, J.; Yamasaki, E. *Mutat. Res.* **31,** 347–379 (1975).
52. Skopek, T.R.; Liber, H.L.; Krolewski, J.J.; Thilly, W.G. *Proc. Natl. Acad. Sci. USA* **75,** 410–414 (1978).
53. Braun, A.G.; Buckner, C.A.; Emerson, D.J.; Nichinson, B.B. *Proc. Natl. Acad. Sci. USA* **79,** 2056–2060 (1982).
54. Autian, J.; Singh, A.R.; Turner, J.E.; et al. *Cancer Res.* **35,** 1591–1596 (1975).
55. Hueper, W.C. *Am. J. Clin. Pathol.* **34,** 328–333 (1960).

56. Oppenheimer, B.S.; Oppenheimer, E.T.; Stout, A.P. *Proc. Soc. Exp. Biol. Med.* **67,** 33–34 (1948).
57. Magnusson, B.; Kligman, A.M. *J. Invest. Derm.* **52,** 268–276 (1969).
58. Laurencin, C.; Domb, A.; Morris, C.; Brown, V.; Chasin, M.; McConell, R.; Lange, N.; Langer, R. *J. Biomed. Mat. Res.* **24,** 1463–1481 (1990).
59. Leong, K.W.; D'Amore, P.; Marletta, M.; Langer, R. *J. Biomed. Mat. Res.* **20,** 51–64 (1986).
60. Jaffe, E.A.; Nachman, R.L.; Becker, C.G.; Minick, C.R. *J. Clin. Invest.* **52,** 2745–2756 (1973).
61. Ross, R. *J. Cell Biol.* **50,** 172–186 (1971).
62. Rock, M.; Green, M.; Fait, C.; Geil, R.; Myer, J.; Maniar, M.; Domb, A. *Polym. Preprint* **32,** 221–222 (1991).
63. Tamargo, R.J.; Epstein, J.I.; Reinhard, C.S.; Chasin, M.; Brem, H. *J. Biomed. Mater. Res.* **23,** 253–266 (1989).
64. Brem, H.; Kader, A.; Epstein, J.I.; Tamargo, R.J.; Domb, A.; Langer, R.; Leong, K.W. *Selective Ther.* **5,** 55 (1989).
65. Brem, H.; Ahn, H.; Tamargo, R.J.; Pinn, M.; Chasin, M. *Am. Assoc. Neurol. Surg.* **349,** (1988).
66. Brem, H.; Tamargo, R.J.; Pinn, M.; Chasin, M. Abstract of the 1988 Annual Meeting of the *Am. Assoc. Neurol. Surg.* **381,** (1988).
67. Domb, A.; Olivi, A.; Judy, K.; Pinn, M.L.; Ewend, M.G.; Goodman, J.H.; Brem, H. ACS Meeting. Philadelphia, PA, (1991).
68. Brem, H.; Domb, A.; Lenartz, D.; Dureza, C.; Olivi, A.; Epstein, J.I. *J. Control. Rel.* 1992 (in press).
69. Brem, H. *Biomaterials* **11,** 699–701 (1990).
70. Langer, R. *J. Control. Rel.* **16,** 53–60 (1991).
71. Domb, A.; Maniar, M.; Bogdansky, S.; Chasin, M. *Critic. Rev. Therap. Drug Carrier Syst.* **8,** 1–17 (1991).
72. Brem, H.; Mahaley, M.S.; Vick, N.; Black,K.; Schold, C.; Burger, P.C.; Friedman, A.H.; Ciric, I.S.; Eller, T.W.; Cozzens, J.W.; Kenealy, J.N. *J. Neurosurg.* **74,** 441–446 (1991).

CHAPTER 4

Bioabsorbable Poly(ester–amides)

Thomas H. Barrows

4.1   Introduction

Research on poly(ester–amides) has been conducted with the objective of incorporating the amide linkage into the backbone of hydrolytically labile polyesters. The amide moiety allows interchain hydrogen bonding which contributes to improved fiber strength and durability at molecular weights that are typically lower than required for most fiber-forming polyesters. Although poly(ester–amides) are not yet available as commercial biomaterials, several examples continue to show promise and are the subject of ongoing development.

Thomas H. Barrows, Life Sciences Research Laboratory, 3M/3M Center, St. Paul, Minnesota 55144, U.S.A.

4.1.1   Background

The terms bioabsorbable, bioresorbable, and absorbable are used interchangeably to describe a polymer that degrades in vivo into products that either are normal metabolites, such as L-lactic acid or succinic acid, or are completely eliminated from the body with or without further metabolic transformations. Synthetic bioabsorbable polymer fibers have largely replaced the use of catgut, a collagen material from ovine or bovine intestinal submucosa, as surgical suture. Synthetic bioabsorbable polymers also are being developed for a variety of implantable surgical devices in addition to suture that are required to perform only a temporary function in vivo [1].

The process of copolymerization is an important way to modify the structure of a polymer to achieve improved properties. Because polyesters and polyamides are two prominent classes of fiber-forming polymers, it is not surprising that research on poly(ester–amide) copolymers has been conducted in the search for unique fiber properties. This approach is exemplified by the heating of polyethylene terephthalate with amine-terminated polycaprolactam to yield a block copolymer with higher modulus than polyamide and better dyeability than polyester [2]. These properties are desirable for textile applications.

Katayama et al. [3,4] synthesized poly(ester–amides) for the purpose of improving undesirable properties of aliphatic polyesters such as low melting points and poor dyeability by copolymerizing these with aliphatic polyamides, which have high melting points and can be intensively dyed. To produce such polymers they first undertook to blend polyesters and polyamides and to copolymerize randomly a mixture of the monomers for polyamides and polyesters. However, the resulting polymer blends and random copolymers had undesirable, worsened properties such as phase separation, coloration, and melting point depression. This problem was solved by the use of a two-step synthesis. The amide bonds first were formed by the synthesis of an amidediol monomer. The amidediol monomer was then polyesterified to give a perfectly regular, alternating copolymer. These steps are illustrated in the following equations. As shown in Eq. (4.1), caprolactone was reacted with 1,6-hexanediamine to form the N,N'-di(6-hydroxycaproyl)hexamethylenediamine amidediol. This amidediol was polymerized with dimethyl adipate using antimony trioxide and calcium acetate as catalysts [Eq. 4.2)] or with adipoyl chloride by heating in nitrobenzene [Eq. 4.3)].

$$\text{NH}_2-(\text{CH}_2)_6-\text{NH}_2 \quad + \quad 2 \quad \text{[caprolactone]} \quad \longrightarrow$$

1, 6-Hexanediamine   Caprolactone

$$\text{HO}-(\text{CH}_2)_5-\overset{\overset{\displaystyle O}{\|}}{\text{C}}-\text{NH}-(\text{CH}_2)_6-\text{NH}-\overset{\overset{\displaystyle O}{\|}}{\text{C}}-(\text{CH}_2)_5-\text{OH} \tag{4.1}$$

Amidediol

$$\text{HO}-(\text{CH}_2)_5-\overset{\overset{\displaystyle O}{\|}}{C}-\text{NH}-(\text{CH}_2)_6-\text{NH}-\overset{\overset{\displaystyle O}{\|}}{C}-(\text{CH}_2)_5-\text{OH} \quad +$$

Amidediol

$$\text{CH}_3\text{O}-\overset{\overset{\displaystyle O}{\|}}{C}-(\text{CH}_2)_4-\overset{\overset{\displaystyle O}{\|}}{C}-\text{OCH}_3 \longrightarrow \qquad (4.2)$$

Dimethyl adipate

$$\left[O-(\text{CH}_2)_5-\overset{\overset{\displaystyle O}{\|}}{C}-\text{NH}-(\text{CH}_2)_6-\text{NH}-\overset{\overset{\displaystyle O}{\|}}{C}-(\text{CH}_2)_5-O-\overset{\overset{\displaystyle O}{\|}}{C}-(\text{CH}_2)_4-\overset{\overset{\displaystyle O}{\|}}{C}\right] + \text{CH}_3\text{OH}$$

Poly(ester–amide)   Methanol

$$\text{HO}-(\text{CH}_2)_5-\overset{\overset{\displaystyle O}{\|}}{C}-\text{NH}-(\text{CH}_2)_6-\text{NH}-\overset{\overset{\displaystyle O}{\|}}{C}-(\text{CH}_2)_5-\text{OH} \quad +$$

Amidediol

$$\text{Cl}-\overset{\overset{\displaystyle O}{\|}}{C}-(\text{CH}_2)_4-\overset{\overset{\displaystyle O}{\|}}{C}-\text{Cl} \longrightarrow \qquad (4.3)$$

Adipoyl chloride

$$\left[O-(\text{CH}_2)_5-\overset{\overset{\displaystyle O}{\|}}{C}-\text{NH}-(\text{CH}_2)_6-\text{NH}-\overset{\overset{\displaystyle O}{\|}}{C}-(\text{CH}_2)_5-O-\overset{\overset{\displaystyle O}{\|}}{C}-(\text{CH}_2)_4-\overset{\overset{\displaystyle O}{\|}}{C}\right] + \text{HCl}$$

Poly(ester–amide)   Hydrogen
                                                         Chloride

Although this type of poly(ester–amide) combined polyester and polyamide properties, it was determined to be unsuitable for bioabsorbable polymer applications due to insufficient hydrolytic susceptibility. This conclusion was based on an experiment that is summarized in the section on α-hydroxyacid amidediols. Other examples of poly(ester–amide) structures not directed toward biomedical applications include the fully aromatic copolymers obtained from the polymerization of aminophenols with benzenedicarbonyl chlorides [5].

An example of a bioabsorbable poly(ester–amide) is provided by Goodman and Kirshenbaum [6], who created hydrolyzable polymers of amino acids and α-hydroxyacids for use as absorbable suture. Since polypeptides and poly(α-hydroxyacids), especially polyglycolic acid (PGA) and polylactic acid (PLA), are well recognized as bioabsorbable, copolymerization of amino acids with α-hydroxyacids was viewed as a plausible route to achieving bioabsorbable polymers with unique properties. The copolymerization of L-alanine and lactic acid was accomplished by reacting L-alanine-N-carboanhydride with L-lactic acid anhydrosulfite as shown in Eq. (4.4). Although an alternating copolymer of this structure would be preferred for optimum fiber properties, this reaction was limited to the synthesis of polymers exhibiting a random sequence of comonomer units.

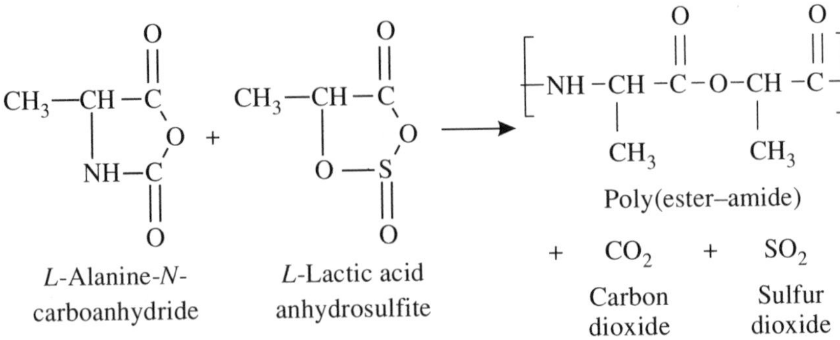

$$
\underset{\substack{\text{L-Alanine-N-}\\\text{carboanhydride}}}{
\begin{array}{c}
\text{O}\\
\text{\textbardbl}\\
\text{CH}_3\text{—CH—C}\\
\mid\quad\quad\text{O}\\
\text{NH—C}\\
\text{\textbardbl}\\
\text{O}
\end{array}}
+
\underset{\substack{\text{L-Lactic acid}\\\text{anhydrosulfite}}}{
\begin{array}{c}
\text{O}\\
\text{\textbardbl}\\
\text{CH}_3\text{—CH—C}\\
\mid\quad\quad\text{O}\\
\text{O—S}\\
\text{\textbardbl}\\
\text{O}
\end{array}}
\longrightarrow
\underset{\text{Poly(ester–amide)}}{
\begin{array}{c}
\text{O}\quad\quad\text{O}\\
\text{\textbardbl}\quad\quad\text{\textbardbl}\\
\left[\!-\text{NH—CH—C–O–CH—C}\!-\right]\\
\mid\quad\quad\quad\mid\\
\text{CH}_3\quad\quad\text{CH}_3
\end{array}}
\quad (4.4)
$$

$$+\quad \underset{\substack{\text{Carbon}\\\text{dioxide}}}{CO_2} \quad + \quad \underset{\substack{\text{Sulfur}\\\text{dioxide}}}{SO_2}$$

Another type of bioabsorbable poly(ester-amide) was formed from an amidediol prepared by the reaction of diethyl oxalate with 6-amino-1-hexanol [7]. This bis-oxamidodiol was then polymerized with diethyl oxalate by transesterification in the presence of a small amount of 1,6-hexanediol to increase the molecular weight of the product. Although poly(ester-amides) obtained from polymerization of bis-oxamido-diols with oxalate esters were sensitive to hydrolysis, polymerization with the use of other diacid esters yielded poly(ester-amides) that were resistant to hydrolysis.

4.2   3M Poly(ester–amides)

Research at 3M has been conducted to develop a new class of bioabsorbable poly (ester–amides) with a high degree of structural versatility while maintaining the objective of utilizing an esterified glycolic acid moiety as the site of hydrolytic instability. This has been accomplished by polyesterification of a series of bis-hydroxyacetamides, also known as amidediols, with diacid chlorides. If lactic acid is used in place of glycolic acid to form the amidediol, then a poly(ester–amide) with greater hydrolytic stability results. The identification codes for this class of polymers are illustrated below where PEA stands for poly(ester–amide) and the x and y stand for the number of methylene units in the diacid and diamine moieties, respectively. S preceding the y indicates that lactic acid with the S configuration (L-lactic acid) rather than glycolic acid has been used to form the amidediol. Likewise, R preceding the y would indicate D-lactic acid and RS would indicate DL-lactic acid. Although virtually any diacid or diamine can be used according to this scheme, linear aliphatic diacids and diamines were preferred to increase the chances of a favorable metabolic fate.

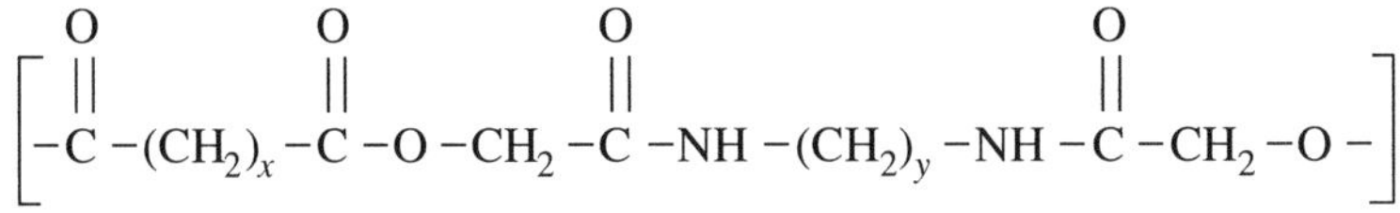

$$\left[-\overset{\overset{\displaystyle O}{\|}}{C}-(CH_2)_x-\overset{\overset{\displaystyle O}{\|}}{C}-O-CH_2-\overset{\overset{\displaystyle O}{\|}}{C}-NH-(CH_2)_y-NH-\overset{\overset{\displaystyle O}{\|}}{C}-CH_2-O-\right]$$

PEA-*x, y*

(4.5)

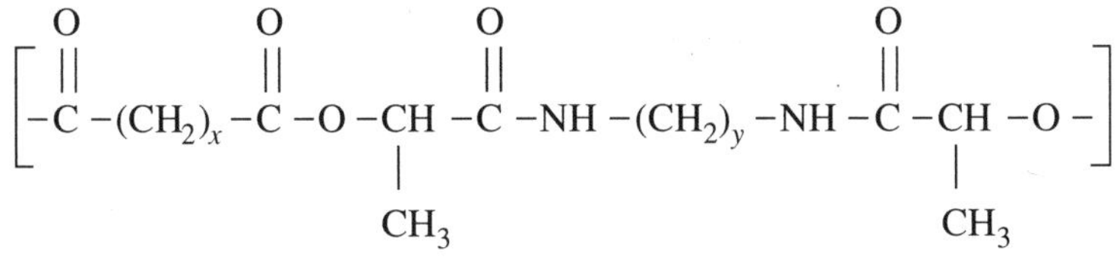

$$\left[-\overset{\overset{\displaystyle O}{\|}}{C}-(CH_2)_x-\overset{\overset{\displaystyle O}{\|}}{C}-O-\underset{\underset{\displaystyle CH_3}{|}}{CH}-\overset{\overset{\displaystyle O}{\|}}{C}-NH-(CH_2)_y-NH-\overset{\overset{\displaystyle O}{\|}}{C}-\underset{\underset{\displaystyle CH_3}{|}}{CH}-O-\right]$$

PEA-*x, Sy*

4.2.1   Amidediol Design

The bioabsorbability of polyglycolic acid is due to the hydrolytic instability inherent in α-hydroxyacid polyesters. This instability results from the electron-withdrawing effect of an electronegative substituent α to the ester carbonyl group. Polylactic acid is less hydrolytically labile than polyglycolic acid due to the electron-donating methyl substituent in the α position. In addition, methyl substitution decreases polymer hydrophilicity which further reduces hydrolytic susceptibility. Polymer crystallinity, however, also plays an important role in the degradation rates of the lactide and glycolide polymers [8–10].

A clear example of the importance of the inductive effect on polymer degradation and absorption time was provided by Wang and Arlitt [11]. These authors implanted various radiolabeled polyesters subcutaneously in rats. The polyesters were prepared by reacting cyclohexane dimethanol with perfluoroadipic, diglycolic, oxoglutaric, and tartaric acids. These acids, which are listed in decreasing order of inductive effect, gave polyesters with a corresponding decrease in the rate of degradation (physical disappearance) and absorption (loss of radiolabel).

In order to utilize the amide moiety in providing an inductive effect for increasing poly(ester–amide) hydrolytic sensitivity, the ester and amide groups ideally should be separated by no more than one methylene group. To demonstrate this requirement, poly(ester–amides) were synthesized by reacting succinyl chloride with each of the following of amidediols:

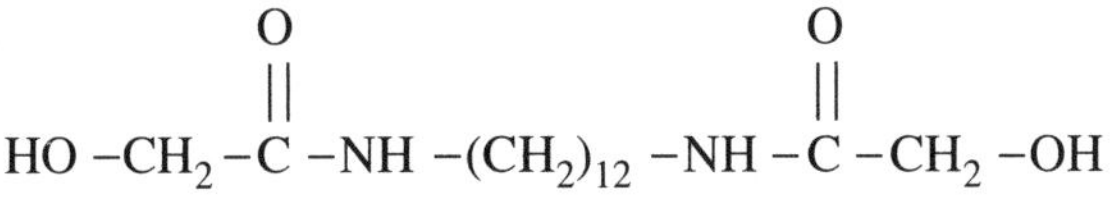

$$HO-CH_2-\overset{\overset{\displaystyle O}{\|}}{C}-NH-(CH_2)_{12}-NH-\overset{\overset{\displaystyle O}{\|}}{C}-CH_2-OH$$

1, 12-Di(hydroxyacetamido)dodecane
C_{12}-Amidediol

$$\text{HO} - (CH_2)_5 - \overset{\overset{\displaystyle O}{\|}}{C} - \text{NH} - (CH_2)_{12} - \text{NH} - \overset{\overset{\displaystyle O}{\|}}{C} - (CH_2)_5 - \text{OH} \qquad (4.6)$$

1, 12-Di(6-hydroxycapramido)dodecane

Accelerated in vitro hydrolysis tests in pH 11 phosphate buffer at 100°C showed that the polymer made from the hydroxyacetamide monomer (C_{12}-amidediol) was fully dissolved after 40 min whereas the polymer made from the above hydroxycapramide monomer was only 30% dissolved after more than 4000 min [12].

4.2.2   Amidediol Synthesis

The synthesis of hydroxyacetamide amidediols has been reported by Skinner and Johansson [13]. These authors prepared amidediols from linear, aliphatic diamines with six through twelve methylene groups and from p-xylylenediamine. A two-step synthesis was employed. First, the diamines were reacted with carbomethoxyacetyl chloride to obtain N,N'-bis(carbomethoxyacetyl)diamines. These intermediates were then treated with sodium in methanol to obtain the corresponding hydroxyacetamide amidediols. The overall yields for this route ranged from 9% (1,6-hexanediamine) to 56% (1,11-undecanediamine). This protected synthesis was employed presumably to avoid the formation of polyester that would occur by direct reaction of hydroxyacid with diamine. It has been discovered, however, that amidediol is a thermodynamically preferred product which can be produced in yields of up to 90% by simply heating the starting materials neat with distillation of water as shown in Eq. (4.7) [14].

$$2 \quad \text{HO} - CH_2 - \overset{\overset{\displaystyle O}{\|}}{C} - \text{OH} \quad + \quad NH_2 - (CH_2)_y - NH_2$$

Glycolic acid   Diamine

$$\downarrow \ -H_2O$$

$$(4.7)$$

$$\text{HO} - CH_2 - \overset{\overset{\displaystyle O}{\|}}{C} - \text{NH} - (CH_2)_y - \text{NH} - \overset{\overset{\displaystyle O}{\|}}{C} - CH_2 - \text{OH}$$

C_y-Amidediol

Table 4.1   bis-Hydroxyacetamide Amidediol Melting Points

Amidediol	Mp, °C	lit. value
1,2-DI-(hydroxyacetamido)ethane	139–141	—
1,4-DI-(hydroxyacetamido)butane	135–138	—
1,6-DI-(hydroxyacetamido)hexane	125–126	126–129
1,8-DI-(hydroxyacetamido)octane	112–114	116–117.5
1,10-DI-(hydroxyacetamido)decane	119–121	122–123.5
1,12-DI-(hydroxyacetamido)dodecane	127–130	129.5–131
1,2-DI-(hydroxyacetamido)-4,9-dioxadodecane	74–77	—
N,N′-DI-(hydroxyacetyl)piperazine	187–190	—
4,4′-METHYLene-bis(hydroxyacetamidocyclohexane)	208–211	—
trans-1,4-cyclohexanebis(hydroxyacetamideomethyl)	182–186	—

4.2.3   Amidediol Properties

A list of amidediols prepared from various diamines and glycolic acid is shown in Table 4.1. A plot of melting point versus methylene chain length for a series of linear aliphatic amidediols is shown in Figure 4.1. The minimum point at $y = 8$ can be interpreted as the result of two trends created by increasing methylene chain length. First, the increase lowers the melting point by reducing the density of amide groups, then it increases the melting point by increasing the molecular weight of the hydrocarbon moiety.

Amidediols synthesized from linear aliphatic diamines were preferred for fiber applications because of their close analogy to nylon. Although side groups could

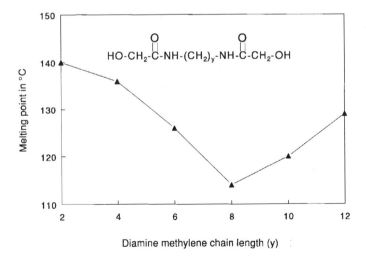

Figure 4.1   bis-Hydroxyacetamide amidediol melting point vs. diamine methylene chain length.

be incorporated in the structure for the purpose of improving sensitivity to proteolytic enzymes [15], the absence of such groups was desired in order to present a more immunologically inert structure and provide optimal polymer orientation and crystallization in fiber fabrication processes. Even numbered diamines were chosen because the existence of well known metabolic pathways for utilization of two carbon units provided some assurance that these compounds would yield relatively innocuous degradation products in vivo [16].

The water solubility of amidediols was of interest because it provided some indication of the hydrophilicity of the corresponding poly(ester–amide). Solubility data for amidediols are shown in Figure 4.2. Although the C_{12}-amidediol appeared insoluble in vitro, this did not necessarily imply that it could not be solubilized in vivo. To investigate in vivo solubility, amidediols with y = 12, 10, 8, and 6 were cast into solid pellets and implanted in the dorsal subcutis of mice, one pellet per animal. The condition of the pellets was inspected visually during necropsy at various times postimplantation. Additional mice were implanted as necessary to obtain an accurate indication of dissolution time. The C_{12}-amidediol pellets dissolved in 140 days and the C_{10}-amidediol pellets dissolved in 30 days. The C_8- and C_6-amidediol pellets dissolved in <24 h [17].

4.3   Polymerization

Polyesterification of amidediol theoretically can be performed by a variety of different condensation reaction methods used in organic synthesis. However, extremely high yield is required if a method is to succeed in producing high molecular weight product. Thus, obtaining high purity monomers and finding the best reaction conditions

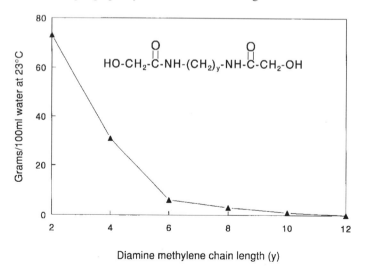

Figure 4.2   Water solubility of bis-hydroxyacetamide amidediols vs. diamine methylene chain length.

is critical to achieving a useful molecular weight. Although low molecular weight polymer can be useful in evaluating some chemical and biological properties, only high molecular weight polymer provides useful information on mechanical properties and end-use performance. Adequate molecular weight of poly(ester–amides) was found to be represented by an inherent viscosity of about 0.60–0.90 (0.5% in 2,2,2-trifluoroethanol at 30°C).

The transesterification method used by Katayama et al, shown in Eq. (4.2) above, did not yield high molecular weight polymer in the case of aliphatic diester monomers. Only dimethyl terephthalate gave product with reasonably high inherent viscosity. Polyesterification of C_{12}-amidediol has been performed both by direct reaction with dicarboxylic acid and by transesterification with dicarboxylic acid dimethyl ester [18]. In the case of direct esterification, water was released as the condensation byproduct which hydrolyzed some of the preformed amide bonds. This resulted in ester–amide interchange which was detected by nuclear magnetic resonance (NMR) in polymer made by the direct method but not in polymer made by the transesterification method [12]. In either case, the molecular weight achieved was less than desired.

In contrast to the use of dimethyl adipate, Katayama et al. obtained high molecular weight polymer with the use of adipoyl chloride as shown in Eq. (4.3). In this case, the adipoyl chloride and amidediol were dissolved in nitrobenzene and heated for 6 h at 140°C under nitrogen. After precipitation of a dimethyl formamide solution of the product into water, poly(ester–amide) with inherent viscosity of 0.95 was obtained in 55% yield ($T_m = 133$°C).

The diacid chloride route was also the preferred method of polymerization for bis-hydroxyacetamide amidediols. Chlorobenzene was used instead of nitrobenzene to prepare PEA-2,12 from succinyl chloride and C_{12}-amidediol [18]. The product separated from solution as the reaction proceeded and formed into a suspension of solid granules under conditions of vigorous mixing. The product was then collected by filtration and residual solvent removed by heating under vacuum. During vacuum heating of the solid, additional increase in molecular weight occurred. The polymer was then ready for melt extrusion or injection molding.

Polymerization in solution would seem to be a feasible alternative. However, solvents that have sufficient polarity to dissolve bis-hydroxyacetamide amidediols at low temperature, are inert to acid chloride, can be easily dried, and are relatively nontoxic were not available. The reaction in solution at elevated temperature, for example in 1,1,2,2-tetrachloroethane, resulted in crosslinked product, presumably due to reaction of acyl chloride end groups with amide nitrogens [12]. Methyl substitution of these nitrogens could block the crosslinking reaction, but would yield polymer that lacks the intermolecular hydrogen bonding capability which provides poly(ester–amide) with its nylon-like fiber properties.

Polymerization of short chain bis-hydroxyacetamides such as C_2-, C_4-, and C_6-amidediol with diacid chlorides was found to be even more problematic than polymerization of C_{12}-amidediol. These problems were solved by using a suspension method of polymerization in which the amidediols remained in the solid state throughout the reaction sequence [19].

4.3.1   Polymer Properties

Poly(ester–amides) can be prepared from a wide variety of different bis-hydroxyacet-amide amidediols and diacids. To begin the evaluation of polymer properties, linear aliphatic diamines and diacids were chosen as starting materials. The C_{12}-amidediol, as mentioned earlier, polymerized most conveniently to high molecular weight. This amidediol was held constant and a series of polymers were prepared using diacids with varying methylene chain lengths. As shown in Figure 4.3, the polymer melting temperature, T_m, was greatest for the shortest chain diacid, that is, succinic acid. There also appeared to be a secondary maximum where diacid and diamine methylene chain lengths were equal. Polymerization of succinyl chloride with amidediols that have shorter methylene chains, for example, $y = 10$, 8, and 6, did not result in a significant decrease or increase in polymer melting temperature. A typical differential scanning calorimeter (DSC) trace for a poly(ester–amide) of this type (PEA-2,6) is shown in Figure 4.4 where the glass transition (T_g) occurred at 35°C, the crystallization exotherm (T_c) occurred at 88°C, and the crystalline melting transition (T_m) occurred at 165°C. This trace was obtained after first melting the polymer and then rapidly cooling it with liquid nitrogen to obtain a relatively amorphous sample before initiation of the heating scan.

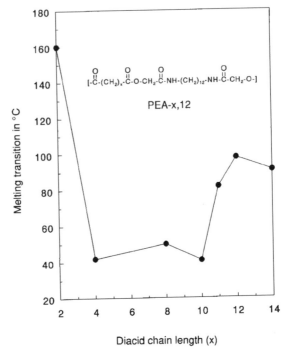

Figure 4.3   Polymer melting transition temperature (T_m) of PEA-x,12 vs. diacid methylene chain length (x).

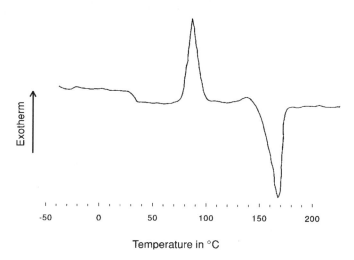

Figure 4.4   Differential scanning calorimetry (DSC) trace of PEA-2,6 after rapid cooling from molten state.

The advantage of using succinic acid in poly(ester–amide) synthesis was appreciated on examination of X-ray diffraction patterns of these polymers before and after being formed into fibers [20]. PEA-2,12 was significantly more crystalline than the poly(ester–amides) synthesized from C_{12}-amidediol and longer chain diacids. The Laue pattern of the succinic acid poly(ester–amide) fiber (PEA-2,12) indicated retention of crystallinity as well as a high degree of orientation. From molecular models it was seen that there is a nearly linear extended chain conformer possible in which alternating C–O groups are arranged as shown in Figure 4.5. This conformer is also the most favorable one for forming a regular H-bonding network with neighboring chains. In a fully extended form, one monomer unit of PEA-2,12 was calculated to be 29 Å long. The diffraction pattern of both the fiber and the original polymer showed a prominent spacing of 25 Å. Thus the chain may not have been fully extended in the crystalline regions. The other prominent lines were consistent with the packing of long molecular chains. The Laue pattern indicated that these chains were probably aligned parallel to the fiber axis. Molecular models showed the succinate segment to be a stiff section with limited conformational freedom, implying that it may be important in inducing polymer crystallization.

The polymerization of C_2-amidediol with long-chain diacids, on the other hand, was also discovered to give poly(ester–amides) with good crystallinity [19]. As shown in Figure 4.5, the PEA-x,2 and PEA-2,y structures provide similarly spaced ester–amide groupings. Polymers made from C_2-amidediol, C_2S-amidediol (made from L-lactic acid and ethylenediamine), and various copolymers made from mixtures of C_2- and C_2S-amidediol and from mixtures of diacid chlorides all gave relatively high crystalline melting transitions (T_m) similar to the PEA-2,y polymers. The thermal transitions for some examples of these PEA-x,2 polymers and copolymers are shown in Table 4.2.

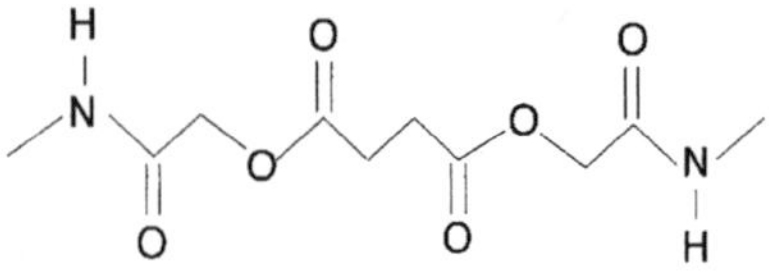

PEA-2,y

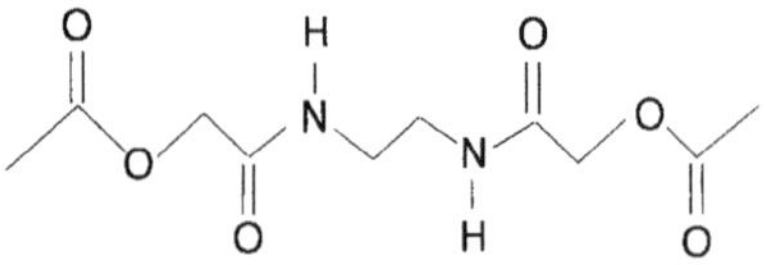

PEA-x,2

Figure 4.5   Arrangement of ester–amide bonds in PEA-2,*y* and PEA-*x*,2 polymers.

4.3.2   Fiber Properties

Melt-spinning of PEA-2,12 has produced monofilament fibers with tensile strength up to 6.5 g/denier. For use as surgical suture, however, monofilament fibers must have good flexibility in addition to strength. Fiber flexural rigidity has been studied by Hoffmann et al. by means of a loop deformation test [21]. These authors found

Table 4.2   DSC Values for Poly(ester–amides) Made from Ethylenediamine

Polymer	T_m (°C)	T_c (°C)
PEA-2,2	190	115
PEA-4,2	167	91
PEA-6,2	145	80
PEA-8,2	155	130
PEA-10,2	159	135
PEA-11,2	159	135
PEA-14,2	185	165
PEA-10,2-co-50%-8,2	150	50
PEA-10,2-co-50%-4,2	120,145,167	55
PEA-10,S2	145	112
PEA-10,RS2	142	110
PEA-10,RS2-co-50%-8,RS2	120	—
PEA-10,RS2-co-50%-2,RS2	125,159	140,65

that the ranking of surgeon preference for various sutures based on handling quality correlated with the ranking of fiber flexural rigidity values, lower being better. To improve upon the flexural rigidity of PEA-2,12 an ether containing copolymer was made by mixing the C_{12}-amidediol with 10 mol% of 1,12-di(hydroxyacetamido)-4,9-dioxadodecane, also referred to as C_{oxy}-amidediol, prior to polymerization with succinyl chloride [14]. Samples of 2-0 U.S.P. suture size monofilament fiber (0.300–0.330 mm dia.) were made from PEA-2,12 and PEA-2,12-co-10%-2,oxy by melt spinning. The fibers were tested in comparison to several commercial U.S.P. 2-0 sutures using a flexural rigidity test [22]. This test was modified from the published procedure by forming the fiber loops around a 100-ml glass beaker rather than a 7.88-mm glass rod. It should be noted that flexural rigidity is highly dependent on fiber processing parameters such as draw ratio and often cannot be maximized without detracting from tensile strength. The incorporation of 10–20% of C_{oxy}-amidediol into the PEA-2,12 structure, however, had a negligible effect on T_m and the improved flexibility was not offset by loss of fiber tenacity. Other mechanical properties for this 10% oxydiol copolymer are compared to properties reported [23] for nylon-12 in Table 4.3.

Fiber properties for PEA-2,6 in comparison to PEA-2,12 and two commercially obtained monofilament absorbable sutures are shown in Table 4.4. The modulus value obtained from the tensile stress–strain curve also provides an indication of fiber flexibility. The modulus of PEA-2,6 was reduced by about 30% by subjecting the fiber to a lower draw ratio. This increased the elongation but decreased the tensile and knot strengths. The U.S.P. knot strength for the 3-0 suture size (1.77 kg), however, was still surpassed under these conditions.

PEA-10,2 was melt extruded into monofilaments and the mechanical properties of the filaments were measured as a function of spinning and drawing conditions [24]. The effect of draw ratio on tensile strength and yield strength is shown in Figure 4.6. The concomitant increase in fiber modulus followed a similar upward inflection from 0.81 to 3.89 GPa as the draw ratio was increased from 3 to 5.

Table 4.3 Mechanical Properties of PEA and Nylon-12

Sample	T_m (°C)	Tensile strength (psi)	Modulus (psi)	Elongation (%)
PEA-2, 12-co-10%-2,oxy				
Extruded Film	165	5400	63,000	410
Injection Molded	165	6300	52,500	470
Nylon-12 poly(lauryl lactam)				
Extruded Film	175	7300	116,000	350

Table 4.4   Fiber Properties

	Commercial Sutures		Experimental Fibers	
	PDS™	*Maxon*™	*PEA-2,6*	*PEA-2,12*
USP size	3-0	3-0	3-0	2-0
Tenacity (g/denier)	5.2	4.5	5.3	6.5
Tensile strength (psi)	96,300	85,000	91,400	—
Modulus (psi)	2,870,000	470,000	1,043,000	550,000
Knot strength (kg)	2.35	—	2.04	3.10
Elongation (%)	49	—	21	35
Strength retention 4 wks. postimplantation (%)	71	59	63	95
Time required for bioabsorption (months)	6	7	8	36 (est.)

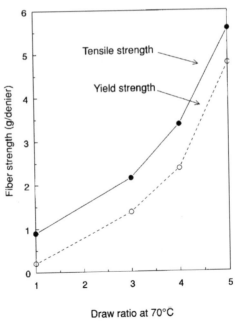

Figure 4.6   Draw ratio vs. tensile strength and yield strength for PEA-10,2 monofilament fibers.

4.4   In Vitro Evaluations

It is well known that the in vivo strength loss profile of synthetic absorbable sutures can be reproduced in vitro due to the simple acid–base catalyzed mechanism of polyester hydrolysis [25]. The accuracy of this correlation probably results from the fact that a relatively small percent change in molecular weight accounts for a large drop in mechanical properties whereas mass loss requires extensive depolymerization. Thus in vitro mass loss testing may be less predictive. Moreover, enzymatic hydrolysis could become an important factor as molecular weight decreases [26]. In the case of bioabsorbable polymers with very long absorption times, however, accelerated in vitro hydrolysis can be useful in ranking the absorbability of various polymers, especially those with close structural similarities [27].

The total hydrolysis of a series of poly(ester–amide) fibers was determined in an accelerated in vitro test by incubating 200-mg samples of monofilament fibers in pH 11 buffer at 80°C and recording the time required for dissolution [28]. As shown in Figure 4.7, PEA-2,6 and PEA-6,2 had a similar rate of dissolution and were fully hydrolyzed in 90 min. PEA-2,12 dissolved more slowly than PEA-10,2 and yielded a 20% residue of insoluble C_{12}-amidediol. PEA-14,2 dissolved only slightly slower than

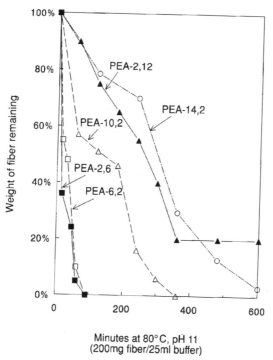

Minutes at 80°C, pH 11
(200mg fiber/25ml buffer)

Figure 4.7   Weight loss of PEA-x,2 and PEA-2,y fibers under conditions for accelerated hydrolysis.

PEA-2,12 but gave a completely clear solution due to the solubility of the long-chain dicarboxylic acid product in the high pH buffer. All three of the PEA-*x*,2 polymers appeared to exhibit a retardation of weight loss relative to the PEA-2,*y* samples prior to 50% hydrolysis. This suggests the possibility that PEA-*x*,2 polymers may possess a more hydrolytically stable crystal structure than the PEA-2,*y* polymers. It has been determined in other bioabsorbable polymers that the crystalline phase is more resistant to hydrolysis than the amorphous phase [29].

In comparing the total hydrolysis of PEA-10,2 multifilament yarn to poly-L-lactic acid (PLA) multifilament yarn under the above conditions, it was found that PEA-10,2 was dissolved in 6 h whereas PLA required 75 h [30]. It was also shown that PEA-10,2 had a rate of strength retention under in vitro physiological conditions that would be useful for the connective tissue healing situations for which PLA has been suggested. Thus PEA-10,2 was concluded to be a superior choice due to the prediction of a more rapid rate of bioabsorption of degraded fiber residues and particles compared to PLA.

The in vitro strength loss profiles of PEA-2,12 and PEA-10,2 are compared to the in vivo strength loss profile of PEA-2,12 in Figure 4.8. The in vivo data [17] were obtained using U.S.P. size 2-0 monofilament suture fiber whereas the in vitro data were obtained using approximately 40-μm diameter single filaments in the case of PEA-10,2 and braids of multifilament yarn composed of 40-μm monofilaments in the case of PEA-2,12. The PEA-2,12 braids were tested in water rather than in physiological buffer for reasons unrelated to this comparison [31]. Because of this the PEA-2,12 in vitro curve is displaced toward longer strength retention relative to the in vivo curve but otherwise has a similar shape. The PEA-10,2 curve, however, exhibits a dramatically longer rate of strength retention which is surprising in view of the faster rate of hydrolysis of PEA-10,2 compared to PEA-2,12 [28]. As mentioned

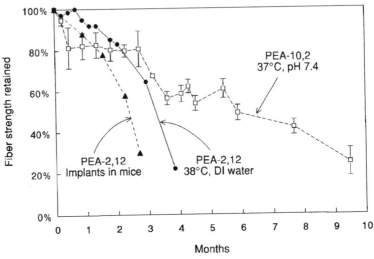

Figure 4.8 Comparison of strength retention in vitro of PEA-2,12 and PEA-10,2 fibers.

above, a more stable crystal structure for the PEA-x,2 polymers relative to the PEA-2,y polymers could account for this observation.

4.5   In Vivo Evaluations

A series of experiments was performed to determine the gross biocompatibility and relative rates of strength loss and absorption of various poly(ester–amide) fibers in vivo [32]. Monofilament fibers (2-0 U.S.P. suture size) prepared by melt spinning were implanted subcutaneously in mice and examined at various times postimplantation for up to 19 months. This initial evaluation showed that low-crystallinity, low-strength polymers had a rapid rate of strength loss. Thus PEA-12,12 retained only 10% of its initial tensile strength after 2 weeks of implantation whereas PEA-2,12 still possessed 98% strength after the same time in vivo. All polymers made from C_{12}-amidediol were bioabsorbed slowly. The thinly encapsulated implants gradually became soft and appeared to have lost considerable mass after 1 year, but residues were still visible after 19 months. All of the poly(ester–amide) fiber implants appeared grossly biocompatible at all time points. This was not the case with chromic catgut suture used as a control, however, as 5 out of 50 implantation sites containing chromic catgut suture exhibited a moderate to severe degree of inflammation.

An in vivo comparison of the degradation characteristics of PEA-2,12 and PEA-2,6 has been conducted with the use of carbon-14 labeled fibers implanted in rats [33]. The metabolism of PEA-2,6 primarily resulted in urinary excretion of mostly unchanged C_6-amidediol whereas the metabolism of PEA-2,12 resulted in both urinary and fecal excretion routes with the recovery of virtually no detectable unchanged C_{12}-amidediol. Although the bioabsorption of PEA-2,6 implants was complete after about 8 months, it was estimated that 3 years would be required for total bioabsorption of PEA-2,12. This study also included the dosing of rats with four different radiolabeled amidediols. As mentioned previously, greater water solubility of amidediol was determined to facilitate more rapid polymer bioabsorption. In the radiolabeling study it was confirmed that water-soluble amidediols injected into rats were rapidly excreted. The greater the water solubility of the amidediol, the faster and more extensively it was excreted unchanged in the urine. Conversely, the more hydrophobic the amidediol, the greater was its fecal excretion and the degree of biotransformation that occurred. In any case, all implanted radiolabeled samples of amidediols were completely eliminated without significant retention in any tissues or organs.

A recent pilot study of the metabolism of radiolabeled PEA-10,2 in 10 rats for 58 days showed that the rate of radiolabel excretion was intermediate between PEA-2,6 and PEA-2,12 [34]. This confirms the validity of the above accelerated in vitro fiber hydrolysis results which also placed the rate of PEA-10,2 hydrolysis as intermediate between PEA-2,6 and PEA-2,12. As expected, radiometric chromatographic analysis identified precursor C_2-amidediol as the major urinary excretion component due to hydrolysis of the subcutaneous PEA-10,2 fiber implants.

The biocompatibility of PEA-10,2 was confirmed by histological examination of subcutaneous implants in rats using PLA as a control material [35]. In this case the implants were prepared by accelerated in vitro degradation of fibers into low molecular weight particles. This provided a simulation of the end-stage of the bioabsorption process with an artificially high concentration of polymer being absorbed into the tissue. Both polymers exhibited an acceptable tissue response under these conditions.

4.6   Toxicology

The acute toxicity of C_6-amidediol was tested in 10 rats by intraperitoneal injection of 5000 mg/kg. Under these conditions there was no evidence of toxic effects for 2 weeks post-dose nor were any visible lesions detected during necropsy. The test was then repeated using PEA-2,6 in powdered form. This result also showed the material to be nontoxic. Histological examination of intramuscular implants of PEA-2,6 in rats revealed a mild cellular reaction which is similar to that observed for commercial synthetic absorbable sutures [36].

Since the PEA-x,2 series has shown promise as an especially useful class of poly(ester–amides), toxicological studies were conducted on C_2-amidediol. Testing of this amidediol provided evidence of toxicological safety for all PEA-x,2 polymers due to the observed release of soluble C_2-amidediol as a product of polymer hydrolysis. These tests included sensitive in vitro genetic toxicology assays such as morphological transformation of Balb/3T3 mouse embryo cells and chromosome aberrations in Chinese hamster ovary cells. Both of these tests were conducted with and without pretreatment of the C_2-amidediol with a rat liver fraction to screen for possible increased toxicity due to metabolic changes. Extracts of PEA-10,2 fiber, in addition to C_2-amidediol, were also tested for cytotoxicity, sensitization, and mutagenicity (Ames Test) and were found to be nontoxic under the conditions of these standard assays [37]. Although C_2-amidediol was not observed to break down further into ethylenediamine, the presence of ethylenediamine as a trace metabolite would not present a problem since this diamine was shown to be noncarcinogenic in a 2-year chronic study in rats [38]. The other products of PEA-x,2 hydrolysis, namely linear aliphatic dicarboxylic acids, are of little concern due to the known pathways for metabolism of these relatively innocuous compounds [39].

4.7   Conclusions

Poly(ester–amides) are a new class of bioabsorbable, biocompatible polymers suitable for fiber, film, and molding applications. Polymer crystallinity, fiber strength retention, and absorption time were optimized by placing a linear hydrocarbon moiety in the diacid monomer and by selecting a short-chain, water-soluble amidediol monomer. Copolymerization of mixtures of amidediols and/or diacids provided a further degree of control over physical properties. The principle mechanism of in vivo degradation appeared to be hydrolysis of the ester bonds. This produced amidediol monomer as a major metabolite which was determined to be nontoxic by various assays.

REFERENCES

1. Barrows, T.H. *Clin. Mater.* **1,** 233–257 (1986).
2. Okazaki, K.; Nakagawa, A.; Nakayama, Y.; Sugii, K. U.S. Patent 3,493,632 (1970).
3. Katayama, S.; Murakami, T.; Takahashi, Y.; Serita, H.; Obuchi, Y.; Ito, T. *J. Appl. Poly. Sci.* **20,** 975–994 (1976).
4. Katayama, S.; Horikawa, H. *Offenlegungsschrift* **2,** 236,949 (1973).
5. East, G.C.; Kalyvas, V.; McIntyre, J.E.; Milburn, A.H. *Polymer* **30,** 558–563 (1989).
6. Goodman, M.; Kirshenbaum, G.S. U.S. Patent 3,773,737 (1973).
7. Shalaby, S.W.; Jamiolkowski, D.D. U.S. Patent 4,209,607 (1980).
8. Carter, B.K.; Wilkes, G.L. *Polym. Prepr.* **24,** 77 (1983).
9. Gilding, D.K.; Reed, A.M. *Polymer* **20,** 1459 (1979).
10. Vert, M.; Chabot, F. *Makromol. Chem. Suppl.* **5,** 30 (1981).
11. Wang, P.Y.; Arlett, B.P. *Polymer Sci. Technol.* **7,** 173 (1973).
12. Barrows, T.H.; Sachi, T.J. Unpublished results.
13. Skinner, W.A.; Johansson, J.G. *J. Med. Chem.* **13,** 319 (1970).
14. Barrows, T.H. U.S. Patent 4,343,931 (1982).
15. Huang, S.J.; Bell, J.P.; Knox, J.R.; Atwood, H.; Bansleben, D.; Bitritto, M.; Borghard, W.; Chapin, T.; Leong, K.W.; Natarjan, K.; Nepumuceno, J.; Roby, M.; Soboslai, J.; Shoemaker, N. In *Proceedings of the Third International Biodegradation Symposium.* Sharpley, J.M.; Daplan, A.M., eds. Applied Science, London, pp. 731–741 (1976).
16. Mahler, H.R.; Cordes, E.H. *Biological Chemistry,* pp. 513–521. Harper & Row, New York (1966).
17. Barrows, T.H.; Johnson, J.D.; Gibson, S.J.; Grussing, D.M. In *Polymers in Medicine II,* pp. 85–80. Chiellini, E.; Giusti, P.; Migliaresi, C.; Nicolais, L., eds. Plenum Press, New York (1986).
18. Barrows, T.H. U.S. Patent 4,529,793 (1985).
19. Barrows, T.H.; Stegink, D.W.; Suszko, P.R.; Truong, M.T. An improved process for increasing stability of poly(esteramides). U.S. Patent pending.
20. Barrows, T.H.; Etter, M.C. Unpublished results.
21. Hoffman, A.H.; Herrmann, J.B.; Lesneski, D.V.; O'Connor, J.L. In *Proceedings of the 5th New England Bioengineering Conference,* pp. 21–25, Cannon, M., ed. Pergamon Press (1977).
22. Barrows, T.H.; Havens, M.A. Unpublished results.
23. Product Information Sheet, Nylon 12—Huels Film Grades, Huels Corp., 750 Third Ave., New York, NY 10017.
24. Spruiell, J.E. Report submitted to 3M, October 28, 1991.
25. Chu, C.C. *J. Biomed. Mater. Res.* **16,** 117–124 (1982).
26. Williams, D.F.; Mort, E. *J. Bioeng.* **1,** 231–238 (1977).
27. Buchholz, B. *The 4th International ITV Conference on Biomaterials: Degradation Phenomena of Polymeric Biomaterials,* Denkendorf, Germany, September 3–5, 1991.
28. Barrows, T.H.; Truong, M.T. Bioabsorbable poly(ester–amide) and method for making same. U.S. Patent pending.

29. Bendix, D.; Entenmann, G. Presented at *The World Congress on Implantology and Bio-materials,* Paris, France, March 8–11, 1989.
30. Barrows, T.H.; Truong, M.T.; Johnson, P.R.; Havens, M.A. *Trans. Fourth World Bioma-terials Congress* **15,** 249 (1992).
31. Bushey, P.R.; May, S.J. Unpublished results.
32. Barrows, T.H.; Grussing, D.M.; Hegdahl, D.W. *Trans. Soc. Biomater.* **6,** 109 (1983).
33. Horton, V.L.; Blegen, P.E.; Barrows, T.H.; Quarfoth, G.J.; Gibson, S.J.; Johnson, J.D.; McQuinn, R.L. In *Progress in Biomedical Polymers,* pp. 263–282. Gebelein, C.G.; Dunn, R.L., eds. Plenum Press, New York 1990.
34. Myhre, P.E. 3M Pharmaceuticals/Drug Metabolism Technical Report, June 8, 1990.
35. Gibbons, D.F.; Barrows, T.H.; Truong, M.T. *Trans. Fourth World Biomaterials Congress* **15,** 408 (1992).
36. Barrows, T.H.; Gibson, S.J.; Johnson, J.D. *Trans. Soc. Biomater.* **7,** 210 (1984).
37. Tests performed by Microbiological Associates, Inc., Bethesda, Maryland 20816 and North American Science Associates, Inc., Northwood, Ohio 43619.
38. Yang, R.S.H.; Garman, R.H.; Maronpot, R.R.; Mirro, E.J.; Woodside, M.D. Bushy Run Research Center Report No. 46–27, May 11, 1984, 547pp, Bushy Run Research Center, Mellon Institute-Union Carbide Corp., Export, PA 15632.
39. Berseth, S.; Poisson, J.P.; Bremer, J. *Biochim. Biophys. Acta.* **1042,** 182–187 (1990).

CHAPTER 5

Amino Acid Derived Polymers

Aruna Nathan and Joachim Kohn

5.1   Introduction: A Historical Perspective

This chapter reviews the medical applications of polymers derived from the naturally occurring α-L-amino acids. The authors have assumed that the reader is familiar with the basic chemical, conformational, and physicomechanical properties of conventional poly(amino acids). Due to the availability of several comprehensive prior reviews (for example Katchalski and Sela [1], Lotan et al. [2], Katchalski [3], Anderson et al. [4], and Fasman [5],) the emphasis of this chapter is on developments that occurred after 1985.

Aruna Nathan and Joachim Kohn, Department of Chemistry, Rutgers University, New Brunswick, New Jersey 08855, U.S.A.

Amino acid derived polymers can be conveniently classified into three groups: conventional poly(amino acids), pseudo-poly(amino acids), and copolymers of amino acids and non-amino acids. Here poly(amino acids) are defined as synthetic polymers composed of α-amino acids linked by *peptide* bonds whereas pseudo-poly(amino acids) are synthetic polymers composed of α-amino acids linked by *nonpeptide* bonds such as ester, carbonate, or urethane linkages. The third category comprises a large number of structurally diverse copolymers that contain both amino acid and non-amino-acid units within the polymer backbone.

Although the first experiments leading to the synthesis of poly(amino acids) can be traced back to the work done by Leuchs in 1906 [6], the intensive exploration of poly(amino acids) did not start until sometime in the 1940s. At that time, procedures had become available for the successful conversion of many amino acids to N-carboxyanhydrides, the necessary monomeric starting materials for the preparation of poly(amino acids) of reasonably high molecular weight (Figure 5.1). The ring-opening polymerization of N-carboxyanhydrides provided a convenient, albeit expensive, route to homopolymers (containing one single amino acid), random copolymers (derived from mixtures of different amino acid N-carboxyanhydrides), and A–B or A–B–A type block copolymers.

Much of the initial work on poly(amino acids) was authoritatively reviewed by Katchalski and Sela [1] in 1958. For all practical purposes, this was the first truly

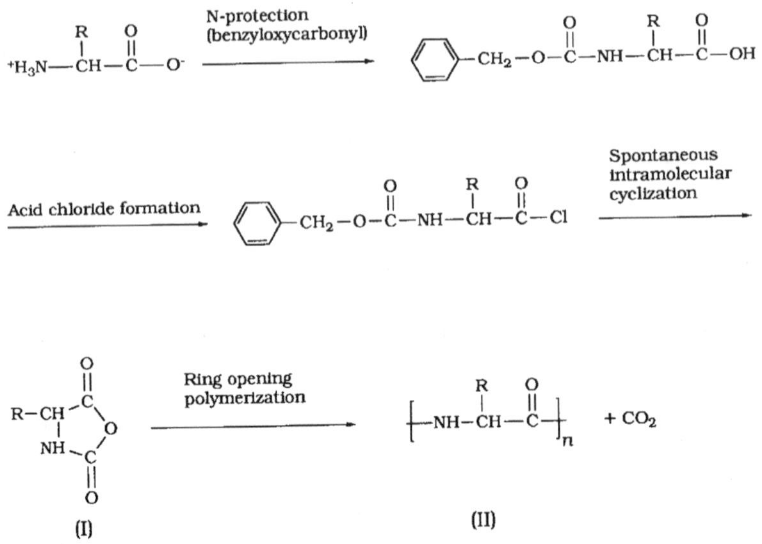

Figure 5.1   General method for the preparation of poly(amino acids) (II) via amino acid N-carboxyanhydrides (I). For the synthesis of the intermediate N-carboxyanhydride, the historic reaction scheme via acid chlorides is shown. Today, procedures for the direct conversion of amino acids to N-carboxyanhydrides by reaction with phosgene are available. The preparation and purification of the labile N-carboxyanhydride is the most expensive step in the reaction sequence.

comprehensive review of the early synthetic work on poly(amino acids). That review has proven to be of lasting value, as the basic reaction schemes used for the chemical synthesis of poly(amino acids) have not changed fundamentally over the last 35 years. In addition, Katchalski and Sela provided a thorough description of the fundamental chemical properties of the poly(amino acids) available at that time. Obviously, those observations are still valid today. For an early, detailed account of the physical properties of poly(amino acids), the reader is referred to the work by Bamford et al. [7].

The driving force behind the exceptionally high interest in poly(amino acids) during the 1950s and 1960s was the early recognition that poly(amino acids) can serve as valuable model systems for natural proteins. Thus, poly(amino acids) played a pivotal role in the elucidation of secondary protein structure and facilitated research aimed at the prediction of protein conformation from the primary amino acid sequence [5]. Poly(amino acids) were also used successfully in the elucidation of enzyme specificities [8], the detection and purification of new proteolytic enzymes [9], and the investigation of the structure–activity correlations governing the antigenicity of natural proteins [10]. Numerous biological activities of poly(amino acids) were discovered and explored before 1960, including possible catalytic properties, the effect of poly(amino acids) on blood clotting, interactions with nucleic acids, interactions with viruses, and antibacterial properties. For detailed coverage of these topics, the reader is referred to an excellent review of the early biological work on poly(amino acids) published in 1959 by Sela and Katchalski [8].

Because of the virtually unlimited number of amino acid sequences, polymer chemists and material scientists were interested in poly(amino acids). The general feeling was that because of their structural diversity, these polymers would be promising candidates for industrial exploitation. Furthermore, poly(amino acids) have well-defined conformations of long-range chain order. These conformations and the associated regularity of structure persist even in solution. These rare properties led many to expect that synthetic poly(amino acids) would provide a basis for unique commercial applications.

In spite of their structural diversity, however, poly(amino acids) share many common traits. Some of these traits turned out to be highly disadvantageous when poly (amino acids) were considered for industrial applications (Table 5.1). Thus, Katchalski and Sela wrote as early as 1958 that "the primary importance of poly(amino acids) has been, and will probably remain, as simple synthetic protein models" [1]. Based on that assessment, the general interest in poly(amino acids) waned during the 1970s as the elucidation of the fundamental properties of proteins came to completion and significant industrial applications of poly(amino acids) appeared unrealistic.

The successful commercial development of the first, degradable surgical sutures between 1966 and 1971 [11–14] led to an intensive search for new, degradable biomaterials. Because poly(amino acids) degrade in vivo to simple amino acids, the application of these polymers in the manufacture of degradable medical implants was an obvious idea. Following the work of Miyamae et al. in 1968 on the use of partially esterified poly(glutamic acid) filaments as surgical sutures [15], several laboratories investigated the use of poly(amino acids) as medical implant materials.

For reasons outlined in Table 5.1, those early investigations did not lead to significant, practical applications. However, two exciting, recent developments may now lead to major, "high tech" industrial and medical applications for poly(amino acids) and other amino acid derived polymers in the near future. The first development relates to the concept of pseudo-poly(amino acids). In those materials many of the unfavorable engineering properties of conventional poly(amino acids) have been eliminated by the replacement of the recurring amide bonds in the polymer backbone by nonamide linkages such as ester, carbonate, or urethane bonds [16–18]. As an additional advantage, pseudo-poly(amino acids) are potentially cheaper to prepare than conventional poly(amino acids), because the expensive amino acid N-carboxyanhydrides are not required. The work on pseudo-poly(amino acids) is reviewed in more detail below.

The second development relates to significant advances made in the preparation of sequential poly(amino acids) and the understanding of the unique properties of these polymers. That progress is exemplified by the ground-breaking work of Tirrell at the University of Massachusetts and Urry at the University of Alabama. Tirrell developed biosynthetic methods for the preparation of sequential poly(amino acids) with an unprecedented degree of control over the sequence and molecular weight distribution of the resulting synthetic polymers. An impressive demonstration of the potential of sequential poly(amino acids) has been provided by Urry, who studied sequential poly(amino acids) built from the repeat sequences of elastin and its analogs. After suitable crosslinking, some of these materials exhibited thermomechanical, electromechanical, or chemomechanical transduction, i.e., the materials carried out mechanical work in response to temperature changes and electrical and chemical stimuli. The potential industrial or medical applications of synthetic, sequential polypeptides are extremely diverse and include novel adhesives, ion-selective transmembrane pumps and channels, miniature motors and contractile systems, materials for the treatment of surgical adhesions, and uniquely designed drug delivery systems. For additional information about the biosynthesis and special properties of sequential poly(amino acids), the reader is referred to a series of recent publications by Urry [19–24] and Tirrell [25–32].

It is a fortunate coincidence that at a time when major technological advances such as the development of medical implants, nonlinear optical devices, electronic information storage devices, or miniature mechanical systems depend critically on the availability of new materials with carefully designed properties, a diverse range of amino acid derived polymers with unique physicomechanical characteristics are becoming available for exploration. It is possible that these developments will ultimately refute the prediction made by Katchalski and Sela in 1958.

5.2   Medical Applications of Conventional Poly(amino acids)

5.2.1   Poly(amino acids) Within the Context of Biomaterials

At the beginning of this century, advances in the material sciences made it possible to use conventional engineering materials in medical implants on a large scale. The use of stainless steel in orthopedic implants represents one of the earliest examples for the successful introduction of a new "biomaterial" into clinical use. Since then, numerous manmade materials have found increasingly sophisticated applications in medicine. Stainless steel, for instance, was replaced in many applications by various titanium alloys and the use of metals in general was augmented by ceramics, composite materials, and by a number of polymeric materials [33,34].

From a practical perspective, it is best to distinguish between long-term and short-term implant applications. Whereas in long-term applications (e.g., permanently implanted prostheses, artificial organ parts, cosmetic implants) exceptionally inert polymers are required, degradable polymers can be applied advantageously in short-term applications such as sutures or drug delivery. Because most poly(amino acids) degrade slowly under physiological conditions, the interest in poly(amino acids) as biomaterials was closely linked to the conceptual and practical advances made in the use of degradable implant materials in general. It is not surprising that degradable sutures were the first widely investigated "biomaterial" application of poly(amino acids) [15].

The intensive exploration of poly(amino acids) as biomaterials started around 1975. The basic rationale for the choice of those materials was nicely summarized in a 1980 publication by Walton, who wrote that

> polypeptides are perhaps a strange choice as biomedical plastics because they are expensive and difficult to produce in large quantities and because fragmentation in vivo could potentially produce immunogenic species. However, the advantages are that with some twenty naturally occurring L-amino acids one can produce a vast array of surface and bulk properties by copolymerizing two or more of these acids. In addition, since the amino acids are components of proteins in native tissues, it is conceivable that certain types of synthetic polypeptide surfaces may mimic native tissue [35].

One of the most widely cited reasons for the use of poly(amino acids) as biomaterials was their perceived structural diversity. Unfortunately, this line of reasoning turned out to be an oversimplification. From an engineering point of view, poly(amino acids) derived from a single amino acid (e.g., homopolymers such as polyalanine, poly-glycine, polyleucine, etc.) can be regarded as "nylon 2" type polyamides. Because most of these polymers are essentially nonprocessible by conventional polymer fabrication techniques, these homopolymers have few practical applications. The most promising plastic materials among the poly(amino acids) are copolymers of two or more amino acids. However, due to the potential immunogenicity of random copolymers of more than two different amino acids [4], the search for biomaterials is essentially limited to a very narrow subset of all available poly(amino acids).

Another widely cited reason for the interest in poly(amino acids) as biomaterials was the perception that poly(amino acids) could be regarded as "analogs" of natural tissue [4]. This, too, was an oversimplification. Although it is well established that most synthetic poly(amino acids) are relatively nontoxic and tissue-compatible materials, poly(amino acids) containing one or two amino acids have very little resemblance to functional natural tissue with regard to their interactions with living cells that are in direct contact with the polymer.

Today, the development of *biologically functional* tissue analogs is one of the most important challenges in biomaterials research. In this context, biological function can refer to the ability of a polymeric surface to support the growth of specific cell types, to prevent attachment and activation of platelets, or to stimulate the growth of healthy tissue at a site of injury. Clearly, simple poly(amino acids) do not usually exhibit these biological activities. However, sequential poly(amino acids) containing specific signal sequences (such as, e.g., the RGD sequence) can probably be designed to mimic specific biological functions of natural tissue. Likewise, pseudo-poly(amino acids) to which biologically active ligands can be readily attached via reactive, pendent side chains may find significant applications as starting materials in the design of functional tissue analogs.

5.2.2   *Immunological Properties*

Within the context of medical implant materials the immunological properties of poly(amino acids) are an important concern. In that context, one has to distinguish between immunogenicity and antigenicity. Here *immunogenicity* is defined as the capacity of a material to provoke an immune response, whereas *antigenicity* relates to the ability of a molecule to interact specifically with an antibody. Although the two terms are often used interchangeably, an antigenic material may not necessarily be immunogenic. Sela noted that one of the most important uses of poly(amino acids) has been in the field of immunology [10]. Indeed, many poly(amino acids) have been shown to be *antigenic*, but it has also been established that certain poly(amino acids) are only very weak *immunogens*.

During the 1950s and 1960s, the exploration of poly(amino acids) as synthetic antigens contributed significantly to the elucidation of the molecular basis of antigenicity. As a "fringe benefit," the work on antigen design provided important insights into the possible use of poly(amino acids) as biomaterials. In essence, homopolymers of amino acids were shown to be very poor immunogens. For example, attempts to raise antibodies against poly-D,L-alanine, polysarcosine, poly-L-aspartic acid, poly-L-lysine, poly-L-hydroxyproline, poly-L-glutamic acid, and poly-L-proline in several animal species were consistently unsuccessful [8,36]. However, the immunogenicity of poly(amino acids) increases with increasing molecular complexity (e.g., branching) and with the number of different amino acid residues present in the copolymer [10].

It is important to realize that the immunogenicity of an amino acid derived polymer is strongly dependent on structure and conformation [37]. For example, a multiresidue, *branched* copolymer of lysine, alanine, tyrosine, and aspartic acid was a powerful synthetic immunogen. That polymer consisted of a poly-L-lysine backbone to which small pendent chains of poly(D,L-alanine) had been attached via the ε-amino groups of lysine. The poly(D,L-alanine) chains were in turn elongated with peptides containing

L-tyrosine and L-glutamic acid [38]. On the other hand, repeated attempts to raise antibodies against an *unbranched* copolymer of L-tryosine and L-aspartic acid consistently failed in several animal species [8].

Those observations indicate that there is no contradiction in the statement that some poly(amino acids) are intensely immunogenic and can serve as immunologically active carriers for haptens and other antigens, whereas other poly(amino acids) can serve as immunologically inactive implant materials. As a general rule, the search for nonimmunogenic implant materials should be limited to unbranched homopolymers and copolymers containing not more than two different amino acid residues in random sequence. Obviously, in view of these observations, the immunological properties of the new, sequential polypeptides investigated by Tirrell or Urry will have to be carefully evaluated prior to their use as biomaterials.

5.2.3   Poly(amino acids) as Medical Implant Materials

In the 1970s, poly(amino acids) were regarded as highly promising candidates for medical applications. However, it rapidly became apparent that in addition to the limitations imposed by immunological considerations, the physicomechanical properties of conventional poly(amino acids) were often highly undesirable from a material engineering point of view (Table 5.1). The need to prepare the expensive amino acid N-carboxyanhydrides as the monomeric starting materials increased the cost of all poly(amino acids) considerably, making poly(amino acids) expensive polymers even if they were derived from (relatively) cheap amino acids. Furthermore, with the exception of some derivatives of poly(glutamic acid), those poly(amino acids) that are insoluble in water are usually also insoluble in common organic solvents. The need for exotic solvent mixtures containing, for example, trifluoroacetic acid, 2,5-dimethyl-2-pyrrolidinone, or o-cresol severely limited the use of solution processing techniques, while the thermal instability of poly(amino acids) in the molten state excluded the use of convenient fabrication techniques such as injection molding or extrusion. For these reasons, poly(amino acids) are generally poor engineering materials and only a very limited number of poly(amino acids) may be commercially viable in a small number of carefully selected specialty applications.

Table 5.1   Undesirable Properties of Many Synthetic Poly(amino acids)

Expensive to produce in larger quantities
Insoluble in common solvents
Thermal degradation on melting
Swelling due to absorption of water
Erratic drug release profiles
Rates of degradation under physiological conditions often too slow

Due to their somewhat more favorable engineering properties, the most promising poly(amino acids) for use in medical implants are the γ-alkyl esters of poly(glutamic acid), the poly(N-alkyl glutamines), and several copolymers of glutamic acid with leucine, methionine, and a small number of additional amino acids. It is therefore not surprising that most laboratories focused their research efforts almost exclusively on those polymers. The development of degradable sutures from poly(γ-benzyl glutamate) was the first widely investigated medical application of a poly(amino acid) [15], followed by detailed studies on the use of poly(amino acids) in the design of drug delivery systems (see below). In addition, the interactions of implanted poly(amino acids) with living tissue were intensively investigated early on [35]. For a comprehensive summary of the exploration of poly(amino acids) as biomaterials up to about 1985, the reader is referred to an excellent review by Anderson et al. [4]. Some of the proposed applications of poly(amino acids) as medical implant materials are summarized in Table 5.2.

The exact mechanism of the in vivo degradation of poly(amino acids) was a widely investigated and somewhat controversial issue. For example, Hayashi et al.

Table 5.2   Applications of Conventional Poly(amino acids) as Implant Materials[a]

Application	Comments	Illustrative reference
Degradable sutures	Filaments drawn from γ-alkyl esters of poly-glutamate. Degradation time could be varied from 2 to 60 days.	Miyamae [15]
Artificial skin substitutes	Polypeptide laminates with nylon–velour. Clinical studies in humans were reported.	Spira [112], Hall [113]
Membranes for artificial kidney	The unusual permeability properties of poly(D,L-methionine–co-leucine) stimulated this research effort.	Martin [114]
Wound dressings	Several laboratories used polyglutamate and/or its derivatives (often in combination with other polymers) in the design of wound dressings or adhesion barriers. A poly-L-leucine sponge is in clinical trials in Japan.	Shiotani [115,116], Shioya [117], Brack [118], Kuroyanagi [119]
Hemostatic agents	A carrier containing thrombin and blood coagulation factor XIII made of a derivative of poly(glutamic acid)	Sakamoto [120]
Hard tissue prostheses	Composite material consisting of calcium phosphate, poly(γ-benzyl L-glutamate), and poly(2-ethyl-2-oxazoline)	Dorman [121]

[a] The use of poly(amino acids) for drug delivery systems and drug carriers is described in a subsequent table.

[39] exposed copolymers consisting of either L-glutamic acid or N-hydroxyalkyl-L-glutamine with various hydrophobic L-amino acids (alanine, leucine, or valine) to pseudo-extracellular fluid containing papain to elucidate the effects of copolymer composition and sequential distributions on the rate of degradation. In all cases, papain degraded those polymers extensively by random chain fracture. The authors made the interesting observation that polymers with uncharged side chains degraded faster than polymers that had a high proportion of negatively charged side chains. The degradation process always followed Michaelis–Menten kinetics. These initial studies were recently augmented by an investigation of the biodegradation of copoly(L-aspartic acid/L-glutamic acid) in vitro using essentially the same methodology [40].

A very detailed study of the enzymatic degradation of poly(N^5-hydroxyalkylglutamines) was published in 1989 by Pytela et al. [41]. They used a series of proteolytic enzymes (papain, cathepsin B, pronase E, pepsin, chymotrypsin A, and elastase) and quantified the degradation by gel permeation chromatography (GPC) in terms of moles of hydrolyzed peptide bonds in the main polymer chain. All proteases investigated in this study degraded the poly(N^5-hydroxyalkylglutamines) by the endopeptidase mechanism, albeit at widely different rates. It is noteworthy that only in some cases did the degradation result in truly low molecular weight products.

Recently, the effects of various surface treatments (glutaraldehyde, carbodiimide, plasma glow) on the rate of biodegradation were explored by Chandy et al. [42]. They monitored the enzymatic and hydrolytic degradation of thin films of poly(L-lactic acid) and poly(γ-benzyl L-glutamate) by weight loss, contact angle, pH changes, and tensile strength studies. The addition of α-chymotrypsin, carboxypeptidase, ficin, esterase, bromelain, and leucine aminopeptidase to the degradation medium tended to increase the rate of strength loss. Leucine aminopeptidase showed the highest enzymic effect on the degradation of glutaraldehyde-treated films of poly(γ-benzyl L-glutamate). Because plasma glow treatment always accelerated the rate of degradation, while the glutaraldehyde and carbodiimide treatments tended to slow degradation to various degrees, the authors proposed that surface modifictaions may provide new ways of controlling the biodegradation of poly(γ-benzyl L-glutamate) and other polymers.

Since 1985, significant advances have been made in the evaluation of the tissue compatibility of polymeric implants. Simple histological evaluations of the implant sites are now being increasingly augmented by detailed studies of dynamic changes that occur within the tissue surrounding the implant. Such studies can be performed both in vitro and in vivo. For example, the interaction of cultured cells with membranes composed of random and block sequences of γ-methyl L-glutamate and γ-benzyl L-glutamate was investigated by Minoura et al. [43], whereas the "cage implant system," allowing the quantitative determination of inflammatory cells and products of cell activation at the implant site, represents a particularly interesting in vivo model system. The cage system was used by Marchant et al. [44] for the in vivo evaluation of poly(2-hydroxyethyl-L-glutamine).

Another focus of modern biomaterials research relates to the absorption of proteins at the polymer surface. Considering that the biomaterial surface is rapidly covered with proteins after implantation or initial contact with body fluids, the importance of the "protein coat" in controlling the overall interaction of the polymer with the surrounding tissue is self-evident. Cho et al. investigated the adsorption of proteins to monolayers of poly(γ-benzyl-L-glutamate)/polyether block copolymer and postulated that the interaction between proteins and the copolymer monolayers may be controlled to a significant extent by hydrophobic bonding [45]. Studies of this kind are particularly important within the context of surfaces that contact blood.

The extensive research effort of many laboratories since about 1970 resulted in the exploration of numerous applications for poly(amino acids) as medical implant materials. A recent computerized literature survey in the Chemical Abstracts database retrieved over 500 publications and patents on the subject and Table 5.2 represents an overview of some of the medical applications that have been suggested for these polymers. In view of this extensive research effort, it is somewhat disappointing that conventional poly(amino acids) have not yet found significant applications as medical implant materials. It appears that the slow and often incomplete biodegradation of many poly(amino acids) in vivo, their marginal engineering properties, and their high cost were the most significant obstacles for the practical application of these polymers as medical implant materials.

5.2.4   Poly(amino acids) as Drug Delivery Devices and Drug Carriers

The most widely investigated medical applications of poly(amino acids) are in the area of drug delivery. In principle, poly(amino acids) can be utilized both as insoluble matrices where the drug is physically adsorbed, entrapped, or dispersed within the polymer (drug delivery devices) or as soluble prodrugs or drug carriers where the drug is chemically bound to the polymer (Table 5.3).

The formulation of drug delivery devices is explored in numerous publications. In a simple, early design, partially esterified poly(glutamic acid) was used in the formulation of a drug delivery system. The initial ratio of glutamic acid to alkyl glutamate in the copolymer controlled the erosion rate of the device and the rate of drug release [46]. In spite of a significant research effort, this simple system has not been used clinically. Later on, the same copolymer system of glutamic acid and ethyl glutamate was used to prepare hydrogel membranes that were permeable to proteins ranging in molecular weight from 12,000 to 69,000 [47]. In addition, copolymers of poly(γ-benzyl-L-glutamate) and poly(N^5-dihydroxyethylaminopropyl-L-glutamine) have been suggested as controlled release matrices for drugs such as 5-fluorouracil, a prednisolone derivative, adriamycin, and pepliomycin [48]. In those systems, the permeability of the copolymer membrane was shown to depend on the molecular weight and chemical structure of the drugs, and on the device characteristics.

A particularly interesting suggestion was made relatively early on by Sparer et al. [49], who exploited the liquid crystalline properties of poly(γ-benzyl-L-glutamate) in the design of a membrane whose permeability could be controlled by an external electric field. Application of an electric or magnetic field caused the liquid crystal

Table 5.3   Illustrative Examples for the Use of Poly(amino acids) as Drug Carriers

Poly(amino acid)	Application	Illustrative references
Poly(L-glutamic acid), poly(3-hydroxypropyl L-glutamine), and copolymers of the above with L-leucine	Controlled release of the narcotic antagonist naltrexone and the antihypertensive minoxidil	Bennett [53], Negeshi [54]
Copolymer of γ-benzyl-L-glutamate with N^5-dihydroxy-ethylaminopropyl L-glutamine	Controlled release of insoluble drugs such as prednisolone, and soluble drugs such as pepliomycin	Seno [48]
Copolymer of glutamic acid with ethyl glutamate	Conrolled release of proteins spanning a molecular weight range of 12,000–69,000	Sidman [47]
Poly(γ-benzyl L-glutamate)	Sustained release of hydrocortisone	Wise [122]
Poly(L-leucine), copolymer of L-leucine with N^5-3-hydroxyethyl L-glutamine	Conrolled release of prednisolone	Shyu [123]
Poly(L-aspartic acid)	Macromolecular prodrug of doxorubicin	Prates [57], Zunino [56,124]
Poly(α-β-aspartic acid)	Macromolecular prodrugs of procaine, histamine, and isoniazid	Giammona [55]
Poly(hydroxyethyl asparagine)	Macromolecular prodrug of 1,2-diaminocyclohexane-trimellitatoplatinum (II)	Fillipova-Voprsalova [125]
Poly(L-glutamic acid)	Macromolecular prodrug of doxorubicin	Kato [126]
Poly(L-lysine), poly(D-lysine)	Macromolecular prodrugs of doxorubicin and methotrexate	Ryser [58], Shen [59,127,128]

to undergo a phase transition from the cholesteric to the nematic phase with a corresponding change in the permeability of the membrane to any solute. When the field was switched off, the membrane reverted back to the cholesteric phase after 4–5 h. Thus, Sparer et al. concluded that the membrane may function as a gate to control solute permeability in drug delivery devices. Very similar results were reported by Bhaskar, who observed a 50–60% increase in the steady state flux through a membrane when poly(γ-benzyl-L-glutamate) underwent a phase transition to the nematic mesophase [50].

Another sophisticated controlled release membrane was suggested by Minaura et al., who used pH-induced reversible conformational changes to vary the permeability of a synthetic membrane. Minaura used an AB–BA type block copoly(amino acid) obtained from L-glutamic acid (A) and L-leucine (B) [51]. This block copolymer membrane had a phase-separated morphology with domains of poly(L-glutamic acid) embedded within a continuous matrix of poly(L-leucine). The reversible conformational change of poly(L-glutamic acid) from an α-helix to a random coil controlled solute permeation through the membrane. In this way, poly(L-glutamic acid) domains in the block copoly(amino acid) membrane simulated the transmembrane channels found in biological membranes.

Poly(amino acids) have also been investigated as macromolecular prodrugs with the drug being covalently bound to the polymer backbone [52]. The carboxylic acid groups of poly(aspartic acid) or poly(glutamic acid), and the amino groups of polylysine served as convenient attachment points for the covalent binding of a variety of drug molecules. Such macromolecular carriers have been formulated both as insoluble microparticles and as injectable drug conjugates.

The attachment of the narcotic antagonist naltrexone and the antihypertensive minoxidil to poly(glutamic acid) and poly(3-hydroxypropyl-L-glutamine) via labile covalent bonds is a representative example for the formulation of a microparticulate drug carrier [53,54]. In that system, the rate of drug release was affected both by the rate of hydrolysis of the drug from the polymer backbone and its rate of diffusion through the polymer matrix. Near constant drug release could be obtained by the incorporation of hydrophobic leucine residues into the polymer backbone.

Numerous publications describe the use of poly(amino acids) as water-soluble macromolecular carriers for a variety of pharmacologically active agents [52]. One of the most obvious designs of such a system involves the attachment of amino group containing drugs to the pendent carboxylic acid groups of poly(glutamic acid) or poly(aspartic acid). For example, procaine, histamine, and isoniazid could be readily linked to poly(α-β-aspartic acid) [55].

Attachment to injectable polymeric carriers is usually suggested for anticancer drugs such as daunomycin, doxorubicin, and methotrexate in an effort to decrease toxicity and increase the duration of antitumor activity. Often promising results obtained in vitro by this approach cannot be readily confirmed in vivo.

Various attachment strategies have been explored. Zunino et al. reported the attachment of doxorubicin to poly(aspartic acid) through degradable ester linkages [56]. Reportedly, the drug conjugate exhibited an improved therapeutic index relative to the free drug [57]. A different approach was explored by Shen and Ryser, who studied the conjugation of methotrexate to poly(L-lysine). This hydrolytically stable conjugate exhibited markedly increased cellular uptake and offered a new way to overcome drug resistance related to deficient transport. Conjugates using poly(D-lysine), on the other hand, failed to inhibit cell growth because the polymer is not degraded in the lysosomes and hence does not release free drug in the cells [58,59].

Those examples illustrate the wide range of applications poly(amino acids) can potentially find as drug carriers. Within the context of drug carriers and prodrugs, the nontoxic, degradable poly(amino acids) have indeed significant advantages over nondegradable polymers. Although no poly(amino acid)-based drug conjugate is currently in routine clinical use in the United States, the synthesis, characterization, and pharmacological evaluation of a large number of poly(amino acid)-based drug conjugates has significantly contributed to our understanding of the pharmacokinetics of soluble drug carriers. It stands to reason that those research efforts will lead to practical applications in the future.

5.3   Pseudo-Poly(amino acids)

5.3.1   Backbone Modification of Poly(amino acids)

Because many of the undesirable physicomechanical properties of conventional poly-(amino acids) can be traced to the recurring amide linkages in the polymer backbone, it was a logical strategy to look for ways to replace the amide (peptide) linkages in conventional poly(amino acids) by a variety of nonamide bonds [60,61].

In peptide chemistry, the term "pseudopeptide" is commonly used to denote a peptide in which some or all of the amino acids are linked by bonds other than the conventional peptide linkage [62]. In analogy to this practice, the term "pseudo-poly(amino acid)" was coined to denote a polymer in which naturally occurring amino acids are linked by nonamide bonds [16–18].

Surprisingly, very few attempts have been made to polymerize amino acids by linkages other than conventional amide bonds. It appears that in 1937 Greenstein was the first to attempt the polymerization of an amino acid via nonamide bonds when he tried unsuccessfully to use the sulfhydryl group of cysteine for the synthesis of a poly-sulfide [63]. A few polymers containing nonamide linkages between individual amino acids were mentioned in the review by Katchalski [1]. Those materials were usually of low molecular weight and their material properties were not described in detail. Later, Fasman attempted to convert poly(serine) to poly(serine ester) by means of the facile N→O acyl shift of serine, but obtained only a structurally ill-defined random copolymer [64]. In 1977, Jarm and Fles prepared several poly(N-benzenesulfonamido serine esters) of low molecular weight by the ring-opening polymerization of N-benzenesulfonamido-L-serine-β-lactones [65]. Because they employed bioincompatible benzenesulfonic acid derivatives for the protection of the serine N-terminus, those polymers were not directly applicable as biomaterials.

The use of pseudo-poly(amino acids) as biomaterials was first suggested by Kohn and Langer in 1984 [60]. In the meantime, a number of these structurally new polymers

Table 5.4   New "Pseudo"-Poly(amino acids)

Polymers	Literature citations
Poly(N-acyl *trans*-4-hydroxy-L-proline ester)	Yu [129], Yu Kwon [73], Kohn [61]
Poly(N-acyl L-serine ester)	Gelbin [70,71], Zhou [68], Fietier [130]
Tyrosine-derived Polyiminocarbonates	Kohn [16,60,61,66,76,77]
Tyrosine-derived Polycarbonates	Kohn [75,78,79,88]

have been prepared and characterized (Table 5.4). The currently available experimental data indicate that the selective replacement of conventional peptide bonds by nonamide linkages within the poly(amino acid) backbone can significantly alter the physical, chemical, and biological properties of the resulting polymer. Furthermore, the modification of the backbone of poly(amino acids) circumvents many of the limitations of conventional poly(amino acids) as biomaterials because pseudo-poly(amino acids) tend to retain the nontoxicity and good biocompatibility often associated with conventional poly(amino acids) while at the same time exhibiting significantly improved material properties. Thus, the approach of backbone modification can lead to a variety of useful biomaterials.

5.3.2   Preparation of Pseudo-poly(amino acids)

5.3.2.1   General Considerations

Because there are no mild, chemical reactions that can be used to transform the amide linkages in the backbone of conventional poly(amino acids) into nonamide linkages such as ester, urethane, or carbonate bonds, it is not possible to simply replace the amide bonds of conventional poly(amino acids) by nonamide linkages. Pseudo-poly(amino acids) must therefore be prepared by suitably designed polymerization reactions.

A synthetically simple approach is based on the use of trifunctional amino acids and dipeptides as monomeric starting materials [66,67]. Such amino acid derivatives can be polymerized by reactions involving the side chain functional group(s). In this way three structurally different types of polymers can be obtained from each trifunctional amino acid [16]. In Figure 5.2a, the polymer backbone consists of linkages formed between the C-terminus and the side chain functional group R; in Figure 5.2b, the polymer backbone consists of linkages formed between the N-terminus and the side-chain functional group; and in Figure 5.2c, the polymer backbone consists of linkages formed between the C-terminus and the N-terminus. If the linking bond in Figure 5.2c is an amide bond, then Figure 5.2c is a schematic representation of a conventional poly(amino acid).

This approach is applicable, among others, to serine, hydroxyproline, threonine, tyrosine, cysteine, glutamic acid, and lysine, and is limited only by the requirement that the nonamide backbone linkages give rise to polymers with desirable material properties.

5.3.2.2   The Preparation of Poly(serine ester)—An Illustrative Example

While the commercially important terephthalic acid derived polyesters are usually prepared by transesterification reactions, the ring-opening polymerization of lactones

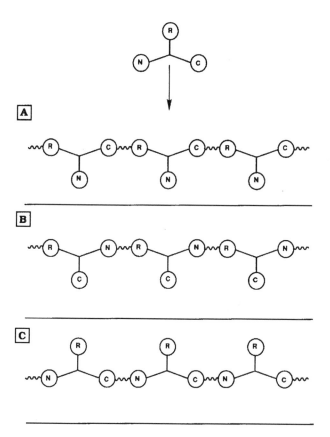

Figure 5.2   The use of trifunctional amino acids for the preparation of pseudo-poly(amino acids). The zigzag line represents a nonamide linkage between individual amino acids. For further details, see text.

is the method of choice for the preparation of several important poly(hydroxyacids) such as poly(lactic acid), poly(glycolic acid), or polycaprolactone. For the preparation of poly(N-acyl- L-serine ester), several approaches were investigated.

The direct esterification of N-protected serine in the presence of a suitable catalyst represents the most straightforward approach to the preparation of poly(N-acyl-L-serine ester). However, this approach failed, due to the facile β-elimination of the serine side chain. Thus, under the dehydrating conditions of the polyesterification, the β-elimination reaction at the serine side chain prevented the formation of polyesters (Figure 5.3a) [68].

The transesterification of N-benzyloxycarbonyl-L-serine methyl ester (Z-Ser-O-Me) was carefully studied and the process was optimized with respect to the reaction conditions and the catalyst used (Figure 5.3b). However, even under optimum conditions, only polymers of low molecular weight were obtained ($M_n \sim 600$) [68].

Those difficulties led to the development of a ring-opening polymerization

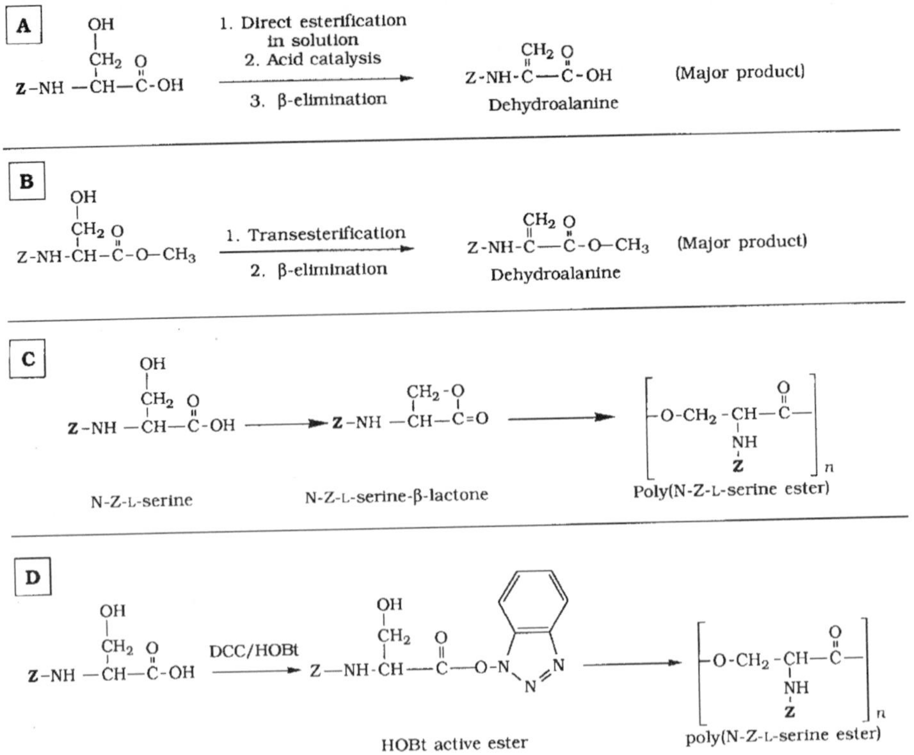

Figure 5.3   For the preparation of poly(N-Z serine ester) (Z = benzyloxycarbonyl) four differ-ent synthetic methods were explored. (A) Direct esterification. (B) Melt transesterification. (C) Ring-opening polymerization of serine-β-lactones. (D) Bulk polymerization of hydroxybenzo-triazole-active esters of serine.

procedure, using N-protected serine-β-lactones [69] as the monomeric starting materials (Figure 5.3c). With tetraethylammonium benzoate (TEAB), an anionic ini-tiator, polymers of relatively high molecular weight ($M_w \sim 40,000$) were obtained under mild conditions (37°C, solution in THF). The disadvantages of that approach are the high cost of preparing the labile serine-β-lactone, and the occurrence of a chain transfer reaction during the polymerization, limiting the maximum molecular weight to about 40,000. Overall, the ring-opening polymerization of serine-β-lactones proved to be an excellent laboratory scale procedure that provided the first samples of poly(*N*-benzyloxycarbonyl L-serine ester). It is particularly noteworthy that the lactonization and subsequent polymerization could be accomplished with less than 0.5% of racemization. The obtained polyesters were therefore optically pure materials [68]. On the other hand, the difficulty of preparing poly(serine ester) in multigram quantities had been one of the main obstacles in the detailed investigation of this new polymer.

Only recently, a method was developed that facilitated the convenient and cost-efficient synthesis of poly(*N*-benzyloxycarbonyl-L-serine ester) via the bulk polymerization of the hydroxybenzotriazole (HOBt) active ester of *N*-benzyloxycarbonyl L-serine (Figure 5.3d) [70,71]. By eliminating the need to isolate and purify the reactive serine-β-lactone that approach is the first practical method for the multigram synthesis of serine-derived polyesters of reasonably high molecular weight.

5.3.3 Polyesters Derived from trans-Hydroxy-L-Proline

Hydroxyproline-derived polyesters represent a specific example of the synthetic concept described in Figure 5.2a. The successful synthesis of poly(*trans*-4-hydroxy-L-proline ester) was first reported by Kohn and Langer in 1987 [61,66]. Thereafter, several additional members of this new family of polyesters were synthesized and investigated in detail by Yu et al. [72,73].

5.3.3.1 Polymer Synthesis

In analogy to the preparation of commercially important polyester fibers such as Dacron, a transesterification reaction was successfully employed in the synthesis of poly(*N*-acyl-trans-4-hydroxy-L-proline esters) (Figure 5.4). To avoid side reactions at the amino group of hydroxyproline, this functionality was protected. Fortunately, a large variety of biocompatible carboxylic acids, ranging from acetic acid to stearic acid, are available for this purpose. Furthermore, the use of temporary blocking groups such as the benzyloxycarbonyl group (Z) or the *tert*-butyloxycarbonyl group (Boc) facilitates the synthesis of unprotected poly(hydroxyproline ester). So far, poly(*N*-palmitoyl *trans*-4-hydroxy-L-proline) has been explored most extensively [74].

5.3.3.2 Properties and Possible Applications of Poly(*N*-palmitoyl *trans*-4-hydroxy-L-proline ester)

Poly(*N*-palmitoyl-*trans*-4-hydroxy-L-proliney-L-proline ester), abbreviated as poly (Pal-Hyp ester), is a wax-like, white material. The polymer is insoluble in water, but readily soluble in most nonpolar organic solvents, including hexane. The glass

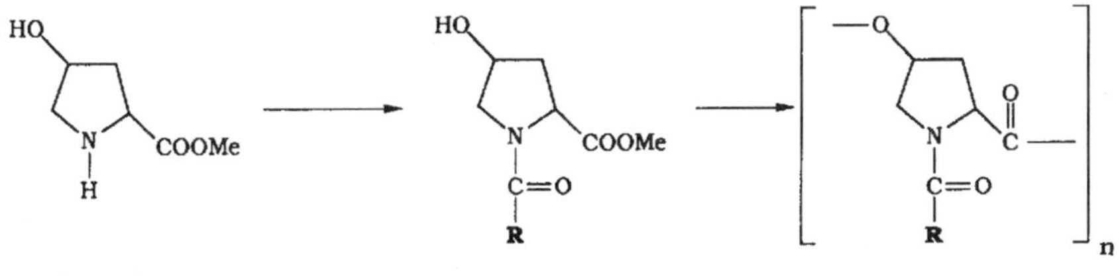

Figure 5.4   General synthetic method for the preparation of hydroxyproline derived polyesters. In the first step, the N-terminus is chemically protected. A wide range of carboxylic acids can be used for this purpose. A polymer particularly useful for drug delivery applications is obtained when R = palmitoyl.

transition occurs around 97°C. The polymer is apparently amorphous, and, like all hydroxyproline derived polyesters, poly(Pal-Hyp ester) can be readily processed by commonly used fabrication techniques (e.g., extrusion, injection molding, compression molding, solvent casting). Furthermore, because the synthesis of poly(Pal-Hyp ester) does not require the expensive hydroxyproline-*N*-carboxyanhydride as monomeric starting material, poly(Pal-Hyp ester) can be prepared at lower cost than the isomeric, conventional poly(amino acid) derived from *O*-palmitoyl *trans*-4-hydroxy-L-proline.

Due to the aliphatic ester linkages in its backbone, poly(Pal-Hyp ester) degrades under physiological conditions. The rate of degradation is slow, probably due to the hydrophobicity of the palmitoyl side chain and the rigidity of the polymer backbone [72]. Hydroxyproline-derived polyesters carrying smaller pendent groups tend to degrade somewhat faster. Overall, long-term drug delivery, such as implantable, multi-year contraceptive formulations, represent an obvious application for poly(Pal-Hyp ester). So far, the release of two different "model dyes," *p*-nitroaniline and Acid Orange, from compression-molded disks of poly(Pal-Hyp ester) has been tested in vitro. Those studies established that the observed release times may vary from several hours to many months, depending on the loading and the hydrophobicity of the released drug. Furthermore, there is no reason evident why poly(Pal-Hyp ester) should not be suitable for the formation of microcapsules or microspheres. For microencapsulated drug formulations the longer degradation time and greater hydrophobicity of poly(Pal-Hyp ester) could be a distinctive advantage over the widely used poly(lactic acid).

5.3.3.3   Biocompatibility

As part of the initial evaluation of hydroxyproline-derived polyesters, poly(Pal-Hyp ester) was subjected to several biocompatibility screening tests [72]. Those studies indicated that poly(Pal-Hyp ester) had little systemic toxicity and elicited a very mild, local tissue response that compared favorably with the responses observed for established biomaterials such as medical grade stainless steel or poly(lactic acid)/poly(glycolic acid) implants.

5.3.4   Polyiminocarbonates and Polycarbonates Derived from L-Tyrosine

Tyrosine-derived polyiminocarbonates and polycarbonates are examples of pseudo-poly-(amino acids) in which a dipeptide (and not a single amino acid) is the basic polymer repeat unit. The observation that aromatic backbone structures can significantly increase the stiffness and mechanical strength of polymers provided the rationale for the use of tyrosine dipeptides in the design of medical implants. Tyrosine is the only major, natural nutrient containing an aromatic hydroxyl group and derivatives of tyrosine dipeptide were used as replacements for industrially used diphenols such

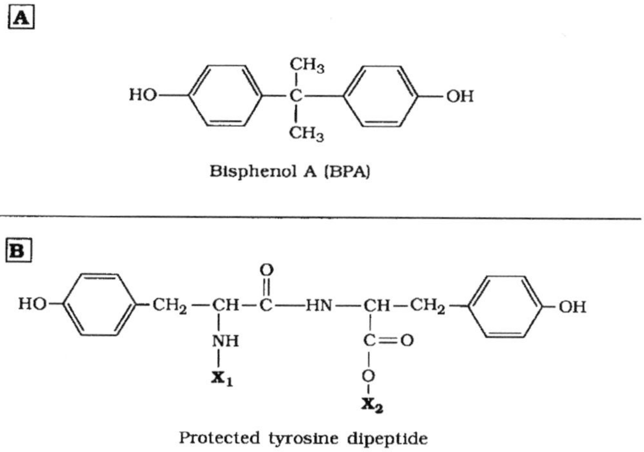

Bisphenol A (BPA)

Protected tyrosine dipeptide

Figure 5.5 Structures of (A) Bisphenol A, a widely used diphenol in the manufacture of commercial polycarbonate resins; (B) tyrosine dipeptide with chemical protecting groups X_1 and X_2 attached to the N- and C-termini, respectively.

as Bisphenol A in the design of medical implant materials (Figure 5.5). The resulting tyrosine-derived polyiminocarbonates and polycarbonates are new polymers that combine favorable engineering properties such as high strength with a high degree of tissue compatibility [16,61,67,75–79].

5.3.4.1   Polymer Synthesis

Polyiminocarbonates are not widely known and are best regarded as the "imine analogs" of the industrially used polycarbonates (Figure 5.6). The exploration of polyiminocarbonates started only in the late 1960s when Hedayatullah reacted aqueous

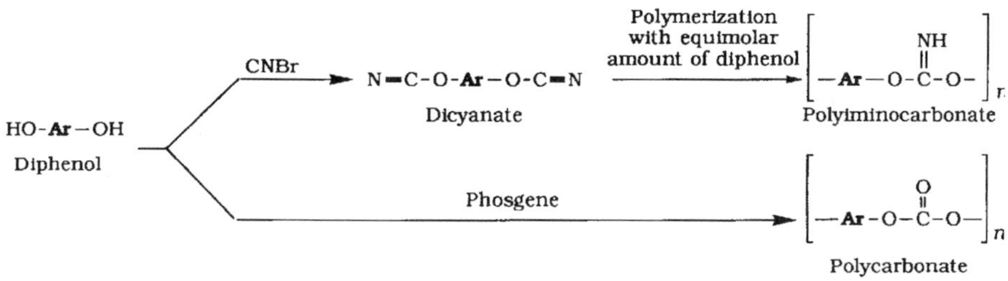

Figure 5.6 General synthetic methods for the preparation of polyiminocarbonates and polycarbonates from either industrial diphenols such as Bisphenol A or tyrosine dipeptide derivatives. The only structural difference between polycarbonates and polyiminocarbonates is the replacement of the carbonyl oxygen in the carbonate linkage by an imino group.

solutions of various chlorinated diphenolate sodium salts with cyanogen bromide [80]. Later, Schminke et al. [81] described (in a now abandoned patent) the synthesis of polyiminocarbonates with molecular weights of about 50,000 by the solution polymerization of a diphenol and a dicyanate. The first comprehensive and systematic study of polyiminocarbonate synthesis in general was published in 1989 [76], followed in 1990 by a detailed exploration of the synthesis and characterization of tyrosine-derived polyiminocarbonates [77]. In essence, tyrosine-derived polyiminocarbonates can be synthesized by the same techniques as Bisphenol A-derived polymers and are best prepared by the reaction of a tyrosine dipeptide with an equimolar quantity of a peptide dicyanate (Figure 5.6). A comprehensive review of interfacial and solution polymerization procedures for the preparation of polyiminocarbonates of high molecular weight has recently been published [16].

Tyrosine-derived polycarbonates are being prepared by the reaction of protected tyrosine dipeptides with either phosgene or the less hazardous triphosgene (Figure 5.6). For laboratory-scale preparations, triphosgene is probably the preferable reagent, whereas for large-scale industrial preparations, the conventional reactors used for the synthesis of poly(Bisphenol A carbonate) would probably require only minor modifications. Under optimized reaction conditions, tyrosine-derived polycarbonates with an apparent weight average molecular weight of up to 400,000 (by GPC, relative to polystyrene standards) were obtained [79].

5.3.4.2   Structure–Property Correlations

The basic tyrosine dipeptide repeat unit facilitated two different approaches for the modification of the polymer structure. First, the protecting groups attached to either the N-terminus or the C-terminus can be varied freely and independently. Second, the polymer backbone linkage itself can be changed because it is possible to prepare both polyiminocarbonates and polycarbonates from the same monomeric repeat unit.

Several detailed investigations of the structure–property correlations have been reported. For example, the effects of the C-terminus protecting groups on the polymer properties were investigated by comparing the properties of the polyiminocarbonates derived from the ethyl, hexyl, and palmityl esters of N-benzyloxycarbonyl-L-tyrosyl-L-tyrosine [82]. These monomers represent a homologous series, differing only in the length of the alkyl group attached to the C-terminus (Figure 5.7).

In a second series of experiments, the tyrosine derivatives desaminotyrosine (Dat) and tyramine (Tym) were incorporated into the monomer, leading to four structurally related dipeptide monomers that carried either no pendent chains at al (Dat-Tym), a N-benzyloxycarbonyl group as the only pendent chain (Z-Tyr-Tym), a hexyl ester group as the only pendent chain (Dat-Tyr-Hex, further abbreviated as DTH), or both types of pendent chains simultaneously (Z-Tyr-Tyr-Hex) (Figure 5.8). After preparation of the corresponding polyiminocarbonates and polycarbonates, the contribution of each type of pendent chain to the polymer properties became evident [77,79]. The most

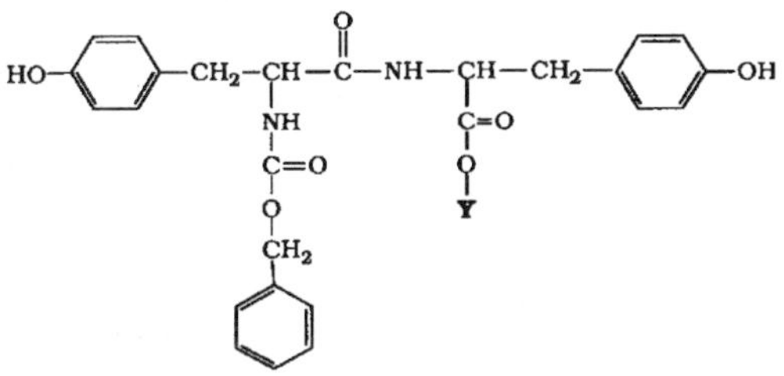

Y = ethyl: poly(Z-Tyr-Tyr-Et iminocarbonate)
Y = hexyl: poly(Z-Tyr-Tyr-Hex iminocarbonate)
Y = palmityl: poly(Z-Tyr-Tyr-Pal iminocarbonate)

Figure 5.7 Structures of three tyrosine derived, protected dipeptides carrying C-terminus protecting groups of increasing length and hydrophobicity. Those dipeptides represent a homologous series of monomers that were used to evaluate the effect of the C-terminus protecting group on the properties of the corresponding polyiminocarbonates and polycarbonates.

significant conclusions of this study were that pendent chains are necessary to obtain processible materials and that the amount of interchain hydrogen bonding has to be minimized. In the absence of any pendent chains, the polymers are sparingly soluble and have softening temperatures that are close to or above their degradation temperature. However, when selecting the pendent chain it is important to minimize possible hydrogen bonding interactions which tend to raise the softening temperature and reduce polymer solubility. Thus, the benzyloxycarbonyl group (containing a urethane linkage which is a strong promoter of hydrogen bonding interactions) did not significantly improve the processibility or solubility of the corresponding polymers whereas the hexyl ester group in Dat-Tyr-Hex (DTH) improved the physicomechanical properties of the resulting polymer significantly. Based on these observations, the desaminotyrosyl-tyrosine alkyl esters were identified as particularly promising monomers for the design of degradable biomaterials [67]. The corresponding polyiminocarbonates and polycarbonates are readily processible materials. Some of these monomers and polymers are now commercially available [74].

The effect of the polymer backbone linkages on the polymer properties was explored by a detailed comparison of poly(DTH iminocarbonate) and poly(DTH carbonate). The replacement of the carbonyl oxygen by an-NH group presents the only molecular difference between those two polymers. Poly(DTH iminocarbonate) and poly(DTH carbonate) are completely amorphous materials and solvent cast films are virtually indistinguishable in appearance. Both polymers exhibit high tensile strength; at a molecular weight of about 100,000, typical values for the tensile strength of poly(DTH iminocarbonate) were about 400–500 kg/cm^2 (40–40 MPa), whereas the corresponding poly(DTH carbonate) had a tensile strength of about 340 kg/cm^2 (34 MPa). Those values compare favorably to the tensile strength values obtained for unoriented, solvent cast film samples of poly(D,L-lactic acid) of similar molecular weight [83]. For comparison, for wet films of poly(γ-methyl-D-glutamate) a much lower tensile

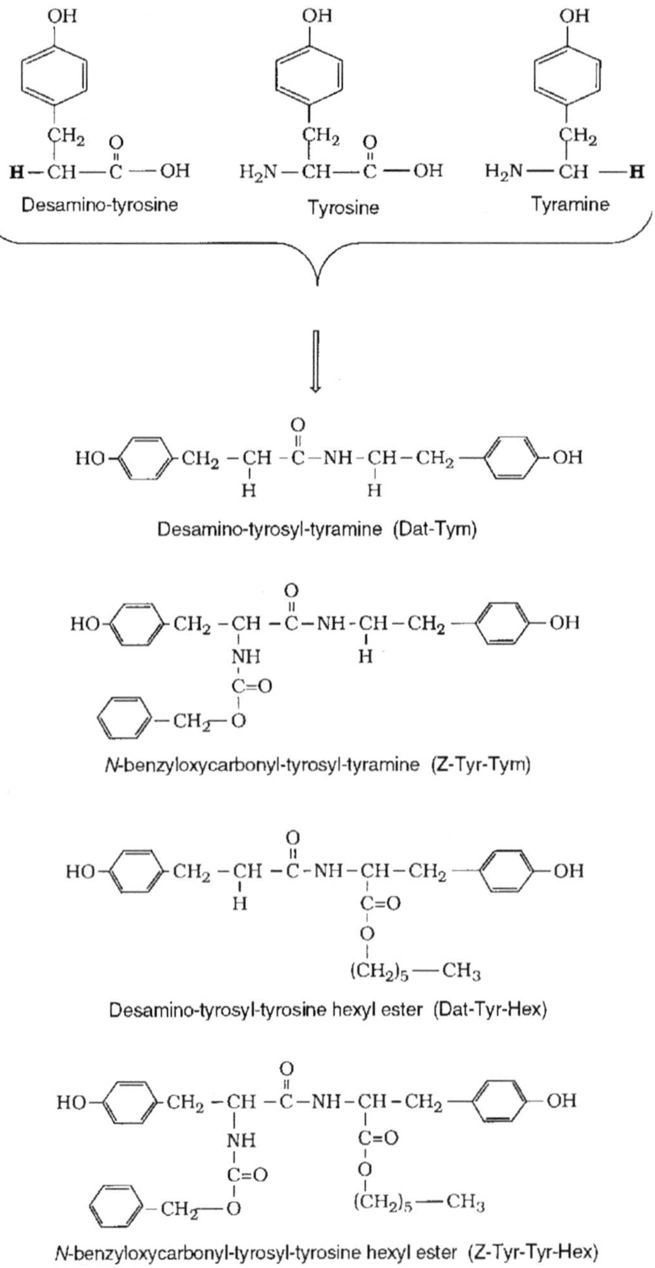

Figure 5.8   Using 3-(*p*-hydroxyphenyl)propionic acid (desaminotyrosine), L-tyrosine, and tyramine as starting materials, four structurally related peptides were prepared that carry various
combinations of protecting groups. The polyiminocarbonate and polycarbonate derived from
Dat-Tyr-Hex (further abbreviated as DTH) were found to have particularly useful engineering
properties.

strength of only 50–100 kg/cm^2 (5–10 MPa) was reported [84]. The tensile moduli of poly(DTH iminocarbonate) and poly(DTH carbonate) were 16,300 kg/cm^2 (1.6 GPa) and 13,900 kg/cm^2 (1.3 GPa), respectively, indicating that the polyiminocarbonate is somewhat stiffer than the polycarbonate.

The most striking difference between poly(DTH iminocarbonate) and poly(DTH carbonate) was in their ductility; whereas poly(DTH iminocarbonate) is a brittle material that failed at about 7% of elongation, poly(DTH carbonate) is a tough material that failed only at about 100% elongation. Because those polymers are derived from identical monomers, this difference in ductility must be attributed to the relatively small change in backbone structure [75].

In general, polycarbonates are more stable than polyiminocarbonates. This general trend is also evident in the tyrosine-derived polymers. When solvent cast films of poly(DTH iminocarbonate) were exposed to aqueous buffer solutions, the films became turbid within a matter of days and swelled noticeably due to the absorption of water. The molecular weight decreased rapidly to about 1000–6000 (weight average) with a concomitant loss of mechanical strength. On the other hand, films of poly(DTH carbonate) absorbed little water and remained intact when immersed in phosphate buffer solution at pH 7.4 (37°C). Overall, a 50% reduction of the initial molecular weight of poly(DTH carbonate) was observed over 6 months. Blending poly(DTH carbonate) with increasing amounts of poly(DTH iminocarbonate) accelerated the degradation of the blends [78]. Although the chemical mechanism of the hydrolytic degradation of poly(Bisphenol A iminocarbonate) has been studied previously [85], detailed evaluations of the degradation behavior of tyrosine derived polymers in vivo have not yet been published.

5.3.4.3   Biocompatibility of Poly(DTH iminocarbonate) and Poly(DTH carbonate)

An evaluation of the tissue compatibility of DTH derived polyiminocarbonate and polycarbonate in rats has recently been concluded [86]. Those studies included medical grade poly(D,L-lactic acid) and medical grade polyethylene as controls. At times ranging from 7 days to 4 months post implantation the subcutaneous implantation sites were histologically evaluated. The general conclusion was that the tissue response elicited by the slowly degrading poly(DTH carbonate) was comparable to the mild response seen for medical grade polyethylene, whereas the fast degrading poly(DTH iminocarbonate) elicited a response that was comparable to the tissue response seen for medical grade poly(D,L-lactic acid).

5.3.4.4   Possible Applications as Medical Implant Materials

Poly(DTH iminocarbonate) and poly(DTH carbonate) are the only tyrosine-derived polymers known so far with engineering properties that appear to be suitable for implant materials. Because those polymers are also tissue compatible, their use in the formulation of drug delivery devices and as orthopedic implants has been explored.

The relatively long degradation time of poly(DTH carbonate) in vivo was exploited in the design of a long-term controlled-release device for the intracranial administration of dopamine [87]. In that application, poly(DTH carbonate) had several potential advantages over other degradable polymers, including the ease with which dopamine could be incorporated into the polymer, an apparent protective action on the dopamine contained within the polymeric matrix, a constant release rate over more than 180 days, and a high degree of compatibility with brain tissue [87].

Due to the relatively high strength of poly(DTH carbonate), this polymer was considered as an orthopedic implant material. Injection-molded bone screws and compression-molded bone pins have been fabricated. Initial evaluations indicate a high degree of strength retention over a 6-month period in vitro and a high degree of compatibility with bone tissue [88].

5.4   Copolymers Derived from Amino Acids and Non-Amino Acids

The basic rationale for the synthesis of copolymers of amino acids and non-amino acids was the expectation that copolymers would exhibit new and/or improved material properties not obtainable by any combination of amino acids alone. Obviously, a virtually unlimited number of structurally different copolymers are theoretically possible and, indeed, a very wide range of amino acid containing copolymers have been described in the literature. The following section (5.4) is a survey of some representative examples of this class of polymers.

5.4.1   Survey of Available, Amino Acid Containing Copolymers

A variety of copolymers of amino acids with non-amino acids have been synthesized, characterized, and evaluated within the context of numerous biomedical applications (Table 5.5). A patent issued in 1972 represents the first specific design of a copolymer of amino acids and hydroxyacids for medical applications [89]. Although these polymers were suggested as possible suture materials, no practical applications emerged.

One of the structural characteristics of amino acid–hydroxyacid copolymers (also referred to as polydepsipeptides) is that they contain both ester and amide bonds in their backbones, the former lacking the ability to act as hydrogen bond donors. Because of this property several publications describe the use of amino acid–hydroxyacid copolymers as model compounds in studies relating to the importance of hydrogen bonding in protein structure and helix to coil transitions (see Goodman and Katakai [90] and references cited therein).

The use of poly(ethylene oxide) (also referred to as poly(ethylene glycol) or PEG) as a copolymer component was widely investigated to increase the solubility of poly(amino acids) in water. That is a particularly useful property in the design of injectable drug carriers. For example, an AB type block copolymer of poly(aspartic

Table 5.5   Copolymers of Amino Acids and Non-Amino Acids and Their Medical Applications

Copolymer system	Suggested applications	Illustrative references
A–B type block copolymers with a poly(amino acid) as the A block and the following B blocks:		
Poly(L-lactide)	Controlled drug delivery	Kim [131]
Poly(ethylene oxide)	Water-soluble drug carriers for anticancer drugs such as doxorubicin	Yokoyama [91–93]
A–B–A type block copolymers with a poly(amino acid) as the A block and the following B blocks:		
Polyoxypropylene	Artificial lung	Kang [94]
Polydimethylsiloxane	Artificial lung	Kumaki [95]
Polyoxyethylene	Controlled drug delivery	Cho [132]
Polybutadiene	hemodialysis membranes	Kugo [96, 101–103]
Segmented multiblock copolymer of		
Poly(ethylene oxide) and Poly(β-benzyl-L-aspartate)	Controlled drug delivery	Yokoyama [104]
Graft copolymer of		
Poly(butyl methacrylate) and poly(glutamic acid)	Controlled drug delivery	Chung [106]
Alternating copolymers of		
Glutamic acid with ethane, butane, and hexane diols and with glycerol	Water-soluble drug carriers	Pramanick [133]

acid) with PEG has been studied as a water-soluble macromolecular prodrug for the anthracycline antibiotic adriamycin [91]. The major advantage of that copolymer over poly(aspartic acid) was the retention of water solubility of the conjugate despite the introduction of a large number of hydrophobic adriamycin residues. The polymer had a micellar structure in aqueous buffer with a hydrophilic outer shell comprised of PEG and a hydrophobic inner core of poly(aspartic acid). After uptake by target cells, the hydrolysis of the poly(aspartic acid) backbone resulted in the release of adriamycin from the conjugate. Monoclonal antibody conjugates of adriamycin with the poly(PEG–aspartic acid) copolymer have been investigated for targeted drug delivery [92,93].

Block and graft copolymers containing two polymer chains with very dissimilar properties are known to undergo phase separations during the casting process to give films with heterogeneous surfaces. Such microphase-separted surfaces can be antithrombogenic [94–98]. A–B–A type block copolymers have been synthesized where the A block was poly(γ-benzyl-L-glutamate) and the B block was either poly-oxypropylene (POP) or polydimethylsiloxane (PDMS), both of which are nonpolar and flexible [94,95]. The POP or PDMS segments contributed to the high gas permeability of the copolymer films.

AB type block copolymers with a polyvinyl or a polysaccharide block and a hydrophobic polypeptide block have been synthesized [99,100]. Such copolymers are interesting from a biological point of view because they form simplified polymeric models of membrane integral proteins. Similarly A–B–A type block copolymers consisting of poly(γ-benzyl-L-glutamate), poly(γ-methyl-D,L-glutamate), or poly-(η-N-z-lysine) as the A block and polybutadiene as the B block have been synthesized and characterized [96,101–103]. The polybutadiene chains have been shown to be in a random coil conformation, whereas the polypeptide chains are in an α-helical conformation. The biological interest in such copolymers is in the preparation of model membranes in which polybutadiene forms amorphous domains surrounded by α-helical polypeptide chains. The mechanical strength and hydraulic permeability of the block copolymer membranes have been shown to be dramatically higher than that of the homopoly(amino acid) [102].

Novel segmented multiblock copolymers containing poly(ethylene oxide) and poly-(β-benzyl-L-aspartate) (PBLA) linked via urethane and urea bonds have been prepared. The water contents of the copolymer membranes could be varied by changing the lengths of the PEO and PBLA segments. Such membranes were found to be suitable for controlled drug release applications [104].

A new type of biomembrane model composed poly(butyl methacrylate)–poly(L-aspartic acid) graft copolymer has been proposed by Maeda et al. [105]. This copolymer has a vinyl backbone and polypeptide branches. The transport of Na^+ ions across the membrane was studied and continuous domains of poly(aspartic acid) were found to function as a transmembrane permeating pathway for Na^+ ions. The ion permeability was controlled by a pH-induced reversible conformational change of the polypeptide segment. Chung et al. have studied the pH induced regulation of permeability of sugars from those graft copolymer membranes [106].

5.4.2   PEG–Lys Copolymers and Their Applications

Copolymers of PEG and L-Lys serve as an illustrative example for the modifiction of the properties of homopoly(amino acids) by copolymerization. In these polymers, the α and η amino groups of lysine were used to connect PEG chains of variable length via urethane bonds (Figure 5.9). The resulting alternating copolymers had pendent carboxylic acid groups (derived from the C-terminus of lysine) that could be used for the attachment of drugs to the polymer backbone and the design of crosslinking reactions for the preparation of degradable and nondegradable hydrogels [107–198].

The PEG–Lys copolymers were synthesized by the reaction of the bis-succinimidyl carbonate derivative of PEG (BSC-PEG) with L-lysine or L-lysine alkyl esters, yielding polymers with weight average molecular weights of up to about 200,000 (Figure 5.9). The pendent carboxylic acid groups were functionalized in a variety of ways to give polymeric derivatives having (among other groups) pendent N-hydroxysuccinimide active esters, hydrazides, amines, alcohols, and aldehydes (Figure 5.10).

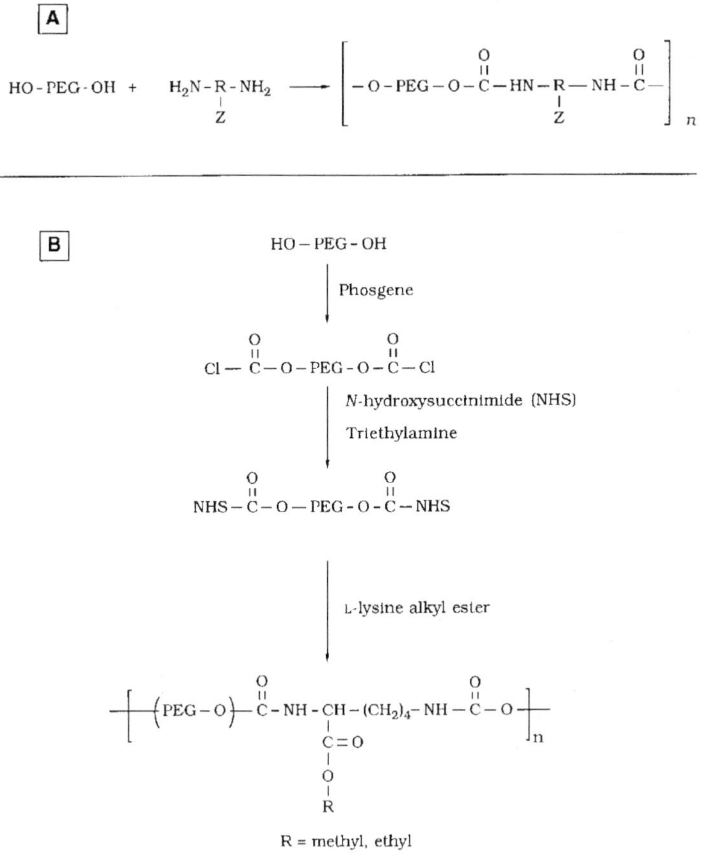

A

$$HO\text{-}PEG\text{-}OH \;+\; \underset{\underset{Z}{\displaystyle |}}{H_2N\text{-}R\text{-}NH_2} \longrightarrow \left[-O\text{-}PEG\text{-}O\text{-}\overset{\overset{O}{\|}}{C}\text{-}HN\text{-}\underset{\underset{Z}{\displaystyle |}}{R}\text{-}NH\text{-}\overset{\overset{O}{\|}}{C}\text{-} \right]_n$$

B

$$HO-PEG-OH$$

| Phosgene

$$Cl-\overset{\overset{O}{\|}}{C}-O-PEG-O-\overset{\overset{O}{\|}}{C}-Cl$$

| N-hydroxysuccinimide (NHS)
| Triethylamine

$$NHS-\overset{\overset{O}{\|}}{C}-O-PEG-O-\overset{\overset{O}{\|}}{C}-NHS$$

| L-lysine alkyl ester

$$-\left[\left(PEG-O \right)-\overset{\overset{O}{\|}}{C}-NH-\underset{\underset{\underset{\underset{\underset{R}{\displaystyle |}}{\displaystyle O}}{\displaystyle |}}{\underset{\underset{C=O}{\displaystyle |}}{CH}}}{CH}-(CH_2)_4-NH-\overset{\overset{O}{\|}}{C}-O \right]_n -$$

R = methyl, ethyl

Figure 5.9 Preparation of poly(PEG–Lys), an alternating copolymer of PEG and L-Lys. (A) Schematic representation of the polymer backbone comprising of PEG and a diamine. (B) Detailed reaction scheme for the synthesis of poly(PEG–Lys) copolymers having alkyl ester pendant chains.

5.4.2.1   Applications of Poly(PEG–Lys) as a Drug Carrier

As an initial model experiment, the attachment of the β-lactam antibiotic Penicillin V to the polymer was studied. When Penicillin V was attached to a poly(PEG–Lys) derivative having pendent alcohol groups through hydrolyzable ester linkages, the polymer–Penicillin V conjugate was as active against a series of clinically important microorganisms as free Penicillin V [110].

For plasma stability and biodistribution studies, radiolabeled poly(PEG–Lys) was prepared using [^{14}C]lysine in the polymerization reaction. Although polyurethanes are used in a variety of biomedical applications (catheters, artificial organs), the use of water-soluble polyurethanes as injectable drug carriers is not known in the

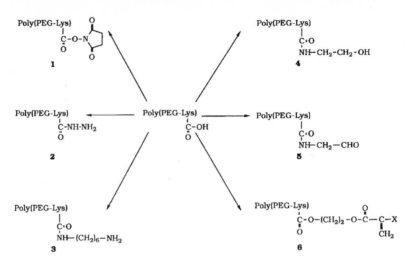

Figure 5.10   Available derivatives of poly(PEG–Lys), having different, chemically reactive pendant side chains: (1) *N*-hydroxysuccinimide (NHS) active ester; (2) hydrazide; (3) amine; (4) hydroxyl; (5) aldehyde; (6) acrylate (X = H) or methacrylate (X = methyl).

literature. Therefore, the stability of the radiolabeled polymer in human plasma was determined by incubation in plasma at 37°C. No degradation of the polymer backbone was observed for over 48 h. The in vivo distribution of the radiolabeled polymer (approximate molecular weight 50,000) in CD1 male mice indicated that most of the polymer was located in the circulating blood and was uniformly distributed in the different tissues. Specifically when the results were calculated in terms of dose per organ, no accumulation of radioactivity by the liver, spleen, or kidney as observed. In acute toxicity studies in mice, no signs of toxicity could be observed up to the maximum possible test dosage of 10 g/kg.

The combination of nontoxicity, stability in plasma, easy excretability, and absence of accumulation in the liver, spleen, or kidney makes poly(PEG–Lys) a promising, injectable drug carrier for the systemic administration of drugs whose activity depends on their concentration in the circulating blood pool.

5.4.2.2   Applications of Poly(PEG–Lys) in the Design of Hydrogels

The active pendent groups of poly(PEG–Lys) derivatives (Figure 5.10) provided convenient attachment points for the reaction with bifunctional crosslinkers. It is particularly noteworthy that those attachment points are evenly distributed along the polymer backbone and that the distance between two neighboring attachment points is controlled by the length of the PEG chain used in the synthesis of the poly(PEG–Lys) backbone. Thus, the uniform and readily controllable spacing of the crosslinks is the major difference between the hydrogels derived from poly(PEG–Lys) and the previously described PEO hydrogels [111].

Because of the wide range of poly(PEG–Lys) derivatives available, numerous crosslinking reactions can be envisioned. For example, the polymer having pendent N-hydroxysuccinimide active ester groups was crosslinked using hexamethylene diamine to obtain membranes with amide crosslinks, the polymer having pendent hydrazide groups was crosslinked with 1,6-hexamethyl diisocyanate to obtain membranes with semicarbazide crosslinks, and the polymers having pendant hydroxyethyl acrylate (HEA) or hydroxyethyl methacrylate (HEMA) groups were photocrosslinked to obtain membranes with crosslinks consisting of poly(acrylic acid) or poly(methacrylic acid), respectively [109].

All poly(PEG–Lys) derived hydrogels swelled extensively. The equilibrium water contents (EWC) ranged from about 64 to 91%, depending on the crosslink chemistry and the crosslink density. In the dry state, all hydrogels formed extremely strong films whose tensile strength exceeded 60 MPa in some cases. Even when fully swollen, poly(PEG–Lys) based hydrogels are significantly stronger than many other hydrogel prepartions of similar EWC. For example, the hydrogels containing poly(methacrylic acid) crosslinks reached a tensile strength of 1.09 MPa at a EWC of 78% and ranked among the mechanically strongest, highly swollen hydrogels reported in the literature so far [109].

Hydrogels crosslinked by semicarbazide linkages lost about 1% of their initial weight per day when exposed to phosphate buffer solutions at 37°C and disintegrated completely after about 40 days when the total cumulative weight loss reached about 50%. Hydrogels containing poly(acrylic acid) or poly(methacrylic acid) crosslinks lost weight at a rate of 1.7% and 1.1% per day, respectively, and completely dissolved when the cumulative weight loss reached about 60%. The hydrogels crosslinked with amide bonds were significantly more stable and did not degrade under physiological conditions.

Possible applications for the degradable hydrogels include implantable hydrogel-based drug delivery systems, membranes for surgical adhesion prevention, and site specific administration of antibiotics. The photocrosslinked membranes with lower EWCs of 50–60% (as compared to 80% for amide and semicarbazide hydrogels) appear to be suitable for diffusional drug release, in particular for macromolecular drugs such as proteins. The nondegradable amide hydrogels have a high EWC and appear suitable for external wound care dressings. Because of their strength and excellent optical clarity, photocrosslinked hydrogels could be considered for contact lens applications, once more stable materials of this type are available.

5.5   Summary

Poly(amino acids) are among the most intensely investigted polymers that have captured the imagination of biologists, biochemists, polymer chemists, theoreticians, and material scientists for over 40 years. The initial and most significant application of those polymers was in biochemistry where they served as readily accessible models for natural polypeptides and proteins. Early attempts to identify large-scale commercial applications for poly(amino acids) were uniformly unsuccessful. Usually, the high cost of the N-carboxyanhydrides required as monomeric starting materials made poly(amino acids) economically unattractive. It is therefore not surprising that the initial excitement and interest waned during the late 1970s when large-scale practical applications appeared unrealistic and poly(amino acids) were no longer needed as model compounds in studies relating to structure and function of proteins.

More recently, the emphasis has been shifting from the conventional poly(amino acids) to *amino acid derived polymers*. This term comprises a large number of polymers in which amino acids are either linked by nonamide bonds [pseudo-poly(amino acids)], or in which amino acids are copolymerized with a truly amazing variety of non-amino acid components. In addition, advances in the chemical and/or biochemical synthesis of sequential poly(amino acids) made new materials available that bear a much closer resemblance to natural polypeptides than the previously available random or block copoly(amino acids).

It is a fortunate coincidence that these new, amino acid derived materials become available at a time when the successful development of a large number of "high tech" applications requires a range of specialty polymers with carefully designed properties. Such high tech applications include, among many others, medical implants, specialty coatings, micromechanical systems, and membranes with controllable permeability. In many of these advanced applications, the key materials are not required in large quantities, and material costs may not be the most decisive factor in the overall commercial viability of the product. Thus, two of the most important obstacles to the early commercialization of conventional poly(amino acids) are declining in significance. One may therefore predict that some of the amino acid derived polymers described in this chapter will find important commercial applications in the future.

REFERENCES

1. Katchalski, E.; Sela, M. In *Advances in Protein Chemistry,* Vol. 13. Anfinsen, C.B.; Anson, M.L.; Bailey, K.; Edsall, J.T., eds. Academic Press, New York, pp. 243–492 (1958).
2. Lotan, N.; Berger, A.; Katchalski, E. In *Annual Review of Biochemistry,* Vol. 41. Snell, E.E.; Boyer, P.D.; Meister, A.; Sinsheimer, R.L., eds. Annual Reviews, Palo Alto, pp. 869–901 (1972).
3. Katchalski, E. In *Peptides, Polypeptides, and Proteins—Proceedings of the Rehovot Symposium on Poly(Amino Acids), Polypeptides, and Proteins and Their Biological Implications.* Blout, E.R.; Bovey, F.A.; Goodman, M.; Lotan, N., eds. John Wiley, New York, pp. 1–13 (1974).
4. Anderson, J.M.; Spilizewski, K.L.; Hiltner, A. In *Biocompatibility of Tissue Analogs,* Vol. 1. Williams, D.F. ed. CRC Press, Boca Raton, FL, pp. 67–88 (1985).
5. Fasman, G.D. *Biopolymers* **26,** S59-S79 (1987).
6. Leuchs, H. *Chem. Ber.* **39,** 857–861 (1906).
7. Bamford, C.H.; Elliot, A.; Hanby, W.E. In *Physical Chemistry—A Series of Monographs.* Hutchinson, E., ed. Academic Press, New York (1956).
8. Sela, M.; Katchalski, E. In *Advances in Protein Chemistry,* Vol. 14. Anfinsen, C.B.;

Anson, M.L.; Bailey, K.; Edsall, J.T., eds. Academic Press, New York, pp. 391–477 (1959).

9. Yaron, A. *Israel J. Chem.* **12,** 651–662 (1974).

10. Sela, M. In *Peptides, Polypeptides, and Proteins—Proceedings of the Rehovot Symposium on Poly(Amino Acids), Polypeptides, and Proteins and Their Biological Implications.* Blout, E.R.; Bovey, F.A.; Goodman, M.; Lotan, N., eds. John Wiley, New York, pp. 495–509 (1974).

11. Kulkarni, R.K; Pani, K.C.; Neuman, C.; Leonard, F. *Arch. Surg.* **93,** 839–843 (1966).

12. Kulkarni, R.K.; Moore, E.G.; Hegyeii, A.F.; Leonard, F. *J. Biomed. Mater. Res.* **5,** 169–181 (1971).

13. Frazza, E.J.; Schmitt, E.E. *J. Biomed. Mater. Res.* **5,** 43–58 (1971).

14. Wasserman, D.; Versfeit, C.C. U.S. Patent 3,839,297, issued 1975.

15. Miyamae, T.; Mori, S.; Takeda, Y. U.S. Patent 3,371,069, issued 1968, assigned to Ajinomoto Co., Inc.

16. Kohn, J. In *Biodegradable polymers in Drug Delivery Systems.* Chasin, M.; Langer, R., eds. Marcel Dekker, New York, pp. 195–229 (1990).

17. Kohn, J. *Drug News Perspectives* **4,** 289–294 (1991).

18. Pulapura, S.; Kohn, J. In *Peptides—Chemistry and Biology (Proceedings of the 12th American Peptide Symposium).* Smith, J.A.; Rivier, J.E., eds. Escom Science Publishers, Leiden (The Netherlands), pp. 539–541 (1992).

19. Urry, D.W. *Polym. Mater. Sci. Eng.* **63,** 329–36 (1990).

20. Urry, D.W. *Polym. Mater. Sci. Eng.* **62,** 587–593 (1990).

21. Urry, D.W.; Parker, T.M.; Reid, M.C.; Gowda, D.C. *J. Bioact. Compat. Polym.* **6,** 262–282 (1991).

22. Urry, D.W.; Hayes, L.C.; Gowda, D.C.; Parker, T.M. *Chem. Phys. Lett.* **182,** 101–106 (1991).

23. Luan, C.H.; Parker, T.M.; Prasad, K.U.; Urry, D.W. *Biopolymers* **31,** 465–75 (1991).

24. Urry, D.W. *Prog. Biophys. Mol. Biol* **57,** 23–57 (1992).

25. McGrath, K.P.; Tirrell, D.A.; Kawai, M.; Mason, T.L.; Fournier, M.J. *Biotechnol. Prog.* **6,** 188–192 (1990).

26. Creel, H.S.; Fournier, M.J.; Mason, T.L.; Tirrell, D.A. *Macromolecules* **24,** 1213–1214 (1991).

27. Tirrell, D.A.; Fournier, M.J.; Mason, T.L. *Polym. Prepr.* **32,** 704–705 (1991).

28. Tirrell, D.A.; Fournier, M.J.; Mason, T.L. *Curr. Opin. Struct. Biol.* **1,** 638–641 (1991).

29. Tirrell, D.A.; Fournier, M.J.; Mason, T.L. *MRS Bull.* **16,** 23–28 (1991).

30. McGrath, K.P.; Fournier, M.J.; Mason, T.L.; Tirrell, D.A. *J. Am. Chem. Soc.* **114,** 727–733 (1992).

31. Krejchi, M.T.; Fournier, M.J.; Mason, T.L.; Tirrell, D.A. *Polym. Prepr.* **32,** 411 (1991).

32. Fournier, M.J.; Creel, H.S.; Krejchi, M.T.; Mason, T.L.; Tirrell, D.A.; McGrath, K.P.; Atkins, E.D.T. *J. Bioact. Compat. Polym.* **6,** 326–338 (1991).

33. Silver, F.; Doillon, C. In *Biocompatibility. Interactions of Biological and Implantable Materials,* Vol. 1. VCH Publishers, New York, 1–24 (1989).

34. Hench, L.L.; Ethridge, E.C. In *Biophysics and Bioengineering Series.* Noordergraaf, A., ed. Academic Press, New York (1982).

35. Walton, A.G. In *Biomedical Polymers. Polymeric Materials and Pharmaceuticals for Biomedical Use.* Goldberg, E.P.; Nakajima, A., eds. Academic Press, New York, pp. 53–83 (1980).

36. Maurer, P.H.; Subrahmanyam, D.; Katchalski, E.; Blout, E.R. *J. Immunol.* **83,** 193–197 (1959).
37. Sela, M.; Katchalski, E.; Olitzki, A.L. *Science* **123,** 1129 (1956).
38. Sela, M.; Fuchs, S.; Arnon, R. *Biochem. J.* **85,** (1962).
39. Hayashi, T.; Tabata, Y.; Nakajima, A. *Polym. J. (Tokyo)* **17,** 463–471 (1984).
40. Hayashi, T.; Iwatsuki, M. *Biopolymers* **29,** 549–557 (1990).
41. Pytela, J.; Saudek, V.; Drobnik J.; Rypacek, F. *J. Control. Rel.* **10,** 17–25 (1989).
42. Chandy T.; Sharma, C.P. *Biomaterials* **12,** 677–682 (1991).
43. Minoura, N.; Aiba, S.; Fujiwara, Y.; Koshizaki, N.; Imai, Y. *J. Biomed. Mater. Res.* **23,** 267–279 (1989).
44. Marchant, R.E.; Sugie, T.; Hiltner, A.; Anderson, J.M. *ASTM Spec. Tech. Publ. (Corros. Degrad. Implant Mater.)* **859,** 251–266 (1985).
45. Cho, C.S.; Kim, H.Y.; Akaike, T. *Pollimo* **14,** (1990).
46. Sidman, K.R.; Schwope, A.D.; Steber, W.D.; Rudolph, S.E.; Poulin, S.B. *J. Membr. Sci.* **7,** 277–291 (1980).
47. Sidman, K.R.; Steber, W.D.; Schwope, A.D.; Schnaper, G.R. *Biopolymers* **22,** 547–556 (1983).
48. Seno, M.; Kuroyanagi, Y. *J. Membr. Sci.* **27,** 241–252 (1986).
49. Sparer, R.V.; Bhaskar, R.K. U.S. Patent 4,513,034, issued 1985, assigned to Merck and Co.
50. Bhaskar. R.K.; Sparer, R.V.; Himmelstein, K.J. *J. Membr. Sci.* **24,** 83–96 (1985).
51. Minoura, N.; Aiba, S.; Fujiwara, Y. *J. Appl. Polym. Sci.* **31,** 1935–1942 (1986).
52. Goldberg, E.P. In *Polymers in Biology and Medicine: A Series of Monographs.* Donaruma, L.G.; Vogl, O., eds. John Wiley, New York (1983).
53. Bennett, D.B.; Adams, N.W.; Li, X.; Feijen, J.; Kim, S.W. *J. Bioact. Compat. Polym.* **3,** 44–52 (1988).
54. Negishi, N.; Bennett, D.B.; Cho, C.S.; Jeong, S.Y.; Van Heeswijk, W.A.R.; Feijen, J.; Kim, S.W. *Pharm. Res.* **4,** 305–310 (1987).
55. Giammona, G.; Giannola, L.I.; Carlisi, B.; Bajardi, M.L. *Chem. Pharm. Bull.* **37,** 2245–2247 (1989).
57. Zunino, F.; Giuliani, F.; Savi, G.; Dasdia, T.; Gambetta, R. *Int. J. Cancer* **30,** 465–470 (1982).
58. Ryser, H.J.P.; Shen, W.C. *Proc. Natl. Acad. Sci. USA* **75,** 3867–3870 (1978). 59. Shen, W.C.; Ryser, J.J.P. *Mol. Pharm.* **16,** 614–622 (1979).
60. Kohn, J.; Langer, R. *Polym. Mater. Sci. Eng.* **51,** 119–121 (1984).
61. Kohn, J.; Langer, R. *J. Am. Chem. Soc.* **109,** 817–820 (1987).
62. Spatola, A.F. In *Chemistry and Biochemistry of Amino Acids, Peptides, and Proteins.* Weinstein, B., ed. Marcel Dekker, New York, ppl 167–357. (1983).
63. Greenstein, J.P. *J. Biol. Chem.* **118,** 321–329 (1937).
64. Fasman, G.D. *Science* **131,** 420–421 (1960).
65. Jarm, V.; Fles, D. *J. Polym. Sci. Polym. Chem. Ed.* **15,** 1061–1071 (1977).
66. Kohn, J.; Langer, R. U.S. Patent 4,638,045, issued 1987, assigned to *Massachusetts Institute of Technology.*
67. Kohn, J.; Pulapura, S. U.S. Patent 5,099,060, issued 1992, assigned to Rutgers University.
68. Zhou, Q.-X.; Kohn, J. *Macromolecules* **23,** 3399–3406 (1990).
69. Arnold, L.D.; Kalantar, T.H.; Vederas, J.C. *J. Am. Chem. Soc.* **107,** 7105–7109 (1985).
70. Gelbin, M.E.; Kohn, J. *Polym. Prepr.* **32,** 241–242 (1991).
71. Gelbin, M.E.; Kohn, J. *J. Am. chem. Soc.* **114,** 3962–3965 (1992).

72. Yu, H. (Ph. D Thesis), Massachusetts Institute of Technology, Cambridge, MA (1988).
73. Yu Kwon, H.; Langer, R. *Macromolecules* **22**, 3250–3255 (1989).
74. A number of pseudo-poly(amino acids) and pseudo-peptide monomers are commercially available from Sigma Chemical Company.
75. Kohn, J. In *Polymeric Drugs and Drug Delivery Systems,* Vol. 469. Dunn, R.L.; Ottenbrite, R.M., eds. (ACS Symposium Series). Comstock, M.J., Series ed. American Chemical Society, Washington, DC, 155–169 (1991).
76. Li, C.; Kohn, J. *Macromolecules* **22**, 2029–2036 (1989).
77. Pulapura, S.; Li, C.; Kohn, J. *Biomaterials* **11**, 666–678 (1990).
78. Pulapura, S.; Kohn, J. In *17th International Symposium for the Controlled Release of Bioactive Materials.* Lee, V.H., ed. Reno, Nevada, Controlled Release Society, Lincolnshire, IL, pp. 154–155 (1990).
79. Pulapura, S.; Kohn, J. *Biopolymers* **32**, 411–417 (1992).
80. Hedayatullah, M. Bull. Soc. Chim. (France) 416–421 (1967).
81. Schminke, H.D.; Grigat, E.; Putter, R. U.S. Patent 3,491,060, issued 1970.
82. Kohn, J.; Langer, R. In *Peptides—Chemistry and Biology: Proceedings of the 10th American Peptide Symposium.* Marshall, G.R., ed. Escom Publishing, Leiden, The Netherlands, pp. 658–661 (1988).
83. Engelberg, I.; Kohn, J. *Biomaterials* **12**, 292–304 (1991).
84. Mohadger, Y.; Wilkes, G.L. *J. Polym. Sci. Polym. Phys. Ed.* **14**, 963–980 (1976).
85. Kohn, J.; Langer, R. *Biomaterials* **7**, 176–181 (1986).
86. Silver, F.H.; Marks, M.; Kato, Y.P.; Li, C.; Pulapura, S.; Kohn, J. *J. Long-Term Effects Med. Implants* **1**, 329–346 (1992).
87. Coffey, D.; Dong, Z.; Goodman, R.; Israni, A.; Kohn, J.; Schwarz, K.O. *Presented at the Symposium on Polymer Delivery Systems at the Spring Meeting of the American Chemical Society,* San Francisco (1992).
88. Lin, S.; Krebs, S.; Kohn, J. In *The 17th Annual Meeting of the Society of Biomaterials.* Scottsdale, AR, Society for Biomaterials, Algonquin, IL, 187 (1991).
89. Goodman, M.; Kirshenbaum, G.S. U.S. Patent 3,773,737, issued 1973, assigned to Sutures, Inc.
90. Katakai, R.; Goodman, M. *Macromolecules* **15**, 25–30 (1982).
91. Yokoyama, M.; Miyauchi, M.; Yamada, N.; Okano, T.; Sakurai, Y.; Kataoka, K.; Inoue, S. *Cancer Res.* **40**, 1693–1700 (1990).
92. Yokoyama, M.; Okano, T.; Sakurai, Y.; Kataoka, K.; Inoue, S. *Biochem. Biophys. Res. Commun.* **164**, 1234–1239 (1989).
93. Yokoyama, M.; Inoue, S.; Kataoka, K.; Yui, N.; Okano, T.; Sakurai, Y. *Makromol. Chem.* **190**, 2041–2054 (1989).
94. Kang, I.; Ito, Y.; Sisido, M.; Imanishi, Y. *Polym. J.* **19**, 1329–1339 (1987).
95. Kumaki, T.; Sisido, M.; Imanishi, Y. *J. Biomed. Mater. Res.* **19**, 785–811 (1985).
96. Kugo, K.; Murashima, M.; Hayashi, T.; Nakajima, A. *Polym. J.* **15**, 267–277 (1983).
97. Kugo, K.; Ohji, A.; Uno, T.; Nishino, J. *Polym. J.* **19**, 375–381 (1987)
98. Cho, C.S,; Kim, S.W.; Komoto, T. *Makromol. Chem.* **191**, 981–991 (1990).
99. Perly, B.; Douy, A.; Gallot, B. *Makromol. Chem.* **177**, 2569–2580 (1976).
100. Douy, A.; Gallot, B. *Polym. Eng. Sci.* **17**, 523–526 (1977).
101. Nakajima, A.; Hayashi, T.; Kugo, K.; Shinoda, K. *Macromolecules* **12**, 840–843 (1979).
102. Nakajima, A.; Kugo, K.; Hayashi, T. *Polym. J.* **11**, 995–1001 (1979).
103. Kugo, K.; Hata, H.; Hayashi, T.; Nakajima, A. *Polym. J.* **14**, 401–410 (1982).

104. Yokoyama, M.; Anazawa, H.; Takahashi, A.; Inoue, S. *Makromol. Chem.* **191**, 301–311 (1990).

105. Maeda, M.; Kimura, M.; Hareyama, Y.; Inoue, S. *J. Am. Chem. Soc.* **106**, 250–251 (1984).

106. Chung, D.; Higuchi, S.; Maeda, M.; Inoue, S. *J. Am. Chem. Soc.* **108**, 5823–5826 (1986).

107. Nathan, A.; Zalipsky, S.; Kohn, J. *Polym. Prepr.* **31**, 213–214 (1990).

108. Kohn, J.; Bolikal, D.; Nathan, A.; Zalipsky, S. In *18th International Symposium on Controlled Release of Bioactive Materials.* Kellaway, I.W., ed. Amsterdam, The Netherlands, Controlled Release Society, Lincolnshire, IL, pp. 329–330 (1991).

109. Nathan, A.; Bolikal, D.; Vyavahare, N.; Zalipsky, S.; Kohn, J. *Macromolecules* **24**, 4476–4484 (199L).

110. Ertel, S.; Nathan, A.; Zalipsky, S.; Agathos, S.; Kohn, J. *Polym. Mater. Sci. Eng.* **66**, 486–487 (1992).

111. Graham, N.B.; McNeill, M.E. *Makromol. Chem. Macromol. Symp.* **19**, 255–263 (1988).

112. Spira, M.; Fissette, J.; Hall, C.W.; Hardy, S.B.; Gerow, F.J. *J. Biomed. Mater. Res.* **3**, 213–234 (1969).

113. Hall, C.W.; Spira, M.; Gerow, F.; Adams, L.; Martin, E.; Hardy, S.B. *Trans. Am. Soc. Artif. Intern. Org.* **16**, 12–16 (1970).

114. Martin, E.C.; May, P.D., McMahon, W.A. *J. Biomed. Mater. Res.* **5**, 53–62 (1971).

115. Shiotani, N.; Kuroyanagi, T.; Koganei, Y.; Miyata, T. Japanese Patent 62246370, issued 1987, assigned to Koken Co., Ltd., Japan.

116. Shiotani, N.; Kuroyanagi, T.; Koganei, Y.; Yoda, R. Japanese Patent 63115564, issued 1988, assigned to Nippon Zeon Co., Ltd., Japan.

117. Shioya, N.; Kuroyanagi, Y.; Koganeo, Y.; Yoda, R. European Patent Application 265906, filed 4 May 1988, assigned to Nippon Zeon Co., Ltd., Japan.

118. Brack, A. French Patent 2625507, issued 1989, assigned to Delalande S. A., France.

119. Kuroyanagi, Y.; Kim, E.; Kenmochi, M.; Ui, K.; Kageyama, H.; Nakamura, M.; Takeda, A.; Shioya, N. *J. Appl. Biomater.* **3**, 153–161 (1992).

120. Sakamoto, I.; Unigame, Takagi K. European Patent Application 162610, filed February 26, 1986, assigned to Unitika Ltd. Japan.

121. Dorman, L.C.; Meyers, P.A. European Patent Application 192068, filed August 27, 1986, assigned to Dow Chemical Co., USA.

122. Wise, D.L.; Midler, O. In *Biopolymers in Controlled Release Systems,* Vol. 2. Wese, D.L., ed. CRC, Boca Raton, FL. pp. 219–229 (1984).

123. Shyu, S.S.; Kuroyanagi, Y.; Seno, M. *Colloid Polym. Sci.* **266**, 587–593 (1988).

124. Zunino, F.; Savi, G.; Giuliani, F.; Gambetta, R.; Supino, R.; Tinelli, S.; Pezzoni, G. *Eur. J. Cancer Clin. Oncol.* **20**, 421–425 (1984).

125. Filipova-Voprsalova, M.; Drobnik, J.; Sramek, B.; Kvetina, J. *J. Control. Rel.* **17**, 89–98 (1991).

126. Kato, Y.; Umemoto, N.; Kayama, Y.; Fukushima, H.; Takeda, Y.; Hara, T.; Tsukada, Y. *J. Med. Chem.* **27**, 1602–1607 (1984).

127. Shen, W.D.; Ryser, H.J.P.; LaManna, L. *J. Biol. Chem.* **260**, 10905–10908 (1985).

128. Shen, W.C.; Du, X.; Feener, E.P.; Ryser, J.J.P. *J. Control. Rel.* **10**, 89–96 (1989).

129. Yu, H.; Lin, J.; Langer, R. In *14th International Symposium on Controlled Release of Bioactive Materials.* Lee, P.I.; Leonhardt, B.A., eds. Toronto, Canada, Controlled Release Society, Lincolnshire, IL, pp. 109–110 (1987).

130. Fietier, I.; Le Borgne, A.; Spassky, N. *Polym. Bull.* **24**, 349–353 (1990).

131. Kim, H.; Sung, Y.K.; Jung, J.; Baik, H.; Min, T.J.; Kim, Y.S. *Taehan Hwahakhoe Chi* **34,** 203–210 (1990).
132. Cho, C.S.; Park, J.W.; Kwon, J.K.; Jo, B.W.; Lee, K.C.; Kim, K.Y.; Sung, Y.K. *Pollimo* **15,** 27–33 (1991).
133. Pramanick, D.; Ray, T.T. *Polym. Bull.* **18,** 311–315 (1987).

CHAPTER 6

Polyphosphazenes as New Biomaterials

A.G. Scopelianos

6.1   Introduction

The emergence of polyphosphazenes as a useful and versatile class of polymers has attracted considerable interest from both academia and industry. The uniqueness of these polymers relates to the fact that they are comprised of an inorganic phosphorus–nitrogen backbone and provide unprecedented synthetic flexibility.

It was this flexibility, coupled with the careful selection and attachment of certain side groups to the phosphazene backbone, that triggered the emergence of multiple potential applications. Historically, early attempts were devoted to the development of hydrolytically stable polymers that displayed excellent thermal and oxidative stability. These early polymers were targeted toward applications that in general were nonbiomedical. During the past two decades, however, this trend has been shifting slowly, as numerous publications have begun to appear suggesting many biomedical uses. These applications range from inert biomaterials for cardiovascular and dental uses, to bioerodable and water-soluble polymers for controlled drug delivery.

As described in subsequent sections, the uniqueness of this system stems from the presence of a nitrogen–phosphorus backbone, as opposed to more commonly employed hydrocarbon-based polymers that depend on such hydrolytically sensitive linkages as

---

A.G. Scopelianos, Ethicon, Inc., A Johnson & Johnson Company, P.O. Box 151, Somerville, New Jersey 08876-0151, U.S.A.

esters (polyglycolides, polylactides, polyhydroxybutyrates, etc.) anhydrides (polyse-bacic acid–bis-carboxyphenoxyalkyl copolymers, etc.), urethane (biomer), activated ethers [poly(ortho esters)], polysaccharides, etc.), amides [poly(amino acids)], etc., for their hydrolytic breakdown. The phosphazene backbone, on the other hand, in the presence of appropriate biocompatible or metabolizable side groups is capable of undergoing hydrolysis to phosphate and ammonium salts with the concomitant release of the side group.

In this chapter, the synthetic aspects, physical properties, and projected biomedical applications of these polymers are discussed. More specifically, the ring-opening polymerization route along with the substitutive mode of synthesis as it relates to each class of substituents is outlined. A description of the relationship between structure and properties along with the flexibility that the synthetic process provides toward tailor-making polymers to meet specific needs is given. Finally, a discussion relating to their potential application as implantable devices, controlled-release drug applications, matrices for cell proliferation, etc., is also included.

6.2   Background

Stokes in 1897 first described the thermal polymerization of hexachlorocyclotriphos-phazene $(NPCl_2)_3(1)$ to a colorless, high molecular weight, insoluble elastomer, later called "inorganic rubber" [1] (2). After prolonged heating at higher temperatures, the elastomer depolymerized to cyclic oligomers. Because of its hydrolytic instability, "inorganic rubber" has no technological use.

Subsequently, and especially during the last 30 years, many attempts have been made to replace the easily hydrolyzable halogen atoms by hydrolytically stable groups, such as alkoxy and aryloxy units. However, these reactions yielded crosslinked, partially substituted products [2–5]. In 1965, Allcock and Kugel prepared the first example of an organic-soluble version of poly(dichlorophosphazene) [6,7]. This was accomplished by the termination of the polymerization of $(NPCl_2)_3$ to high molecular weight polymer before 50% of trimer was consumed. This linear, uncrosslinked material readily dissolves in organic solvents and reacts with amines [8], alkoxides [8,9–11], or organometallic reagents [12] to form hydrolytically stable, high molecular weight poly(organophosphazenes). These reactions are shown in Scheme 1, where n ≈ 10,000–15,000.

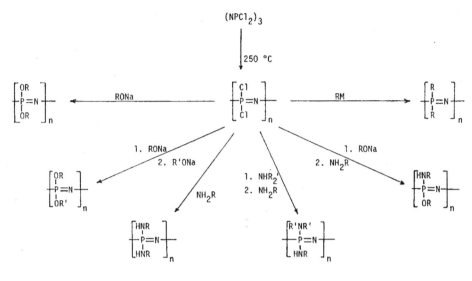

Scheme 1

Similar nucleophilic reactions have also been carried out with the substrate $(NPF_2)_n$ [13,14]. Mixed substituent copolymers have also been made [11,15]. Thus, the polyphosphazenes offer an unprecedented synthetic versatility.

The properties found for poly(organophosphazenes) are dependent on the nature of the substituent introduced. By controlled variations in the type and the ratio of side groups, new polymers can be made that are tailored to suit their intended purpose.

For polymers not intended for biomedical application, some of the properties found include resistance to weathering; high elasticity [8,11,16]; low temperature flexibility; resistance to petroleum-based solvents, heat, and high-intensity radiation [16–20]; and ability to act as insulators [21] and flame- and heat-resistant hydraulic fluids and lubricants [22]. Properties such as flame retardance [11] and the ability to retain colors [17] have also been described. The general ultraviolet-visible transparency of the phosphorus–nitrogen skeleton has created an interest in the potential use of polyphosphazenes as ultraviolet-resistant coatings [23]. Polymers with fluoroalkoxy or aryloxy substituent groups are at present manufactured for use as seals in automotive and aircraft engines as heat, vibration, and electrical insulation and as flame-retardant foams [24].

Recently, poly(organophosphazenes) have been synthesized with alkyl ether alkoxy side groups attached to the phosphorus atoms of the skeleton [25–33]. These species are water stable and either water-soluble or hydrophilic polymers. Specific members of this series form weak complexes with metal salts, which are excellent solid electrolyte materials. Polyphosphazenes have also been suggested as electronic or electrooptical materials. For example, introduction of ferrocenyl side groups provides sites for electrochemical oxidation and reduction [50–54]. Highly polarizable aromatic side groups have also been introduced and were shown to generate second-order nonlinear optical properties [34,35].

6.3   Biomedical Applications

Although the primary objective of this chapter is to review the development and future directions of bioerodible polyphosphazenes, it is also important to review other potential biomedical applications, which include hydrolytically stable and biocompatible polymers. This is especially important in order to gain true appreciation for the flexibility of this polymeric system and the breadth of its potential. An excellent review on this topic was recently published by Allcock [36].

6.3.1   Biostable and Bioactive Polymers

It is well established that most currently employed biomedical polymers were originally developed for industrial applications. Likewise, polyphosphazenes that might be considered suitable for biomedical applications have progressed through a similar sequence.

For example, the most extensively studied polyphosphazenes are hydrophobic, bearing fluoroalkoxy side groups (3) and (4) [37,38]. Polymer (3) was the first fully substituted and uncrosslinked polyphosphazene prepared. Undoubtedly these materials were selected because of their high fluorine content and expected minimal tissue response, as had been previously demonstrated with highly fluorinated materials such as Teflon. Indeed these materials were shown to be biocompatible when implanted subcutaneously in rats [38]. W.M. Reichert and coworkers [39] have also shown that the hemocompatibility of such materials is dependent on the molecular motions of the polymer interface. In an attempt to control molecular motion through irradiation and crosslinking they were able to show that increased levels of bound water, as a result of higher crosslinking, resulted in an increased thrombogenic response.

$$\left[\begin{array}{c} OCH_2CF_3 \\ | \\ -N=P- \\ | \\ OCH_2CF_3 \end{array} \right]_n \qquad \left[\begin{array}{c} OCH_2CF_3 \\ | \\ -N=P- \\ | \\ OCH_2(CF_2)_xCF_3 \end{array} \right]_n$$

$$\qquad 3 \qquad\qquad\qquad\qquad 4$$

Perhaps because of their availability and ease of preparation the next most studied series of polymers is the aryloxyphosphazenes (5) and closely related derivatives (6), (7), (8), (9), (10), and (11) (Scheme 2).

For example, polymer (6), because of its polyelectrolyte nature, can be crosslinked with dissolved cations such as calcium to form a hydrogel matrix [40]. In fact, this

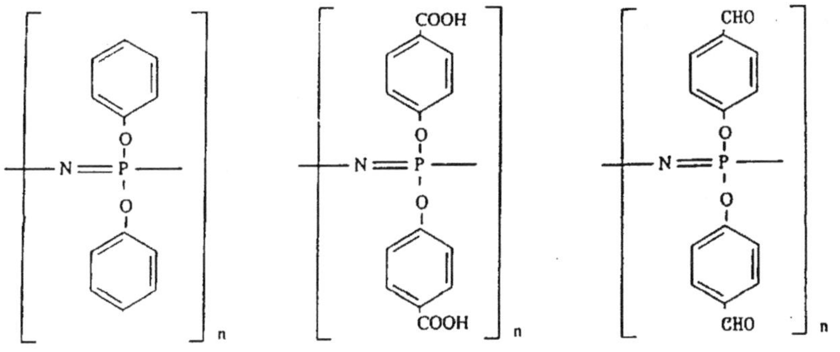

Scheme 2

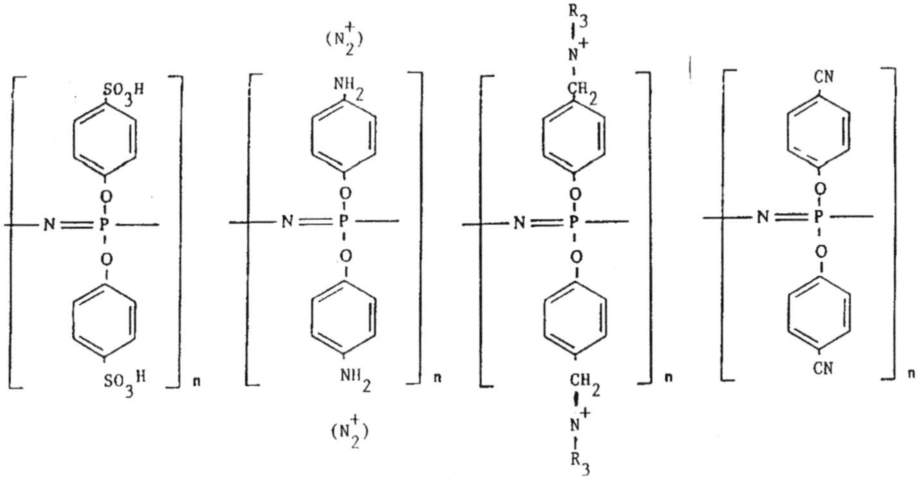

method is very similar to the one employed for the preparation of alginate micro-spheres. Two lines of hybridoma cells were encapsulated within calcium crosslinked polyphosphazene gel microbeads without affecting their viability or their capacity to produce antibodies. Interaction with poly(L-lysine) produced a semipermeable mem-brane that was capable of retaining the cell-secreted antibodies inside the beads.

Schiff's base coupling techniques were employed for the covalent binding of sulfadi-azine (antibiotic), dopamine, amino-4-picoline (analgesic), and hydralazine (antihyper-tensive) to polymer (7) [41]. Further studies suggested that in an aqueous environment the Schiff's base hydrolyzed to release the bioactive agents. The potential of using such systems as depot drug delivery devices is evident. It should be realized, however, that the aryloxy phosphazenes are hydrolytically stable materials and therefore complete breakdown of this polymer may not be possible.

Fully and partially sulfonated polymers (8) both as free acid and sodium salt were evaluated for their antibacterial, anticoagulant, and cell adhesion properties [42]. The polymers showed antibiotic activity against *Salmonella typhimurium* (TA 100) and *Bacillus subtilis* but not against *Salmonella pullorum, Escherichia coli,* or *Streptococcus faecalis.* It was postulated that the activity may be due to the fact that *Salmonella typhimurium* is a genetically engineered strain without a cell coating and with an inferior DNA repair mechanism. Most likely, the polymers disrupt the bacterial cell membrane, and this inhibits the cell's ability to control diffusion through the membrane. It was reported that there appears to be no inherent antibiotic character associated with the phosphazene structure itself, but only with the side group structure.

In blood clotting experiments, the unsulfonated, sulfonated, and protamine-immobilized films were tested in a Lindholm cell with EDTA-treated human blood. The unsulfonated films had an average clotting time of 12 min. The sulfonated and protamine-immobilized films both had average clotting times of > 60 min. Thus, these materials exhibited enhanced blood compatibility with respect to the untreated films.

In additional experiments, the immobilization of mammalian cells on these polymers was also evaluated [42]. All the sulfonated films had a complete monolayer coverage of cells after 3 days of incubation. By contrast, the unsulfonated polymer surfaces were colonized only in small regions of the surface. The ionic and hydrogel character of the sulfonated films appeared to enhance their suitability for cell attachment and growth. Unlike sulfonated polystyrene, these films are flexible after sulfonation and cell attachment.

In an attempt to prepare novel antithrombogenic materials, polymer (10) was complexed with heparin [43]. The method utilizes a tertiary ammonium binding site connected to the polyphosphazene backbone via an aryloxy side group. Preliminary experiments have shown that mean clotting times in a Lindholm cell and using bovine blood were several-fold longer than in unheparinized controls. Some leaching of heparin from the polymeric surface was also seen as evidenced by decreased clotting times.

Polymer-bound catecholamines were also prepared by the diazotization of the aminophenoxy-bearing polyphosphazene (9) followed by coupling with dopamine, DL-epinephrine, and DL-norepinephrine [44]. The aminophenoxy units were generated by reduction of 4-nitrophenoxy groups over PtO_2 and hydrogen. The phosphazene skeleton was unaffected by the reduction, diazotization, and diazo coupling. The polymeric derivatives were found to elicit inhibition of prolactin released from rat pituitary cells in culture for 48 h.

Glucose-6-phosphate dehydrogenase (G-6-PDH) and trypsin have been linked covalently to a surface-modified poly[bis(aryloxy)phosphazene] supported on porous alumina particles [45]. Poly(diphenoxyphosphazene) was surface nitrated and then reduced to the aminophenoxy derivative. The aminophenoxy sites were then activated by reaction with cyanogen bromide, nitrous acid, or glutaric dialdehyde, followed by treatment with the enzymes in aqueous buffer solutions. The maximum immobilization yields (8–20% for G-6-PDH and 50–60% for trypsin) were achieved when the glutaric dialdehyde activated system was used. For both proteins, the immobilized enzyme retained its activity and remained linked to the support through numerous cycles. By contrast, the same enzymes absorbed on uncoated alumina particles were displaced readily by washing. The storage stabilities of both immobilized enzymes were much greater than those of the free enzymes in solution. For the G-6-PDH system, enzyme activity and scanning electron microscopy were used to examine the relationship between the substrate topography/surface area and the concentration of immobilized protein molecules.

With use of similar aminophenoxy-bound phosphazenes, chemotherapeutic or herbicidal agents were linked by peptide-coupling techniques [46]. The synthetic procedure calls for the preparation of 4-cyanophenoxy-bound polyphosphazene (11) followed by reduction of the cyano groups to 4-(aminomethyl)phenoxy units with the use of BH_3-THF. Condensation of the pendant amino groups with N-acetylglycine, N-acetyl-DL-penicillamine, p-(dipropylsulfamoyl)benzoic acid, and 2,4-dichlorophenoxyacetic acid was accomplished with the use of dicyclohexylcarbodiimide. In preliminary studies, the polymers along with the coupled bioactive agents appeared to be hydrolytically stable at 37°C in a buffered aqueous medium at pH 7.4. This is expected in view of the hydrophobic character of the phenoxy cosubstituent groups and their well established hydrolytic stability. It was postulated that the induction of biological activity, either of the phosphazene-bound acids or the free carboxylic acids released by hydrolysis, would probably be enhanced by the use of more hydrophilic and hydrolytically unstable cosubstituent groups.

6.3.2   Water-Soluble Polymers

It is important to note that the polymeric systems described so far were based on the employment of hydrophobic and hydrolytically stable and insoluble side groups. Furthermore, as a result of side group functionalization, they also served as sites for the covalent attachment of bioactive agents. Interestingly, however, some of the early approaches were based on the employment of both water-stable and water-soluble polyphosphazenes as opposed to the insoluble materials described above. Those polymers were then employed as prospective plasma extenders or carriers of bioactive agents. The following two examples illustrate this point.

First, the synthesis of high molecular weight polymeric water-soluble carriers (12) and (13) for the finding of metalloporphyrins was described [47,48] as well as the behavior of these polymers as coordination liquids for hemin and heme in aqueous solution and in the solid state.

In the presence of aqueous polymer (12) hemin or heme exists as hematin–hemin hydroxide or bis-aquoheme, respectively, with only weak (probably acid–base) binding between the polymer and metalloporphyrin. However, polymer (13) binds strongly to both heme and hemin via the imidazole residues. In aqueous solution or as solid films, the presence of the polymeric ligands did not, however, prevent the irreversible oxidation of heme on contact with oxygen.

$$\left[\begin{array}{c} NHMe \\ | \\ -N=P- \\ | \\ NHMe \end{array} \right]_n \qquad \left[\left(\begin{array}{c} NHMe \\ | \\ N=P- \\ | \\ NHMe \end{array} \right)_{0.x} \left(\begin{array}{c} NHMe \\ | \\ N=P- \\ | \\ NH \\ | \\ (CH_2)_4 \end{array} \right)_x \right]_n$$

12   13

In the late 1970s, polymeric and oligomeric, water-soluble anticancer derivatives were synthesized by the reaction of [NP(NHCH$_3$)$_2$]$_3$ or [NP(NHCH$_3$)$_2$]$_4$ with K$_2$PtCl$_4$ [49,50]. The anticancer properties of square planar platinum compounds have shown that the marked chemotherapeutic effects are accompanied by serious toxic side effects, including bone marrow depletion and damage to the intestinal mucosa and kidney. It was reasoned that many of these side effects may result from rapid excretion through the semipermeable membranes of the kidney system. However, because high molecular weight polymers cannot pass through semipermeable membranes, it was assumed that the binding of a platinum complex to a polymer should retard the rate of excretion.

Reaction products obtained were shown to contain a square planar platinum atom bonded to the nitrogen atoms of the phosphazene skeleton for both the polymeric and cyclic derivatives. The compounds showed tumor inhibitory activity in initial anticancer screening tests.

Most recently, the water-soluble polymer (14) was prepared by reacting the sodium salt of 2-(2-methoxyethoxy)ethanol with polydichlorophosphazene [51–53]. Films were solution cast from water and then crosslinked by exposure to γ-irradiation to form hydrophilic membranes and hydrogels capable of swelling with water. Potential applications for such materials include soft tissue replacement, contact lenses, artificial skin, dialysis membranes, blood vessel replacement, etc. However, it has been found in preliminary experiments that facile diffusion of small molecules occurs from these hydrogels. Water-soluble and ultraviolet-detectable small molecules with different molecular structural features and basicities were found to diffuse rapidly from hydrogel discs at approximately the same rate (~ 70% loss from the matrix

$$\left[\begin{array}{c} OCH_2CH_2OCH_2CH_2OCH_3 \\ | \\ -N=P- \\ | \\ OCH_2CH_2OCH_2CH_2OCH_3 \end{array} \right]_n$$

14

in 1 h). Moreover, the rate of diffusion was essentially unaffected by crosslink density. Thus, hydrogels derived from polymer (14) provide a facile pathway for small molecule diffusion, an effect that can be attributed to the ease of chain and side group reorientation in this polymer, and the fact that the minimum degree of crosslinking achieved was ~ 1 crosslink/25 repeating units. It was postulated that this would provide ample free volume (water molecules) to allow almost unrestricted diffusion of small-molecule species.

6.3.3   Bioerodible and Bioactive Polymers

Although more than 300 different types of polyphosphazenes have been synthesized over the years, only few appear to meet the requirements necessary for bioerosion and biocompatibility. In general, it appears that these polymers fall into two categories: first, polyphosphazenes that are substituted with amines of low pK_a and second, activated alcohols. The following sections deal with these two areas in some detail.

6.3.3.1   Amine-Substituted Polyphosphazenes

The most comprehensive studies aimed at understanding the effect of different amino side groups on the hydrolytic behavior of phosphazenes were undertaken in the early 1970s and 1980s [54–56]. In these studies, the general hydrolytic characteristics of several aminophosphazene high polymers and analogous small-molecule cyclic analogues were examined. The latter have been used as models in an attempt to understand the behavior of the high molecular weight polymers [57]. The hexa(amino)cyclotriphosphazenes (15–26) (Scheme 3) were prepared by the reactions between hexachlorocyclotriphosphazene and ammonia, methylamine, ethyl glycinate, imidazole, benzylamine, trifluoroethylamine, aniline, potassium pyrrolide, pyrrolidine, piperidine, or morpholine. The compound [NP(NHCH$_2$CONHCH$_3$)$_2$]$_3$ (18) was prepared by the reaction between (17) and methylamine.

Compound (19) was the only aminocyclophosphazene that hydrolyzed at a detectable rate in water or aqueous dioxane at 25°C. At 100°C, compounds (15)–(19) hydrolyzed (as evidenced by changes in the ^{13}P NMR spectra) but species (20)–(26) did not. Hydrolysis of (15)–(19) yielded ammonia, phosphate, and the appropriate side group amine or amino acid. Under these reaction conditions the sensitivity to hydrolysis decreased in the order (19) > (17) > (18) > (15) > (16) > (22) > (20), (21), (23)–(26) on the basis of the speed of changes in ^{31}P NMR spectra and recovered starting materials.

In acidic media at 25°C, compounds (15)–(19) degraded to phosphoric acid, ammonium ion, and the appropriate amine salts. The sensitivity to acidic hydrolysis in 1 N or weaker hydrochloric acid decreased in the order (19) > (15) > (17) ≥ (18) > (16). Hydrolysis was faster in 5 N hydrochloric acid than in 1 N acid. In the stronger acid medium the sensitivity to hydrolysis decreased in the order (19) > (18) > (15) > (16).

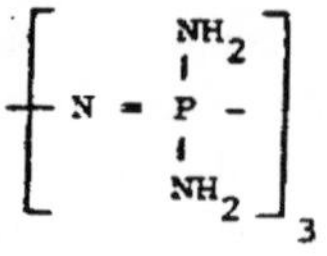

$$\left[\!\!\begin{array}{c} NH_2 \\ | \\ N\!-\!P\!- \\ | \\ NH_2 \end{array}\!\!\right]_3 \qquad \left[\!\!\begin{array}{c} HNCH_3 \\ | \\ N\!-\!P\!- \\ | \\ HNCH_3 \end{array}\!\!\right]_3 \qquad \left[\!\!\begin{array}{c} O \\ || \\ HNCH_2COCH_2CH_3 \\ | \\ N\!-\!P\!- \\ | \\ HNCH_2COCH_2CH_3 \\ || \\ O \end{array}\!\!\right]_3$$

15 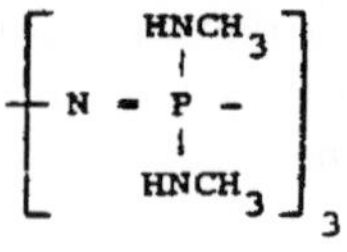   16 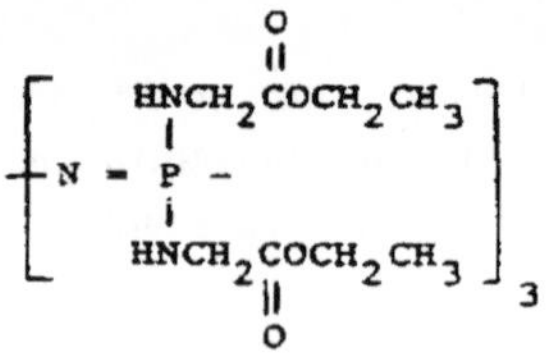   17

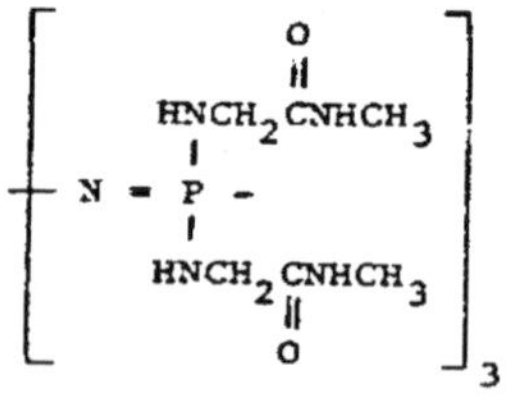

$$\left[\!\!\begin{array}{c} O \\ || \\ HNCH_2CNHCH_3 \\ | \\ N\!-\!P\!- \\ | \\ HNCH_2CNHCH_3 \\ || \\ O \end{array}\!\!\right]_3 \qquad \left[\!\!\begin{array}{c} N \\ | \\ N\!-\!P\!- \\ | \\ N \end{array}\!\!\right]_3 \qquad \left[\!\!\begin{array}{c} HNCH_2\text{—Ph} \\ | \\ N\!-\!P\!- \\ | \\ HNCH_2\text{—Ph} \end{array}\!\!\right]_3$$

18 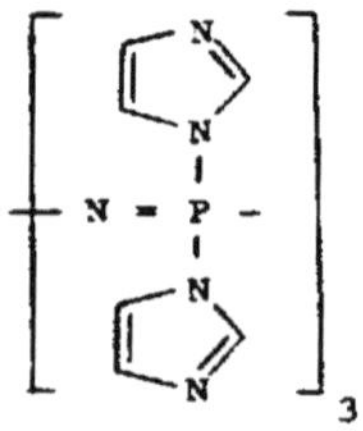   19 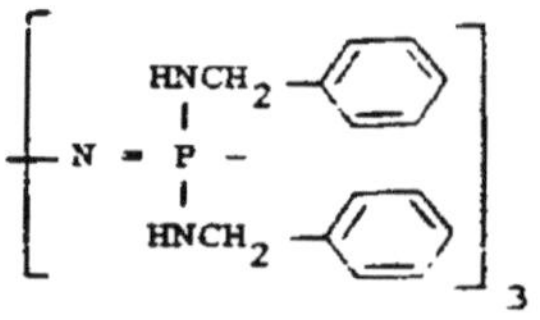   20

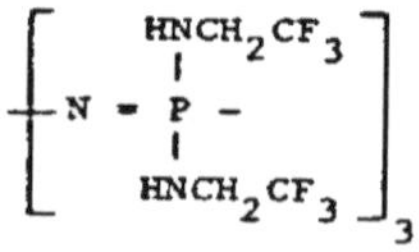

$$\left[\!\!\begin{array}{c} HNCH_2CF_3 \\ | \\ N\!-\!P\!- \\ | \\ HNCH_2CF_3 \end{array}\!\!\right]_3 \qquad \left[\!\!\begin{array}{c} HN\text{—Ph} \\ | \\ N\!-\!P\!- \\ | \\ HN\text{—Ph} \end{array}\!\!\right]_3 \qquad \left[\!\!\begin{array}{c} N \\ | \\ N\!-\!P\!- \\ | \\ N \end{array}\!\!\right]_3$$

21 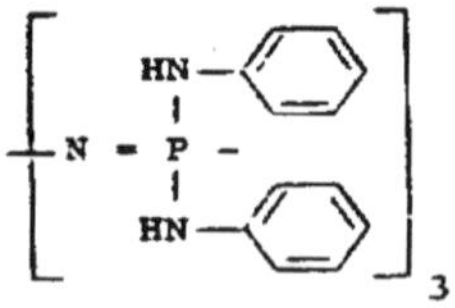   22 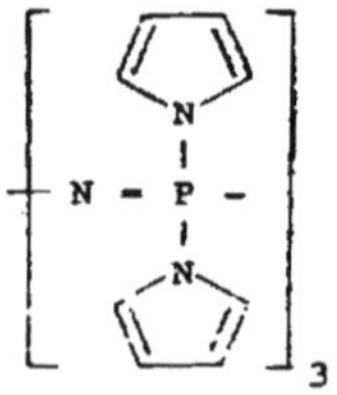   23

$$\left[\!\!\begin{array}{c} N \\ | \\ N\!-\!P\!- \\ | \\ N \end{array}\!\!\right]_3 \qquad \left[\!\!\begin{array}{c} N \\ | \\ N\!-\!P\!- \\ | \\ N \end{array}\!\!\right]_3 \qquad \left[\!\!\begin{array}{c} O \\ N \\ | \\ N\!-\!P\!- \\ | \\ N \\ O \end{array}\!\!\right]_3$$

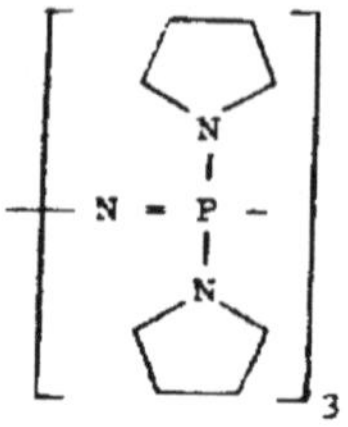

24 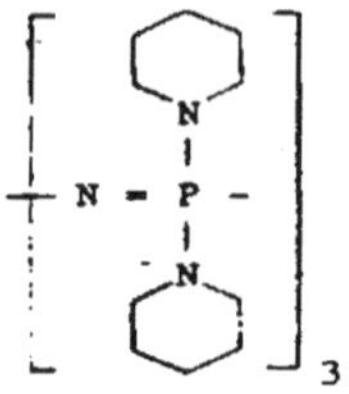   25 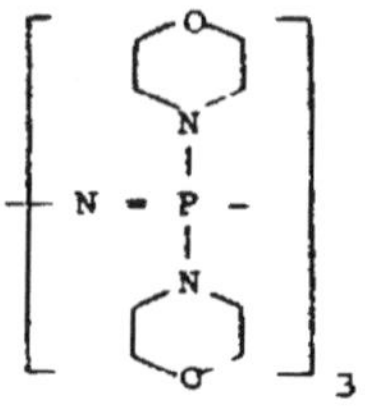   26

Scheme 3

The overall speed of hydrolysis in 1 N or stronger acid was higher than in water. The behavior of (20)–(26) could not be followed because of the insolubility of these species in acidic aqueous dioxane.

In basic media containing 1 N or stronger sodium hydroxide, the rates of hydrolysis were very slow, with the order of decreasing sensitivity to hydrolysis being (19) > (15), (18) > (16), (17), (20)–(26). Species (16), (17), and (20)–(26) were essentially stable to strong base for several weeks. Only slight hydrolysis of (15) and (18) occurred in

1 N sodium hydroxide solution during 24 h but (19) was unstable during the same treatment. In weaker alkaline solutions (between 0.5 and 1 N), the hydrolysis of (15)–(18) was still a slow process. However, (17), (18), and (19) degraded within 24 h when the concentration of sodium hydroxide was reduced to between 0.03 and 0.5 N.

Two different but interconnected mechanistic pathways appear to take place. In the first, hydrolytic removal of one amino residue from phosphorus occurs to yield species of type $N_3P_3R_5OH$ before cleavage of the phosphazene ring takes place. In the second, cleavage of the phosphazene ring is a faster reaction following protonation of the ring nitrogen atoms. Those compounds that contained amino acid ester or amide side groups hydrolyzed only after prior initial conversion of the ester or amide units to free carboxylic acid groups.

In a separate study designed to establish the hydrolytic tendencies of polyamino-phosphazenes and to draw correlations with the previous study, polymers (27)–(34) were synthesized and their properties studied [58] (Scheme 4).

The monomethylamino group was chosen as a cosubstituent because of its small dimensions and because it is a water-solubilizing substituent group for polyphosphazenes. The choice of an amino acid ester as a substitute group, rather than an amino acid, was dictated by the expectation that free carboxylic acid groups would participate in chlorine substitution (to yield crosslinks) and could also cause degradation of the polymer backbone.

All the amino acid ester containing polymers synthesized in this work underwent a slow decrease in molecular weight in solution and in the solid state at 25°C. For example, some samples of $[NP(NHCH_2COOC_2H_5)_2]_n$ (27) underwent a decrease in chain length from ~ 10,000 repeating units to fewer than 1000 repeating units in 15 days in the solid state at 25°C. The related mixed substituent polymer (28) underwent a chain shortening from about 8000 to 1000 repeating units in 25–30 days. The other polymers behaved similarly. Cooling of the polymers to −32°C considerably reduced the rate of decomposition.

Several mechanistic possibilities have been proposed. These are: (a) a mechanism involving prior hydrolysis of ester linkages to yield carboxylic acid groups; (b) an attack on the skeleton by the carbonyl groups of the substituent units; and (c) a hydrolytic cleavage of the bond connecting the side group to phosphorus.

Reaction (a) is plausible for all the amino acid ester polymers synthesized in this work. It is well known that amino acid esters are susceptible to hydrolysis to the free amino acid. Moreover, commercial amino acid ester hydrochlorides may contain traces of the free amino acid which could be carried through the synthetic sequences. Hence, the possibility exists that the final number of polymer fragments depends on the number of carboxylic acid residues in the polymer. Added benzoic acid or benzoyl peroxide markedly enhanced the rate of chain cleavage. Moreover, the polymers that appeared to be the most resistant to chain cleavage were those that contained phenylalanine ester, and these might be expected to be more hydrophobic than, for example, polymers that contained ethyl glycinate residues. However, protection of the ester grouping against hydrolytic attacks by the synthesis of a *tert*-butyl glycino-substituted polymer (35) (Scheme 4) had no perceptible effect on the rate of molecular weight

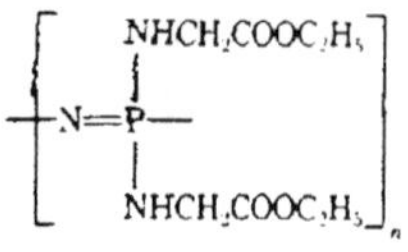

$$
\left[\begin{array}{c} NHCH_2COOC_2H_5 \\ -N=P- \\ NHCH_2COOC_2H_5 \end{array} \right]_n
$$

27

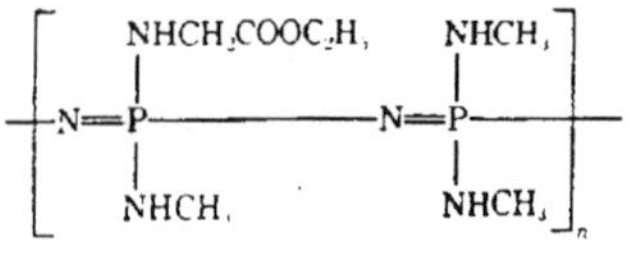

$$
\left[\begin{array}{cc} NHCH_2COOC_2H_5 & NHCH_3 \\ -N=P- \quad -N=P- \\ NHCH_3 & NHCH_3 \end{array} \right]_n
$$

28

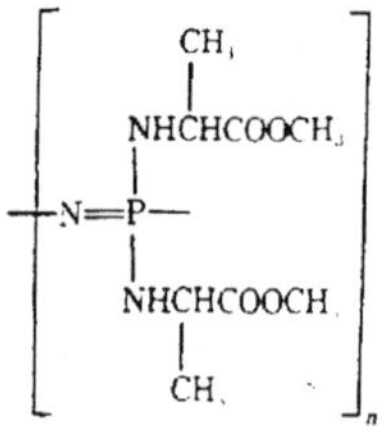

$$
\left[\begin{array}{c} CH_3 \\ NHCHCOOCH_3 \\ -N=P- \\ NHCHCOOCH_3 \\ CH_3 \end{array} \right]_n
$$

29

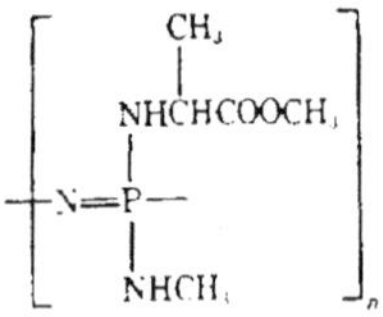

$$
\left[\begin{array}{c} CH_3 \\ NHCHCOOCH_3 \\ -N=P- \\ NHCH_3 \end{array} \right]_n
$$

30

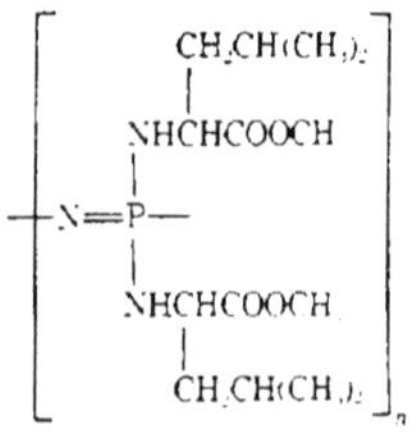

$$
\left[\begin{array}{c} CH_2CH(CH_3)_2 \\ NHCHCOOCH_3 \\ -N=P- \\ NHCHCOOCH_3 \\ CH_2CH(CH_3)_2 \end{array} \right]_n
$$

31

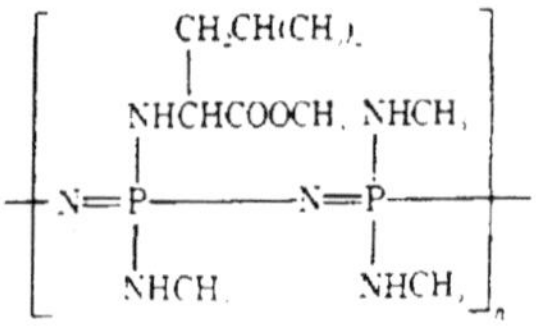

$$
\left[\begin{array}{cc} CH_2CH(CH_3)_2 & \\ NHCHCOOCH_3 & NHCH_3 \\ -N=P- \quad -N=P- \\ NHCH_3 & NHCH_3 \end{array} \right]_n
$$

32

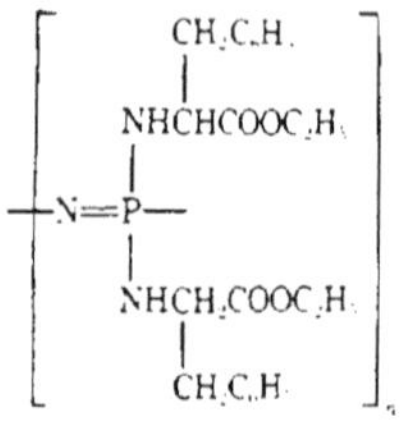

$$
\left[\begin{array}{c} CH_2C_6H_5 \\ NHCHCOOC_2H_5 \\ -N=P- \\ NHCH_2COOC_2H_5 \\ CH_2C_6H_5 \end{array} \right]_n
$$

33

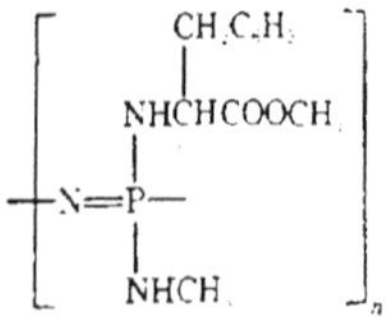

$$
\left[\begin{array}{c} CH_2C_6H_5 \\ NHCHCOOCH_3 \\ -N=P- \\ NHCH_3 \end{array} \right]_n
$$

34

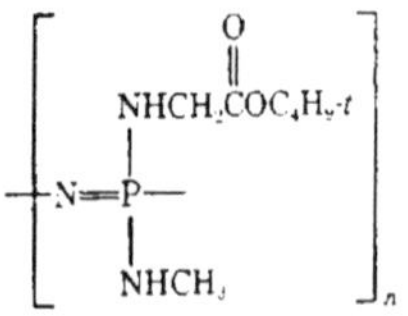

$$
\left[\begin{array}{c} O \\ \| \\ NHCH_2COC_4H_9\text{-}t \\ -N=P- \\ NHCH_3 \end{array} \right]_n
$$

35

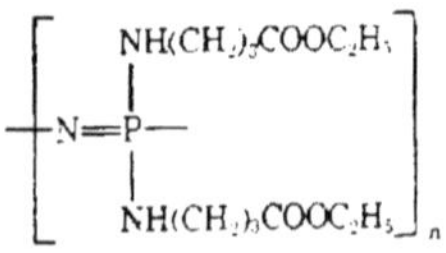

$$
\left[\begin{array}{c} NH(CH_2)_2COOC_2H_5 \\ -N=P- \\ NH(CH_2)_2COOC_2H_5 \end{array} \right]_n
$$

36

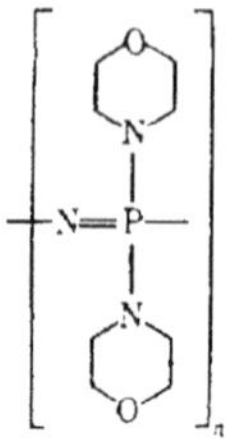

37 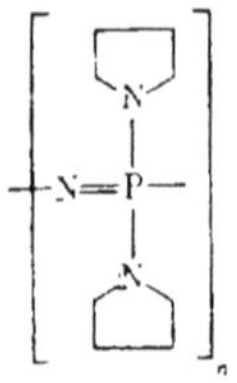   38

$$
\left[\begin{array}{c} NHCH_2CH_2OC_2H_5 \\ -N=P- \\ NHCH_2CH_2OC_2H_5 \end{array} \right]_n
$$

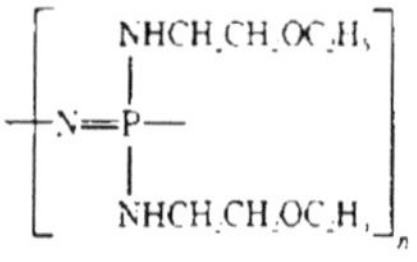

39 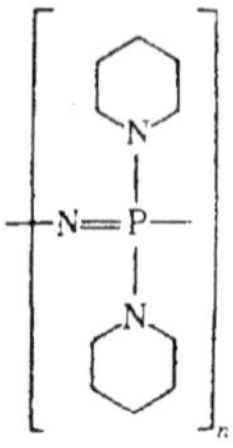   40

Scheme 4

decline. Although side group ester hydrolysis provides a plausible mechanism for the initiation of chain cleavage, the experimental results indicate that other mechanisms are probably involved as well.

Attack by a side group carboxyl unit (mechanism b) is compatible with the experimental data. Butyramide causes extensive decomposition and chain cleavage during reactions with $(NPCl_2)_n$. Separation of the ester function from the amino function [as in (36)] did not prevent a spontaneous decline in molecular weight.

Evidence was also found to support mechanism c. The hydrolytic cleavage of a P–N side group bond would be facilitated by a high leaving group ability of this side group. Thus, ethyl glycinate has a pK_a at 25°C of 7.75, whereas butylamine has a pK_a value of 10.78. Poly[bis(butylamino)phosphazene] is an especially stable polymer. Similarly, poly[bis(morpholino)phosphazene] (37) underwent a skeletal cleavage reaction in the atmosphere, but poly[bis(pyrolidino)phosphazene] (38) and poly[bis(piperidino)phosphazenel] (40) did not. The pK_a value for morpholine at 25°C is 8.33, whereas the values for pyrrolidine and piperidine are 11.27 and 11.12, respectively. Polymer (39) underwent chain cleavage, a result that can be related to a pK_a value of 9–10 for $H_2NCH_2CH_2OC_2H_5$. This rough correlation of polymer instability with leaving group ability provides valuable guidance for the tailored synthesis of hydrolytically degradable polymers.

Since the discovery that amino acid ester substituted polyphosphazenes degrade slowly at physiological pH, approaches involving cosubstitution with bioactive agents have also emerged. It was hoped here that the concomitant breakdown of the polymer along with the hydrolytic detachment of the side groups would allow for sustained release of the bioactive agent. One such system (41) comprised of 65–90% of glycine ethyl ester with 10–35% of a spacer(s) group such as L-lysine ethyl ester linked to a bioactive agent (D), in this case naproxen, was developed and the breakdown kinetics of the polymer and release profiles for drug established [59].

$$
\begin{array}{ll}
D_{(m)} & \\
| & D \quad = \text{Drug} \\
S & D_m \quad = \text{Model compound} \\
| & \\
\left[N{=}P \right]_n & S \quad = \text{Spacer} \\
| & \\
\text{Glyet} & \text{Glyet} \quad = \text{Glycine ethyl ester}
\end{array}
$$

41

Although a nearly zero-order release was observed, the amount released was only 1.5% and ≈0.4% at 20 days from the 10% and 35% naproxen-substituted polymers, respectively. When phenylacetic acid was used as the drug, about 9% and 1.5% release at 20 days was seen for the 10% and 60% substituted polymers, respectively.

In a study that also employs a similar approach steroidal residues derived from desoxoestrone, estrone, 17β-estradiol, 17α-ethynylestradiol, estradiol, 3-methyl ether, and 1,4-dihydroestradiol 3-methyl ether have been linked to a polyphosphazene chain via the sodium salt of the steroidal hydroxy function [60]. The degree of replacement of P–C bonds by P–OR units was in the range of 0.5–40%, depending on the reaction conditions. The residual chlorine atoms were removed by reaction with methylamine, ethyl glycinate, or n-butylamine. Stable polymers were obtained when the steroidal units were linked to phosphorus through an aryloxy residue. However, linkage through an alkoxy residue led to instability and chain cleavage. The use of ethyl glycinato residues as cosubstituent groups yielded hydrolytically degradable polymers. A typical structure for the methylamine cosubstituent derivative is shown in (42).

$$\left[\left(\begin{array}{c} OR \\ | \\ -N=P- \\ | \\ NHCH_3 \end{array}\right)_x \left(\begin{array}{c} NHCH_3 \\ | \\ -N=P- \\ | \\ NHCH_3 \end{array}\right)_y\right]_n$$

42

All the polymers were flexible or brittle film-forming materials. The solution properties depended on the ratio of the amino residues. For species of formula (42), the polymers were soluble in water or ethanol when fewer than 5% of the side groups were steroidal residues. These same products were insoluble in most organic media. Polymers with 5–20% of steroidal substituents were only slightly soluble in water and formed emulsions in aqueous media. However, they were soluble in tetrahydrofuran. Higher ratios of steroidoxy to methylamino units yielded materials that were soluble only in tetrahydrofuran.

The use of ethyl glycinato cosubstituent groups in place of methylamino yielded polymers that underwent chain cleavage in aqueous media. The effect was not observed when methylamino units were present as side groups. More extensive studies to measure breakdown kinetics and release profiles of the attached steroids were not reported.

In an interesting set of reactions, the local anesthetics procaine, benzocaine, chloroprocaine, etc., were attached to the polyphosphazene backbone by direct aminolysis [61]. Here, the objective was to modify the duration of biological activity by linkage of the active molecule via the primary amino function to a polyphosphazene skeleton. Both cyclic trimeric and high-polymeric phosphazene skeletal systems were used. The cyclic trimeric derivatives served as small-molecule models for the preliminary examination of reaction conditions, monitoring of side reactions, and the development of analytical techniques. Because of the relatively small size of these molecules, complete halogen replacement was achieved yielding fully substituted polymers. All the polymers were soluble in organic solvents.

The gel permeation chromatography (GPC) average molecular weights were in the range of 4×10^5 to 5×10^5, values that are somewhat lower than those normally found for poly[(arylamino)phosphazenes]. This may reflect a tendency for depolymerization as a consequence of the forcing reaction conditions needed for complete halogen replacement.

Glass transition temperatures were in the range 27–58°C and can be compared to the value of 91°C for $[NP(NHC_6H_5)_2]^n$. Hydrogen bonding undoubtedly plays a part in reducing the torsional mobility of polyphosphazenes of this type, compared to say $[NP(OC_6H_5)]_n$ ($T_g = -8°C$).

The hydrolysis behavior and bioactivity of the cyclic and polymeric derivatives have not yet been examined in detail. However, preliminary experiments indicate that the procaino-substituted high polymers undergo a slow hydrolysis in buffered aqueous media at pH 7. Clearly, more studies are needed.

6.3.3.2   Alkoxy-Substituted Polyphosphazenes

Although most of the early work was focused on the development of amino substituted and potentially biodegradable phosphazenes, most recently Allcock and co-workers have introduced what appears to be the first alkoxy-substituted and hydrolyticaly unstable macromolecules. More specifically, glyceryl-substituted polyphosphazenes (43) were prepared to provide water solubility and possible sites for either crosslinking or attachment of bioactive agents through the free hydroxyls [62]. Furthermore, hydrolytic breakdown was expected to generate glycerol, phosphates, and ammonium salts.

$$\left[\begin{array}{c} \overset{\displaystyle OH \quad OH}{\underset{\displaystyle |\quad\quad|}{O-CH_2-CH-CH_2}} \\ | \\ -P=N- \\ | \\ \underset{\displaystyle |\quad\quad|}{O-CH_2-CH-CH_2} \\ \overset{\displaystyle}{\underset{\displaystyle OH \quad OH}{}} \end{array}\right]_n$$

43

Because glycerol is a trifunctional agent, it was necessary to protect two of the hydroxyl groups while allowing the third one, in the form of sodium salt, to nucleophilically replace the chlorine atoms in poly(dichlorophosphazene). Deprotection of the isopropylidene and methoxymethylene groups was achieved with aqueous acetic acid.

The protected polymers were found to be noncrystalline elastomeric materials with glass transition temperatures ranging from -19 to $-35°C$ based on the protecting groups employed. GPC molecular weights were estimated at 1×10^6. The deprotected polymer had a T_g of $-19°C$ and was water soluble. Although attempts to crosslink the polymer under γ-irradiation (2 Mrad) failed, treatment with adipoyl chloride or hexamethylene diisocyanate in dimethylformamide (DMF) resulted in extensive crosslinking and generation of stable hydrogels. Entrapment of tripelennamine hydrochloride within the hydrogel followed by measurement of the release kinetics of the agent in water achieved complete release within 26 h.

The deprotected polymer was found to hydrolyze completely to glycerol, phosphate, and ammonia in water after 150 h at $100°C$ and after 720 h at $37°C$.

6.3.3.3   Mixed Substituent Polyphosphazenes

As discussed previously, it is possible to introduce more than one substituent onto the polyphosphazene backbone through the sequential addition of selected nucleophiles. The degree of substitution is easily controlled by introducing the first nucleophile at a mole ratio suitable to achieve a certain level of attachment followed by the second and, if desirable, the third nucleophile to replace all chlorine atoms in poly(dichlorophosphazene). Using this strategy four new types of polyphosphazenes were synthesized that bear α-D-glucosyl side groups together with methylamino (44), trifluoroethoxy (45), phenoxy (46), and (methoxyethoxy) ethoxy (47) cosubstituted groups [63].

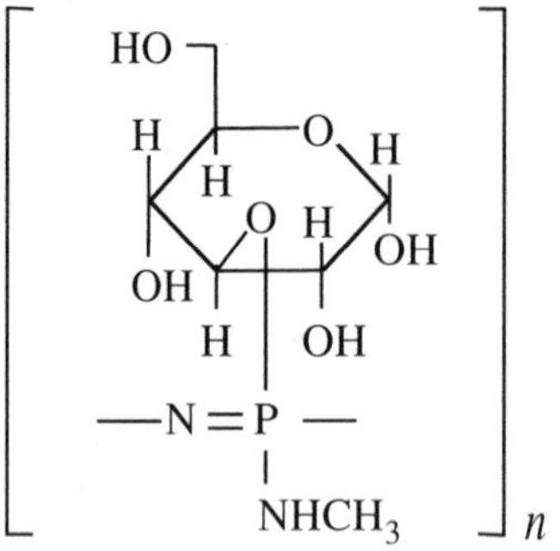

44

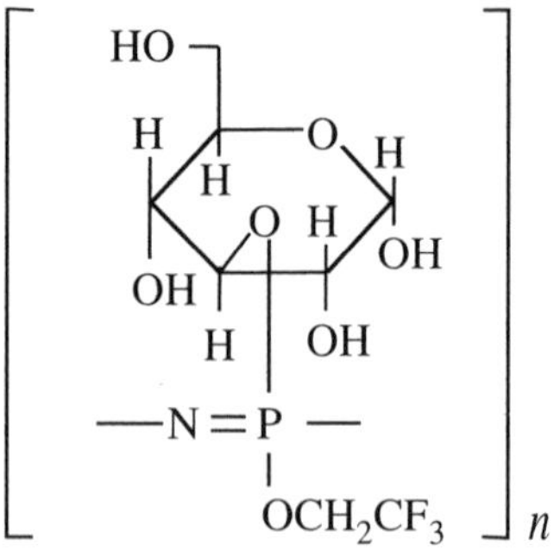

45

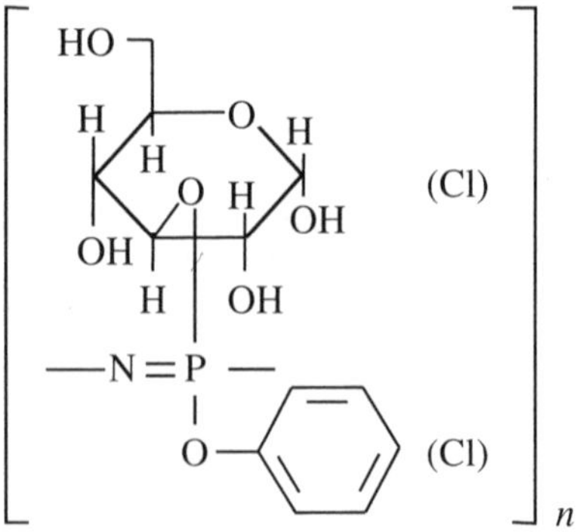

46

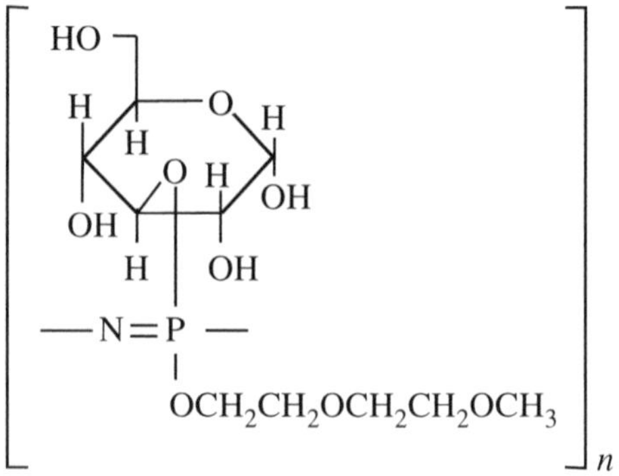

47

The syntheses were accomplished by first reacting the sodium salt of diacetone D-glucose with poly(dichlorophosphazene) followed by the introduction of methylamine or the appropriate alkoxide of aryl oxide to replace the remaining chlorines. Finally, the protecting groups on the glucosyl were removed by treatment with acid.

On the basis of hydrolysis experiments, it appears that only the α-D-glucosyl/methylamino cosubstituted polyphosphazene is hydrolytically unstable. For example, at 100°C in aqueous buffered solutions (pH 6–8) the polymers degraded within 24–96 h whereas at 37°C and pH 7.4, a $t_{1/2}$ of 165–175 h was estimated. Clearly, more work is needed to elucidate the breakdown kinetics and establish whether these materials represent viable biodegradable polymeric candidates.

Perhaps the most comprehensive study conducted with hydrolytically unstable polyphosphazenes involves the development of a bioerodible monolithic controlled delivery device [64]. A polyphosphazene bearing various ratios of imidazole and methylphenoxy substituents (48) was selected for the study because it has been established that imidazole side groups are hydrolytically labile whereas the methylphenoxy groups impart stability. It was postulated that the breakdown kinetics as well as the hydrophilic–hydrophobic character of the device could be modulated through the ratios of the cosubstituents.

48

Indeed, it was shown that degradation of a 20% imidazole substituted polyphosphazene was extremely slow (4% in 600 h) whereas the 80% imidazole substituted polymer degraded on the order of hours at room humidity. At 45% imidazole substitution about 30% degraded in approximately 300 h.

When p-nitroaniline was loaded at 1–5% complete release in vitro at 37°C, pH 7.4, was achieved over a 250-h period with the 20% imidazole-substituted polymer. At 5–10% loadings of progesterone complete release was seen after 1000 h in vitro. In the studies involving polypeptides, release of 25% loaded bovine serum albumin from coated and uncoated 20% imidazole-substituted polymer was on the order of ~ 50% and ~ 40%, respectively, after 500 h in vitro.

REFERENCES

1. Stokes, H.N. *Am. Chem. J.* **17**, 275 (1895); *Ber;* **28**, 437 (1895); *Am. Chem. J.* **19**, 782, (1897).
2. Goldschmidt, G.; Dishon, B. *J. Polym. Sci.* **3**, 481 (1948).
3. Lenton, M.V.; Lewis, B.; Pearce, C.A. *Chem. Ind.* (London), p. 1387 (1964).
4. Mirhej, M.E.; Henderson, J.F. *J. Macromol. Chem.* **1**, 187 (1966).
5. Evans, R.L. U.S. Patent 3,271,330 (1966) (to 3M Company).
6. Allcock, H.R.; Kugel, R.L. *J. Am. Chem. Soc.* **87**, 4212 (1965).
7. Allcock, H.R.; Kugel, R.L.; Valan, K.J. *Inorg. Chem.* **5**, 1709 (1966).
8. Allcock, H.R. *Phosphorus–Nitrogen Compounds* Academic Press, New York (1972).
9. Tate, D.P. *J. Polym. Sci. Polym. Symp.*, **No. 48**, 33 (1974).
10. Singler, R.E.; Schneider, N.S.; Hagnauer, G.L. *Polym. Eng. Sci.* **15**, 321 (1975).
11. Chattapadhyay, A.K.; Hinrichs, R.L.; Rose, S.H. *J. Coatings Technol.* **51**, 87 (1979).
12. Allcock, H.R.; Chu, C.T.-W. *Macromolecules* **12**, 551 (1979).
13. Allcock, H.R.; Patterson, D.B.; Evans, T.L. *J. Am. Chem. Soc.* **99**, 6095 (1977).
14. Allcock, H.R.; Patterson, D.B.; Evans, T.L. *Macromolecules* **12**, 172 (1979).

15. Rose, S.H. *J. Polym. Sci.* **B6,** 837 (1968).
16. Allcock, H.R. *Chem. Rev.* **72,** 315 (1972); *Chem. Britain,* **10,** 118 (1974); *Sci. Am.* **230,** 66 (1974); *Chemtech.* **5,** 552 (1975); *Science* **193,** 1214 (1976); *Agnew Chem. Int. Ed. in Eng.* **16,** 147 (1977); *Acc. Chem. Res.* **11,** 81 (1978); *Acc. Chem. Res.* **12,** 351 (1979).
17. Allcock, H.R. *Acc. Chem. Res.* **12,** 351 (1979).
18. Singler, R.E.; Schneider, N.S.; Hagnauer, G.L. *Polym. Eng. Sci.* **15,** 321 (1975).
19. Kyker, G.S.; Antowiak, T.A. *Rub. Chem. Tech.* **47,** 32 (1974).
20. Sekata, K.; Magill, J.H.; Alarie, Y.C. *J. Fire Flammability* **9,** 50 (1978).
21. Kajiwara, M.; Saito, H. *Polymer* **17,** 1013 (1976).
22. Audrieth, L.F.; Steinman, R.; Toy, A.D.F. *Chem. Rev.* **32,** 109, (1943).
23. Lund, L.G.; Paddock, N.L.; Proctor, J.E.; Searle, H.T. *J. Chem. Soc.* 2542 (1960).
24. Kyker, G.S. *AMMR Technol. Assess. Plan. Conf.* (May, 1975).
25. Allcock, H.R.; Austin, P.E.; Neenan, T.X.; Sisko, J.T.; Blonsky, P.M.; Shriver, D.F. *Macromolecules* **19,** 1508 (1986).
26. Blonsky, P.M.; Shriver, D.F.; Austin, P.E.; Allcock, H.R. *Polym. Mater. Sci.* **53,** 118 (1985).
27. Blonsky, P.M.; Shriver, D.F.; Austin, P.E.; Allcock, H.R. *Solid State Ionics* **18, 19,** 258 (1986).
28. Allcock, H.R.; Kwon, S.; Riding, G.H.; Fitzpatrick, R.J.; Bennett, J.L. *Biomaterials* **19,** 509, (1988).
29. Allcock, H.R.; Lavin, K.D.; Riding, G.H. *Macromolecules* **18,** 1340 (1985).
30. Allcock, H.R.; Riding, G.H.; Lavin, K.D. *Macromolecules* **20,** 6 (1987).
31. Saraceno, R.A; Riding, G.H.; Allcock, H.R.; Ewing, A.G. *J. Am. Chem. Soc.* **110,** 980 (1988).
32. Saraceno, R.A.; Riding, G.H.; Allcock, H.R.; Ewing, A.G. *J. Am. Chem. Soc.* **110,** 7254 (1988).
33. Manners, I.; Riding, G.H.; Dodge, J.A.; Allcock, H.R. *J. Am. Chem. Soc.* **111,** 3067 (1989). 34. Dembeck, A.A.; Kim, C.; Allcock, H.R.; Devine, R.L.S.; Steier, W.H.; Spangler, C.W. *Chem. Mater.* **2,** 97 (1990).
35. Allcock, H.R.; Dembeck, A.A.; Kim, C.; Devine, R.L.S.; Shi, Y.; Steier, W.H.; Spanger, C.W. *Macromolecules* **24,** 1000 (1991).
36. Allcock, H.R. In *Biodegradable Polymers as Drug Delivery Systems.* Langer, R.; Chassin, M., eds. Mercel Dekker, New York (1990).
37. Allcock, H.R.; Kugel, R.L. *J. Am. Chem. Soc.* **87,** 4216 (1965).
38. Wade, C.W.R.; Gourlay, S.; Rice, R.; Hegyeli, A.; Singler, R.; White, J. In *Organometallic Polymers.* Carraher, C.E.; Sheats, J.E.; Pittman, C.V., eds. Academic Press, New York, p. 289. (1978).
39. Reichart, W.M.; Filisko, F.E.; Barenberg, S.A. *J. Biom. Mat. Res.* **16,** 301–312 (1982).
40. Allcock, H.R.; Kwon, S. *Macromolecules* **22,** 75 (1978).
41. Allcock, H.R.; Austin, P.E. *Macromolecules* **14,** 1616 (1981).
42. Allcock, H.R.; Fitzpatrick, R.J.; Salvatti, L. *Chem. Mater.* (in press).
43. Neeman, T.X.; Allcock, H.R. *Biomaterials* **3,** 78 (1982).
44. Allcock, H.R.; Hymen, W.C.; Austin, P.E. *Macromolecules* **16,** 1401 (1983).
45. Allcock, H.R.; Kwon, S. *Macromolecules* **19,** 1502 (1986).
46. Allcock, H.R.; Neeman, T.X.; Kossa, W.C. *Macromolecules* **15,** 693 (1982).
47. Allcock, H.R.; Greigger, P.P.; Gardner, J.E.; Schmutz, J.L. *J. Am. Chem. Soc.* **101,** 606 (1970).
48. Allcock, H.R.; Neeman, T.X.; Boso, B. *Inorg. Chem.* **24,** 2656 (1985).

49. Allcock, H.R.; Allen, R.W.; O'Brien, J.P. *J. Am. Chem. Soc.* **99**, 3984 (1977).
50. Allen, R.W.; O'Brien, P.J.; Allcock, H.R. *J. Am. Chem. Soc.* **99**, 3987 (1977).
51. Allcock, H.R.; Austin, P.E.; Neenan, T.X.; Sisko, J.T.; Blonsky, P.M.; Shriver, D.F. *Macromolecules* **19**, 1508 (1986).
52. Blonsky, P.M.; Shriver, D.F.; Austin, P.E.; Allcock, H.R. *J. Am. Chem. Soc.* **106**, 6854 (1984).
53. Allcock, H.R.; Kwon, S.; Riding, G.H.; Fitzpatrick, R.J.; Bennett, J.L. *Biomaterials* **19**, 509 (1988).
54. Topelmann, W.; Kroschwitz, H.; Schroter, D.; Patzig, D.; Lehmann, H.A. *Z. Chem.* **19**, 273 (1979).
55. Riesel, L.; Hermann, E.; Patzmann, H.; Somieski, R.; Kroschivitz, R.; Schroter, D.; Lehmann, H.A. *Z. Chem.* **10**, 466 (1970).
56. Allcock, H.R.; Fuller, T.J.; Matsummura, K. *Inorg. Chem.* **21**, 515 (1982).
57. Allcock, H.R. *Acc. Chem. Res.* **12**, 351 (1979).
58. Allcock, H.R.; Fuller, T.J.; Mack, D.P.; Matsummura, K.; Smeltz, K.M. *Macromolecules* **10**, 824 (1977).
59. Grolleman, C.W.J.; deVisser, A.C.; Walke, J.G.C.; Van der Groot, H.; Timmerman, H. *J. Controlled Rel.* **3**, 143 (1986) and **4**, 133 (1986).
60. Allcock, H.R.; Fuller, T.J. *J. Am. Chem. Soc.* **103**, 2250 (1981).
61. Allcock, H.R.; Austin, P.E.; Neeman, T.X. *Macromolecules* **15**, 689 (1982).
62. Allcock, H.R.; Kwon, S. *Macromolecules* **21**, 1980 (1988).
63. Allcock, H.R.; Pucher, S.R. *Macromolecules* **24**, 23 (1991).
64. Laurencin, C.T.; Koh, H.J.; Neeman, T.X.; Allcock, H.R.; Langer, R. *J. Biom. Mat. Res.* **21**, 1231 (1987).

CHAPTER 7

Bacterial Polyesters: Structural Variability in Microbial Synthesis

Richard A. Gross

7.1   Introduction

Bacterial polyesters, commonly referred to as poly(hydroxyalkanoates), PHAs, have generated great interest over the past 10 years. The structure of β- or 3-hydroxyalkanoate polymers containing a variety of *n*-alkyl side chain substituents is shown in Figure 7.1. The polymers are stereochemically pure and, therefore, are isotactic. The polymeric repeat units are found in the (*R*) stereochemical configuration.

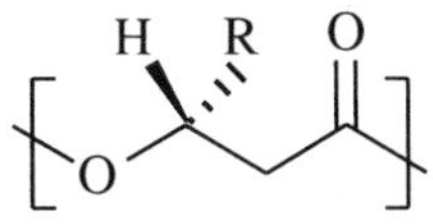

Poly(3-hydroxyalkanoates)

Figure 7.1   The structure of 3-hydroxyalkanoate polymers containing various chain length *n*-alkyl substituent groups. Abbreviations used for repeat unit structures for *n*-alkyl side group containing polymers: 3HB, 3-hydroxybutyrate; 3HV, 3-hyroxyvalerate; 3HC, 3-hydroxycaproate; 3HH, 3-hydroxyheptanoate; 3HO, 3-hydroxyoctanoate; 3HN, 3-hydroxynonanoate; 3HD, 3-hydroxydecanoate; 3HUD, 3-hydroxyundecanoate; 3HDD, 3-hydroxydodecanoate.

Richard A. Gross, Department of Chemistry, University of Massachusetts Lowell, Lowell, Massachusetts 01854, U.S.A.

The most studied members of this family are poly(3-hydroxybutyrate), P3B, and poly(3-hydroxybutyrate-co-3-hydroxyvalerate), P(3HB-co-3HV). The copolyesters poly(3HB-co-3HV) from *Alcaligenes eutrophus* are marketed as Biopol by Zeneca Bio Products in the U.K. Primary factors that have greatly accelerated the interest in, and research on, PHAs are the following:

- Biotechnological advances in the production and isolation of P3HB and P(3HB-co-3HV) so that large quantities became available from Zeneca for material studies.
- Thermoplasticity of these polymers, allowing extrusion and injection molding using conventional processing equipment.
- Useful physical properties, most notably observed for the P(3HB-co-3HV) copolyesters.
- Increased interest in the development of biodegradable plastic materials for disposable plastic applications.
- The biodegradability of P3HB and P(3HB-co-3HV) in a number of biologically active disposal sites such as compost and marine environments.
- Flexibility of biocatalyst systems to incorporate a large range of repeat unit structural entities into the polymers.
- The relatively slow chemical hydrolysis of P(3HB-co-3HV) compared to that of other biomaterials (for example, polylactide) used for medical applications.
- Encouraging preliminary results found for bacterial polyester during in vivo investigations.

PHAs alone as well as blended with other polymers, other polyesters such as poly(η-caprolactone), poly(vinylalcohol) and related copolyesters, and starch based materials currently occupy leading positions in the development of environmentally degradable polymeric materials. The ubiquitous formation of PHAs in many microbial types appears to have provided a rich array of microorganisms and environmental conditions that lead to PHA depolymerization and subsequent mineralization of the degradation products formed. When it is considered that poly(3HB-co-3HV) is produced from glucose and propionic acid in an all aqueous fermentation process, and that the polymer when disposed can be completely converted to biogas, water, biomass, and humic materials, it becomes readily apparent that poly(3HB-co-3HV) represents a truly "environmentally friendly" plastic from nature.

This review of PHA research will emphasize the metabolic flexibility exhibited in PHA microbial polymer formation. It is written from the perspective of a polymer synthetic chemist to display the exciting potential of the PHA biochemical machinery for the production of new tailored polymeric materials. Other areas of interest in bacterial polyester research such as PHA synthetic analogues, blends, processing, and physical and biological properties will be the subject for another review [1].

7.2   Biosynthesis of Bacterial Polyesters

7.2.1   Background Information

In 1925, a microbiologist named Lemoigne at the Pasteur Institute in Paris first described P3HB [2–4]. This was indeed a rather wonderful gift presented by the microbiology community to polymer scientists. Lemoigne probably had no idea that 67 years after his discovery, extensive, exciting, and original research would still be ongoing in a range of interdisciplinary fields. The productivity in this field is surely evident by the numerous significant and informative reviews that have been published to date covering various aspects of PHA research [5–17].

Compilations of microorganisms that accumulate PHAs, their classification according to Bergey's manual, the maximum PHA content found to date, PHA-producing substrates, limiting compounds leading to PHA production, and categorization of the type of studies carried out on the various PHA producing strains can be found in reviews by Brandl [7] and Steinbuchel [9].

The research reviewed below on polyester biosynthesis focuses mainly on the variety of structural entities that have been incorporated into PHAs using different microorganisms and physiological conditions. In addition, methods for altering the copolymer composition are discussed as well as studies that address other questions such as the comonomer sequence distribution of bacterial copolyesters.

7.2.2   Aliphatic Side Groups of Poly(β-hydroxyalkanoates)

P3HB was the first member of the PHA family to be discovered and was regarded for many years as the only structural type of bacterial polyester. An important experiment by Wallen and Rohwedder in 1974 [18] was carried out where PHAs were isolated by chloroform extraction of activated sludges. Unexpectedly, a large percentage of the sludge dry weight (1.3%) was PHA. Furthermore, analysis by gas chromatography–mass spectroscopy revealed that the polymer contained four different repeat units, specifically 3-hydroxybutyrate (3HB), 3-hydroxyvalerate (3HV), 3-hydroxycaproate (3HC), and 3-hydroxyheptanoate (3HH), where the first two were predominant. In addition, Findlay and White [19] isolated PHA from marine sediment and showed that the PHA sample contained a number of different repeat unit structural types. Also, Capon et al. [20] isolated PHAs from freshwater blue-green alga which had predominantly 3HB and 3HV repeat units. The molar ratios of 3HB to 3HV were as high as 1:2 for one isolate. The research carried out by these workers above clearly showed that the family of bacterial polyesters extended beyond simple homopolymers of 3HB.

Copolyesters of 3HB with a repeat unit containing a hydrogen in place of the methyl pendant group (3-hydroxypropionate, 3HP) were prepared by Nakamura et al. [21].

These workers used *A. eutrophus* as the biocatalyst and 3-hydroxypropionic acid, 1,5-pentanediol, and 1,7-heptanediol (diols with an *even* number of carbon atoms were found to produce 4-hydroxybutyrate, 4HB, repeat units, see below) to produce P(3HB–co-3HP) copolyesters containing up to 7 mol% 3HP repeat units (see Figure 7.2). Analysis of these copolymers by ^{13}C nuclear magnetic resonance (NMR) spectroscopy indicated that they have a random comonomer sequence distribution. The molecular weights as determined by gel permeation chromatography ranged from Mn values of 1,400,000 to 36,000 g/mol, depending on the carbon source used and its concentration in the polymer producing media.

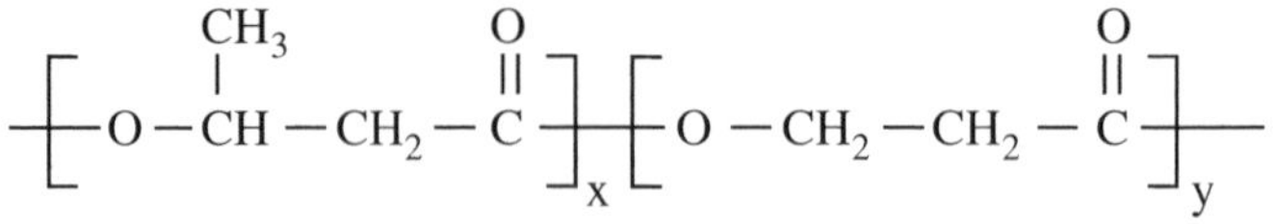

3HB   3HP

Figure 7.2   Structure of P(3HB–co-3HP) synthesized by *A. eutrophus*. This copolyester has been formed with up to 7 mol% 3HP repeat units.

The work discussed above which demonstrated that bacterial polyesters may be isolated from sludge [18], marine sediment [19], and freshwater alga [20] clearly established that it should not be difficult to obtain many microorganisms that have the metabolic flexibility for the incorporation of 3HV repeat units. Further evidence of the abundance of microbial strains capable of producing P(3HB–co-3HV) was presented by Haywood et al. [22], who reported that many P3HB producers were capable of forming this copolymer. In another study, Liebergesell et al. [23] found that cultivation of 11 chemolithotrophic and 15 nonsulfur purple bacterial strains, in the presence of propionate, valerate, or heptanoate, produced a polymer that contained 3HV. In addition, some of the nonsulfur purple bacteria investigated in this study produced PHA containing up to 7 mol% 3HV when cultured on acetate. Therefore, it became rather apparent that the structure of PHAs could vary significantly from that of the simple homopolymer P3HB and that the inclusion of 3HV repeat units into bacterial polyesters using various microbial species should not be problematic. It is important to mention that some microbial species do not readily incorporate 3HV even when cultured with the preferred odd chain carbon sources that lead to 3HV incorporation. Specific examples that demonstrate this point may be found in the work of Liebergesell et al. [23], who found that most sulfur purple bacteria did not incorporate 3HV under the preferred culture conditions. Another example of extremely limited microbial flexibility of a microbial species was found with *Rhodobacter sphaeroides* [24] which, when cultured on propionate/malate and pyruvate/malate, formed PHA consisting of approximately 98 mol% 3HB and, at best, 2 mol% 3HV.

An important advance was made in PHA research when ICI developed technology for the commercial development of PHAs using *Alcaligenes eutrophus H16* [25]. The commercial production of a series of P(3HB-co-3HV) copolyesters (see Figure 7.3) was accomplished by feeding *A. eutrophus* propionic acid and glucose in the second or polymer-producing stage [14,15]. This was an exciting development, as the addition of 3HV units into the polymer chains had beneficial effects on the properties of the materials derived from these polymers. Currently, these copolymers are available from Zeneca Bio Products in a variety of copolymer compositions and are marketed under the trade name Biopol.

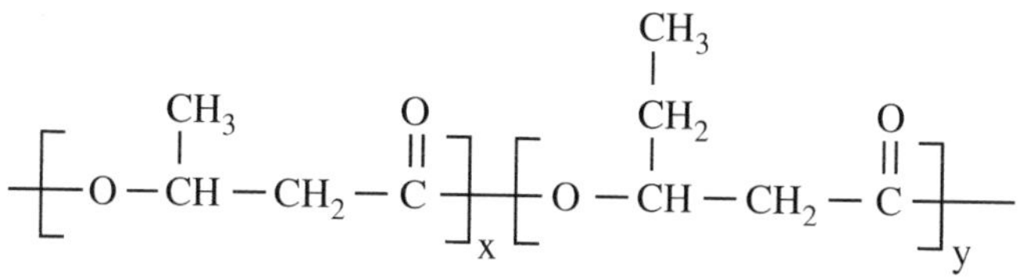

$$\left[O-\underset{\underset{CH_3}{|}}{CH}-CH_2-\underset{\underset{}{\overset{O}{\|}}}{C} \right]_x \left[O-\underset{\underset{\underset{CH_3}{|}}{CH_2}}{CH}-CH_2-\underset{\underset{}{\overset{O}{\|}}}{C} \right]_y$$

3HB   3HV

Figure 7.3   Structure of P(3HB–co-3HV) which is currently marketed as Biopol by Zeneca. Biopol is commercially available from Zeneca with 3HV contents ranging from approximately 0 to 16 mol%.

Of commercial importance is the demonstration that P(3HB-co-3HV) can be formed by metabolism of inexpensive carbohydrate feed sources. Particularly interesting examples of carbon sources used for the formation of P(3HB-co-3HV) are fructose, glucose, gluconate, and starch. A tabulation of various literature references for the formation of P(3HB-co-3HV) by what has been termed as "unrelated" substrates has been prepared by Steinbuchel [26].

Studies carried out by Bluhm et al. on bacterial P(3HB-co-3HV) samples (3HV content from 0 to 47 mol%) produced by *Alcaligenes eutrophus* were shown to be statistically random copolymers by ^{13}C NMR analysis of the respective carbonyl carbon dyad signal intensities [27]. Further studies were also carried out by Doi et al. that are in agreement with this conclusion [28]. Because the effects of the PHA comonomer sequence distribution as well as the side group structure on the relaxation time of the respective PHA carbonyl carbons have not as yet been determined, the ability to carry out an accurate quantitative analysis of the repeat unit sequence distribution by observation of the corresponding ^{13}C NMR dyad signal intensities becomes questionable. The P(3HB-co-3HV) repeat unit sequence distribution was also studied by fast atom bombardment mass spectrometry (FAB-MS) analysis carried out on oligomers of P(3HB-co-3HV) separated by high performance liquid chromatography (HPLC) [29]. The FAB-MS analysis was in agreement with the measurements made by ^{13}C NMR indicating that the copolyesters contain statistically random repeat unit sequence distributions as well as trace amounts of P3HB. The fact that random copolymers are produced is extremely significant since it indicates that the polymerase enzyme has a high degree of substrate flexibility and will accept either the D-3-hydroxyvalerate-SCoA or the D-3-hydroxybutyrate-SCoA monomer units [5,6,8,9] with little preference as to the nature of the incoming monomer or the growing polymer chain end structure.

Fuller and co-workers [30] showed that by growth of the phototrophic bacterium *Rhodospirillum rubrum* on 3-hydroxypentanoate as the sole carbon source, it was possible to obtain P(3HB-co-3HV) with up to 90mol% 3HV content. Furthermore, when this microorganism was grown on 3-hydroxyhexanoate and 3-hydroxyheptanoate as sole carbon sources, it was possible to obtain up to 26 and 4 mol% of the respective C-6 (propyl side group) and C-7 (butyl side group) repeat units into the resulting copolymers. This study further extended the spectrum of 3-hydroxyalkanoate repeat units of increased side chain length that could be incorporated into bacterial polyesters formed by a microorganism that clearly had a preference for relatively shorter (methyl and ethyl) n-alkyl side group chain lengths. Similarly, Doi and co-workers produced a range of P(3HB-co-3HV) copolyesters containing between 0 and 90% HV using various mixtures of butyric and pentanoic acids as carbon sources [31].

Pioneering work by Witholt and co-workers has established that *Pseudomonas oleovorans* is capable of growth and subsequent metabolism of n-alkanes to PHAs in two-phase liquid–liquid fermentations [32–34]. When *P. oleovorans* was grown under ammonium-limiting conditions on the substrates hexane through dodecane, PHAs were formed that, depending on the growth substrate used, contained variable amounts of repeat units with significantly longer (greater than ethyl) side chain n-alkyl substituents [33]. Witholt and co-workers reported that when the organism was grown on octane and nonane, PHAs were formed that contained predominantly 3HO and 3HN, respectively [32,33]. Further analysis of these products showed that the polymer repeat units have an [R] enantiomeric excess of at least 90% [35]. Therefore, the stereochemical configuration of these relatively longer side chain length repeat units is consistent with that found, in every case, for PHAs synthesized of shorter side chain length (such as P3HB and P(3HB-co-3HV). The discovery and investigations on polymers formed by *P. oleovorans* are fascinating when it is considered that other microbial strains such as *A. eutrophus* and *R. rubrum* showed a rather poor ability to incorporate n-alkyl substituents such as n-propyl (3HC) and n-butyl (3HH). Drastically different PHA side group structures containing various functional groups were then envisioned and some interesting examples have been demonstrated (see the section below on functionalized PHAs for additional information).

Some of the technical difficulties involved in working with a biphasic alkane–water medium for *P. oleovorans* growth and PHA production were overcome by using a homogeneous medium that contained the sodium salt of n-alkanoic acid carbon sources [36,37]. In this work, PHAs were produced that contained at least two major (> 5 mol%) repeat unit types. That is, depending on the carbon source used, the PHAs can have n-alkyl pendant groups with chain lengths from methyl to nonyl and copolymers containing as many as six different types of β-hydroxyalkanoate repeat units [36,37]. In most cases, the major repeating unit in the polymer had the same chain length as the n-alkanoic acid used for growth, but units with two carbon atoms less or more than the acid used as a carbon source were also generally present in the polyesters formed [36,37]. The maximum cellular yield and polymer content (in percent of the cellular dry weight) measured were 1.5 g/L and 49%, respectively, using nonanoate as the carbon source [37]. The Mw and Mw/Mn values of these PHAs measured by gel permeation chromotography (GPC) (relative to polystyrene molecular weight standards) ranged between 160,000 and 360,000 g/mol and 2.4 to 3.0, respectively

Table 7.1   Composition of PHA Samples Produced by *P. oleovorans* Grown on *n*-Alkanoates

Carbon source	Repeating units found in PHA, mol%								
	3HB	3HV	3HC	3HH	3HO	3HN	3HD	3HUD	3HDD
Caproate	3	<1	72		22		3		
Heptanoate		7	<1	86	<1	7			
Octanoate	<1	1	6		75		17		
Nonanoate		3	<1	20	5	72			
Decanoate	<1	1	7		44		47		<1

[36,37]. Structural analysis of these polymers by ^{13}C NMR spectroscopy was carried out and the peaks corresponding to the respective repeat units were assigned [37]. Unfortunately, in this case, NMR analysis did not provide the information required to carry out an analysis of the comonomer sequence distribution.

Ballistreri et al. carried out investigations on PHAs formed by *P. oleovorans* where mixtures of nonanoic and octanoic acid were used as the carbon sources for growth and polymer production [38]. It was shown that the PHA products contained all of the repeat units expected for the polymers produced from either alkanoic acid alone. In other words, the composition of PHAs was controlled by mixing various alkanoic acids of different chain lengths. Thus, predictable polymer repeat unit compositions can be obtained in this manner. In addition, studies of the comonomer sequence distribution were also carried out by the separation of oligomeric polymer degradation products (obtained after controlled methanolysis) by HPLC and analysis of the fractions by FAB-MS. It was determined that the oligomer composition and that calculated assuming Bernoullian statistics based on a random comonomer sequence distribution were in agreement. Therefore, FAB-MS analysis of oligomeric degradation fractions showed that random copolyesters are produced by *P. oleovorans* when grown on *n*-alkanoic acids. It was noted above [37] that NMR did not prove sensitive to the comonomer sequence distribution and, therefore, failed to provide an answer to this problem. Another approach that proved successful in confirming the random nature of *n*-alkyl side chain copolyesters produced by *P. oleovorans* was the introduction of a chlorinated functionality into the PHA side chains by growth of the microorganism on a mixture of octane and 1-chlorooctane [39]. Methylene carbon resonances on the polymer side chains were now split due to comonomer sequence effects.

The idea of mixing a substrate that is used for both growth and polymer formation with a substrate that promotes cell growth but is not metabolized to form PHA was used for the formation of PHAs by *Pseudomonas oleovorans* that contain branched side chain groups [40]. Cultivation of *P. oleovorans* using 7-methyloctanoate as the sole carbon source supported cell growth and the formation of PHAs with methyl branched side chain groups (see Figure 7.4). In contrast, 6- and 5-methyloctanoate used as sole carbon sources resulted in relatively slower growth and the absence of microbial PHA formation. However, when 6- and 5-methyloctanoate were used in combination with nonanoic acid (NA) as cosubstrates, the branched aliphatic molecules were metabolized to form PHA repeat units with methyl branches in the pendant

group. Also of interest was the observation by ^{13}C NMR that the methyl branched
5-methyloctanoate repeat units contained a significantly higher content of one of the
two possible diastereomers while little to no stereoselection was seen in the formation
of PHA containing 6-methyloctanoate repeat units.

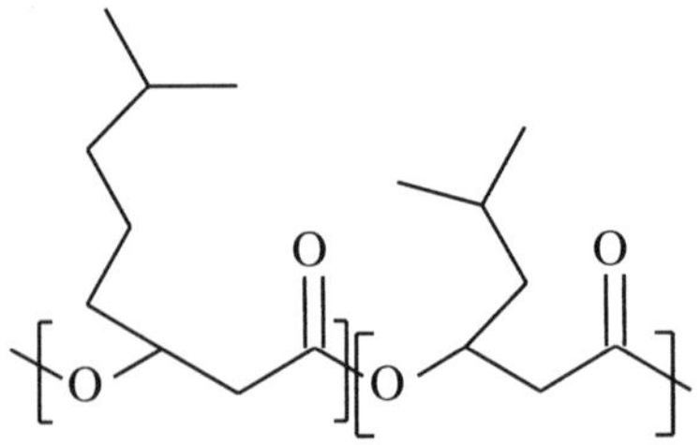

Figure 7.4 A copolymer of 3-hydroxy-7-methyloctanoate and the two carbon atom shorter
3-hydroxy-5-methylhexanoate formed by *P. oleovorans* cultivated on 7-methyloctanoate.

Haywood et al. [22] investigated a range of *Pseudomonads [putida, oleovorans,
aeruginosa* (NCIB 9904 and 8626, and 2F32) *fluorescens* and *testosteronii*] for growth
and PHA formation on various *n*-alkanes, *n*-alcohols, and *n*-alkanoic acids. These
microorganisms produced PHAs with side chain groups ranging from *n*-propyl to *n*-
heptyl, where the PHA composition was dependent on the microorganism and the
carbon source used. Furthermore, the most predominant repeat unit carbon chain
length corresponded to the carbon source structure (octanol would give predominantly 3OA repeat units) whereas less predominant repeat unit carbon chain lengths
contained either two less or two more carbon atoms as was previously observed for *P.
oleovorans* [35–37]. Therefore, it appears that a number of different *Pseudomonads*
similarly metabolize hydrocarbon substrates, which may or may not contain an oxidized terminal carbon, to form PHA copolyesters with *n*-alkyl side chain substituents
ranging from *n*-propyl to *n*-heptyl (C$_6$-C$_{10}$). Further confirmation of this assertion
was presented in work carried out by Huisman et al. [41] where it was shown
that fluorescent *Pseudomonads*, specifically *Pseudomonas aeruginosa*, *Pseudomonas
putida*, and *Pseudomonas fluorescens*, readily accumulate PHAs with pentyl and heptyl
side chain groups.

An important characteristic of a PHA biosynthetic method is the ability of the microbial biocatalyst to produce desirable PHA structures from inexpensive carbohydrate
(termed "unrelated" [26,42]) carbon sources. The successful formation of P(3HB-co-
3HV) copolymers has been carried out from "unrelated" carbon sources [26], and
was discussed above. Research to produce PHAs containing relatively longer side
chain groups from carbohydrate feed stocks has also been carried out. Haywood et al.
[42] have identified a number of *Pseudomonas* species that form PHAs that contain
as the predominant repeat unit 3-hydroxydecanoate (C$_{10}$ or 3HD). For one specific
strain investigated, PHAs with predominantly 3HD repeat units were formed from
acetate, glycerol, lactate, succinate, glucose, gluconate, and fructose. Interestingly,
when octanoate was used as the carbon source for the microorganism, a polymer with
95 mol% 3HO repeat units was accumulated. Timm et al. [43] have also reported

the formation of PHAs that contain mainly 3HD repeat units (55–76 mol%) by *Pseudomonas aeruginosa* and 15 other strains of this species when the cells were cultured using gluconate as the carbon source under nitrogen-limiting conditions. For some of the strains investigated in this work, PHA formed represented 70% of the cellular dry weight. These workers found that 41 of 55 *Pseudomonas* species studied accumulated this type of polymer from gluconate. The formation of PHA structures containing primarily 3HD or 3HB/3HV repeat units, as described above, provides microbial products with widely different properties from inexpensive carbohydrate feedstocks. These findings have obvious important commercial and ecological implications that will surely stimulate a great deal of future research.

7.2.3 Bacterial Polyesters That Contain Linkages Other than Strictly β- or 3-Hydroxypropioesters

Doi and co-workers discovered that bacterial polyesters can be formed that have linkages other than between the 3- or β-hydroxyl functionality of one repeat and the carboxylic acid functionality of its neighbor [44]. Copolyesters of 3HB and 4-hydroxybutyrate, 4HB (see Figure 7.5) were formed by *A. eutrophus* cultured using a range of carbon sources including 4HB, γ-butyrolactone, 4-chlorobutyric acid, 1,4-butanediol, and 1,6-hexanediol [45–48]. The effect of the concentration and choice of the above carbon sources for copolyester production gave a range of copolyester compositions with up to 36 mol% 4HB [8]. When *A. eutrophus* was cultured on a mixture of 10 mM fructose and 10 mM γ-butyrolactone, a P(3HB-co-8% 4HB) copolyester was formed and the percent PHA of the cellular dry weight was 48%. P(3HB-co-4HB) copolymers produced under a range of physiological conditions had Mn and Mw/Mn values from 100,000 to 480,000 and 1.7 to 2.9, respectively [8].

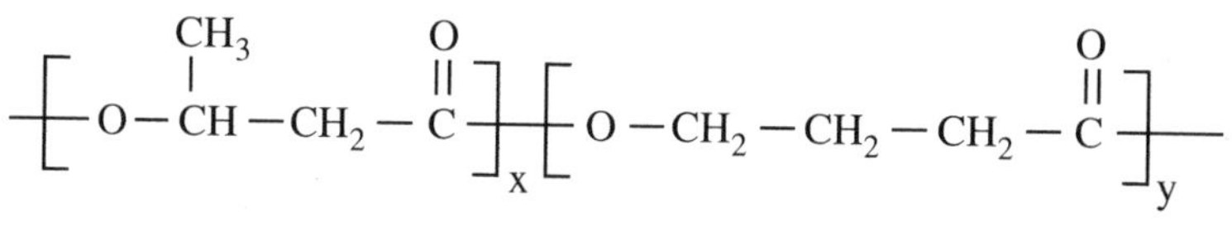

3HB   4HB

Figure 7.5   Copolyesters of 3HB and 4HB produced by the biocatalyst system *A. eutrophus*.

The comonomer sequence distribution of P(3HB-co-4HB) formed from 4-chlorobutyric acid, γ-butyrolactone, 4-hydroxybutyric acid, and γ-butyrolactone/fructose as carbon sources were analyzed by using ^{13}C NMR spectroscopy [45]. Four peaks with additional splitting due to tetrad effects were observed in the carbonyl region which provided quantitative information on the dyad comonomer sequence distribution (assuming that the delay time used in these experiments was adequate). These authors then compared the experimental dyad comonomer sequence distribution with that expected for a random copolymer and found good agreement. Interestingly, one sample formed using γ-butyrolactone and butyric acid as co-carbon sources yielded a polymer that, by ^{13}C NMR analysis, showed a large deviation from a random comonomer distribution. Fractionation of this product into boiling acetone-soluble and -insoluble fractions resulted in the isolation of two copolyesters having random copolymer structures and different comonomer mole fractions (86 and 7 mol% 4HB, respectively) [49].

The idea that polyesters which are not exclusively poly(β-esters) can be formed by biological methods was extended further by Doi et al., who showed that 5-hydroxyvalerate (5HV) can also be incorporated into the polymer structure [50]. The formation of terpolyesters containing 3HB, 3HV, and 5HV by *A. eutrophus* (see Figure 6, below) was claimed by using various mixtures of 5-chloropentanoic acid with pentanoic acid as carbon sources.

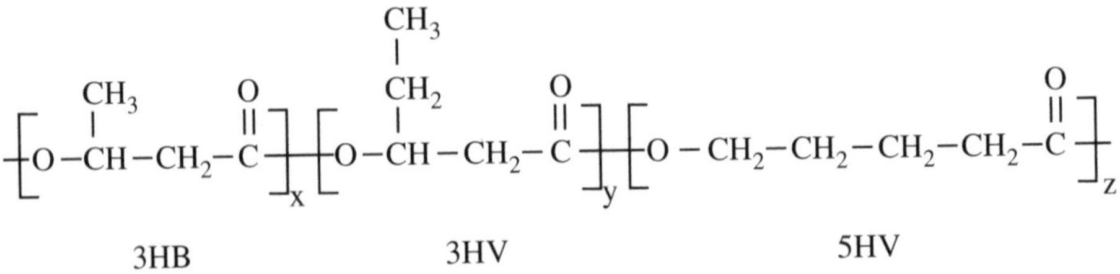

Figure 7.6 Bacterial polyester formation by *A. eutrophus* of a terpolymer containing 3HB, 3HV, and 5HV repeat units.

It was noted by these workers that poor polymer production (1% of the cellular dry weight, ~50 mg polyester/L) resulted when 5-chloropentanoic acid was used as the sole carbon source. However, when 5-chloropentanoic acid/pentanoic acid mixtures (0.5/1.5 g for 100 mL culture volumes) were used, 19% of the cellular dry weight (0.8 g polyester/L) was a polyester that contained 26 mol% 3HB, 65 mol% 3HV, and 9 mol% 5HV as determined by ^{1}H NMR spectral integration.

7.2.4   Bacterial Polyesters with Functional Pendant Groups

Studies were carried out by this author where 4-pentenoate was used as the sole carbon source for growth and polymer production by *R. rubrum* and *A. eutrophus* [51].

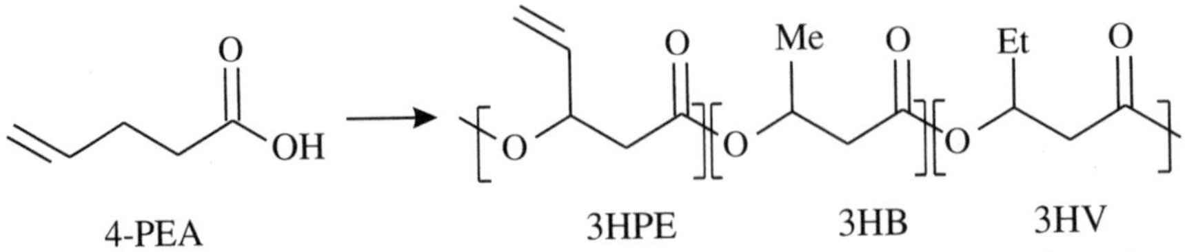

Figure 7.7 Terpolyesters of 3HPE, 3HB, and 3HV formed by the phototrophic bacterium *R. rubrum* when cultivated on 4-pentenoate.

Interestingly, it was found that *R. rubrum* formed a product that contained the repeat units 3HB (11 mol%), 3HV (59 mol%), and 3-hydroxy-4-pentenoate (3HPE, 30 mol%). Thus, P(3HB-co-3HV) copolyesters that also contain vinyl pendant 3HPE repeat units (see Figure 7.7) were formed [51]. In contrast, *A. eutophus* metabolized the vinyl groups to saturated monomeric structures such that no vinyl pendant groups were formed. This work demonstrates that the metabolic flexibility desired in the production of different PHAs may not be observed if a researcher limits his or her work to only a selected few PHA producing microbial strains. In addition, thermal analysis indicated that the terpolymer formed by *R. rubrum* had two distinctly different melting transitions. Therefore, it may be that the vinyl containing terpolymer product from *R. rubrum* consists of a mixture of two different copolyesters [52].

Vinyl side chain containing polymers have also been synthesized using *P. oleovorans* [35]. Growth of this microorganism on 1-octene showed that a minimum of 46mol% of the polymer repeat unit side chains contain a terminal unsaturated functionality. Therefore, by changing the ratio of 1-octene to octane, the degree of side chain unsaturation may be altered [35]. Fritzsche et al. [53] studied growth and polymer production on the carbon sources 3-hydroxy-6-octenoic acid (*cis–trans* mixture) and 3-hydroxy-7-octenoic acid. Polymers obtained from cultures grown on 3-hydroxy-6-octenoic acid as the carbon source had primarily 3-hydroxy-6-octenoate repeat units (both *cis* and *trans* isomers) and only a small amount of the corresponding 3-hydroxy-4-hexenoate. It was determined that the total amount of saturated repeat units was < 6mol%. The polyester formed from 3-hydroxy-7-octenoate had a repeat unit composition consisting of predominantly the unsaturated C-8 unit, where only approximately 1mol% of the repeat units were found to be saturated and approximately 2mol% of the repeat units were the corresponding unsaturated C-6 unit. Therefore, it was concluded that the metabolic pathway involving hydrogenation of the growth substrate double bonds was not important for *P. oleovorans*. The use of unsaturated 3-hydroxy acids relative to alkenes resulted in higher degrees of unsaturation on the polymer side chain groups.

Recently, Huijberts et al. [54] have shown that PHAs with unsaturated side chains can be formed by cultivation of *P. putida* on glucose. In one experiment, *P. putida* accumulated 21% of its dry cell weight as PHA which contained 17% and 4% (wt/wt) of the repeat units 3-hydroxy-5-*cis*-dodecenoate (C12:1) and 3-hydroxyl-7-*cis*-tetradecenoate (C14:1), respectively (see Figure 7.8). Similar results were obtained in cultures carried out using fructose and glycerol as carbon sources for PHA formation. The authors believe that the corresponding monomers for PHA biosynthesis are formed from intermediates of de novo fatty acid biosynthesis. This hypothesis would clearly explain how PHAs with side group unsaturation may be formed from glucose metabolism. These observations are, of course, very exciting as they create the opportunity to obtain functionalized PHAs directly from carbohydrate feed stocks.

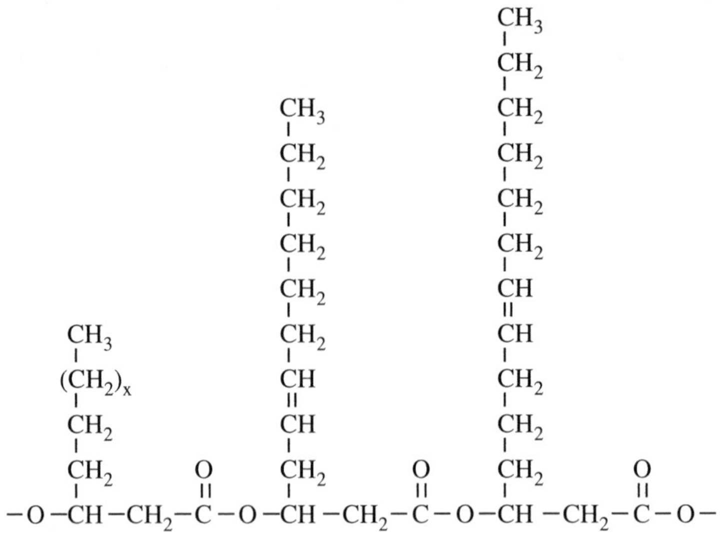

CH$_3$
|
CH$_2$
|
CH$_2$
|
CH$_3$ CH$_2$
| |
CH$_2$ CH$_2$
| |
CH$_2$ CH$_2$
| |
CH$_2$ CH$_2$
| |
CH$_2$ CH
| ||
CH$_3$ CH$_2$ CH
| | |
(CH$_2$)$_x$ CH   CH$_2$
| || |
CH$_2$ CH   CH$_2$
| | |
CH$_2$ O   CH$_2$ O   CH$_2$ O
| || | || | ||
−O−CH−CH$_2$−C−O−CH−CH$_2$−C−O−CH−CH$_2$−C−O−

Figure 7.8   Vinyl side group containing PHA formed by *P. putida* cultivated on glucose.

It has been reported by Fritszche et al. [55] that when *P. oleovorans* is cultured on 5-phenylvaleric acid (PA) as the sole carbon source, a homopolymer of 3-hydroxy-5-phenylvalerate (3HPV) repeat units was produced (see Figure 7.9). In other words, bacterial polyesters with phenyl side chain substituents were accumulated when *P. oleovorans* was cultivated on PA.

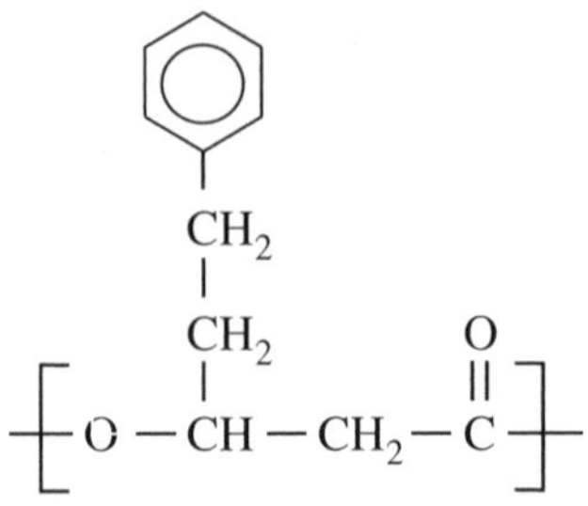

P(3HPV)

Figure 7.9   The formation of PHAs with phenyl side group substituents by cultivation of *P. oleovorans* on 5-phenylvaleric acid.

Furthermore, *P. oleovorans* was cultured using different mixtures of PA and either *n*-nonanoic acid (NA) or *n*-octanoic acid (OA) [56]. The polymers formed contained repeat units expected from metabolism of NA or OA as well as HPV repeat units. In one example, a PHA product formed using a 1:2 (mol/mol) molar ratio of NA and PA, respectively, for growth and polymer production contained 0.6 mol% 3HV, 16.0 mol% 3HH, 41.1 mol% 3HN, 1.7 mol% 3HUD, and 40.6 mol% HPV. Interestingly, from measurements of the rate at which the substrates were consumed and PHA was produced, as well as analysis of the PHA composition as a function of the culture time, it appeared that *P. oleovorans* consumed NA and PA independently for energy and polymer production. This information, in combination with the observation that the product showed two distinct glass transition temperatures and could be fractionated into components of relatively higher and lower HPV content, makes it likely that the PHA produced is indeed a mixture at least two copolyesters. The authors believed their results substantiate that two different polymerase enzymes are present in *P. oleovorans* [57] which can simultaneously produce two different polymers.

Kim et al. [58] successfully carried out the formation of PHAs by *Pseudomonas oleovorans* which have brominated side chain substituents. In this work, the brominated substrates 6-bromohexanoic acid (6BRHA), 8-bromooctanoic acid (8BROA), and 11-bromoundecanoic acid (11BRUA) were evaluated for PHA production. It was found that these carbon sources do support cell growth but when used alone are not metabolized to form PHA. However, when these bromoalkanoic acids were used in combination with octanoic or nonanoic acids, PHAs with between 2 and 37 mol% of brominated side chain groups were formed. The relatively constant PHA molecular weight and mol% of brominated repeat units noted at various culture times, along with the observation by differential scanning calorimetric analysis that the PHA products have a single glass transition temperature and melting transition temperature, all indicate that the brominated PHAs formed were random copolymers. The authors also compared the incorporation of 11BRUA into PHA by introducing a 1:1 molar ratio of NA and 11BRUA as cosubstrates at the initiation of the culture as opposed to the sequential addition of NA followed by the addition of 11BRUA after the growth reached the deceleration phase. In both cases, the same total molar quantities of the two carbon sources were added to the culture media. It was observed that the biomass formation was similar for the two cultures whereas the PHA yield almost doubled when the two carbon sources were fed in sequence. Of greatest interest is that the sequential feeding led to a relatively much lower mol% of brominated side chain groups (3.7 vs. 37.5 mol%). This experiment clearly shows the increase in metabolism of 11BRUA to form PHA when it was used along with NA as a cosubstrate. It is clearly envisioned by these workers that the bromo functionalities in the side chains of these PHAs may readily be modified through a variety of well established chemical transformations to provide many other important side chain groups.

7.2.5   Isotopic Labeling of Poly(hydroxyalkanoates)

It was demonstrated by this author that the biodeuteration of P3HB may be accomplished using the microorganisms *Rhodobacter sphaeroides* and *A. eutophus* [59]. The formation of deuterated P3HB was studied using the following carbon sources and solvents: (1) acetate in H_2O; (2) D_3-acetate in H_2O; (3) acetate in 90–92% D_2O; and (4), D_3-acetate in 90–92% D_2O. In all cases, the P3HB produced under deuterium-enriched conditions was of high molecular weight and was partially deuterated. Fourier transform infrared spectroscopy (FT-IR), pyrolysis gas chromatography–mass spectrometry

(PGC-MS), and NMR were used in this study to establish the extent and distribution of deuterium in the P3HB samples produced. Considerable differences in the extent and distribution of deuterium were found between these two microorganisms and the three deuterium-enriched culture conditions investigated. Using the culture conditions where both the solvent and acetate were deuterated led to the formation of P3HB by *Rb. sphaeroides* and *A. eutrophus* with 69 and 42%, respectively, of the repeat units which contained five of six possible positions deuterated. Interestingly, high levels of biodeuteration at the P3HB methine hydrogen position (the hydrogen attached to the P3HB main chain β-carbon) was observed when *Rb. sphaeroides* was grown in D_2O/H_2O (92/8) on H_3-acetate. The deuteration of the nicotinamide nucleotide-linked acetoacetyl-CoA reductase by D-^{1}H exchange with D_2O was postulated as being responsible for this occurrence.

A study was carried out by Doi et al. to investigate the biosynthesis of copolyesters from ^{13}C-labeled acetate and propionate [60]. Specific labeling of either P3HB or P(3HB-co-3HV) was obtained in this manner.

Therefore, the use of either specifically labeled carbon sources or solvent for the formation of bacterial polyesters represents a general method for the formation of labeled PHA structures. The labeled polymers thus obtained may be depolymerized to their respective repeat units to provide a variety of labeled chiral synthons. This approach can be exploited as a preparative route for a large range of high value specialty chemicals.

Acknowledgments

Research carried out in my laboratory which is directed towards the development of new approaches to polymer synthesis using biocatalyst systems is supported by the National Science Foundation, Division of Materials Research, under a Presidential Young Investigator Award (Grant DMR-9057233). I am also grateful to Dr. Satish Pulapura who assisted in the preparation of this manuscript.

REFERENCES AND NOTES

1. A review article entitled "Bacterial polyesters and synthetic analogues: synthetic methods, physical and biological properties" is currently in preparation and will appear in *Progress in Polymer Science,* A Pergamon Publication.
2. Lemoigne, M. *Ann. Inst. Pasteur (Paris)* **39,** 144 (1925).
3. Lemoigne, M. *Bull. Soc. Chim. Biol.* **8,** 770 (1926).
4. Lemoigne, M. *Ann. Inst. Pasteur (Paris)* **41,** 148 (1927).
5. Dawes, E.A.; Senior, P.J. *Adv. Microbiol. Physiol.* **10,** 135 (1973).
6. Dawes, E.A. *Microbial Energetics.* Blackie, Glasgow, London (1986).
7. Brandl, H.; Gross, R.A.; Lenz, R.W.; Fuller, R.C. Plastics from bacteria and for bacteria: Poly(3-hydroxyalkanoates) as natural, biocompatible, and biodegradable polyesters. In *Advances in Biochemical Engr./Biotechnology,* **vol. 41** Ghose, T.K.; Fiechter, A.; eds. Springer: Berlin (1990).
8. Doi, Y. *Microbial Polyesters,* VCH, New York (1990).

9. Steinbuchel, A. Polyhydroxyalkanoic acids. In *Biomaterials: Novel Materials from Biological Sources,* Byrom, D., ed. Stockton Press, New York, pp. 123–213 (1991).
10. Steinbuchel, A.; Schlegel, H.G. *Mol. Microbiol.* **5,** 535 (1991).
11. Tomita, K.; Saito, T.; Fukui, T. Bacterial metabolism of poly(3-hydroxybutyrate). In *Biochemistry of the Metabolic Process,* Lennon, D.L.F.; Stratman, F.W.; Zahlten, R.N., eds. pp. 353–366. Elsevier, New York (1983).
12. Anderson, A.J.; Dawes, E.A. *Microbiol. Rev.* **54,** 450 (1990).
13. Marchessault, R.H. *Polym. Prepr. Am. Chem. Soc.* **29,** 594 (1988).
14. Byrom, D. *Trends Biotechnol.* **5,** 246 (1987).
15. Holmes, P.A. *Phys. Technol.* **16,** 32 (1985). 16. Holmes, P.A. In *Developments in Crystalline Polymers,* **Vol. 2.** Bassett, D.C., ed. pp. 1–65, Elsevier, London (1988).
17. See ref. 9 for other reviews on PHAs.
18. Wallen, L.L.; Rohwedder, W.K. *Environ. Sci. Technol.* **8,** 576 (1974).
19. Findlay, R.H.; White, D.C. *Appl. Environ. Microbiol.* **45,** 71 (1983).
20. Capon, R.J.; Dunlop, R.W.; Ghisalberti, E.; Jefferies, P.R. *Phytochemistry* **22,** 1181 (1983).
21. Nakamura, S.; Kunioka, M.; Doi, Y. *Macromol. Reports* **A28(Suppl. 1),** 15 (1991).
22. Haywood, G.W.; Anderson, A.J.; Dawes, E.A. *Biotechnol. Lett.* **11,** 471 (1989).
23. Liebergesell, M.; Hustede, E.; Timm, A.; Steinbuchel, A.; Fuller, R.C.; Lenz, R.W.; Schlegel, H.G. *Arch. Microbiol.* **155,** 415 (1991).
24. Brandl, H.; Gross, R.A.; Lenz, R.W.; Lloyd, R.; Fuller, R.C. *Arch. Microbiol.* **155,** 337 (1991).
25. Holmes, P.A.; Wright, L.F.; Collins, S.H. ICI Eur. Pat. Appl. 0052459 (1981); Eur. Pat. Appl. 0069497 (1983).
26. See Ref. 9, pp. 169–170
27. Bluhm, T.L.; Hamer, G.K.; Marchessault, R.M.; Fyfe, C.A.; Veregin, R.P. *Macromolecules* **19,** 2871 (1986).
28. Doi, Y.; Kunioka, M.; Nakamura, Y.; Soga, K. *Macromolecules* **19,** 2860 (1986).
29. See Nedea, M.E.; Morin, F.G.; Marchessault, R.H. *Polym. Bull.* **26,** 549 (1991) and references therein.
30. Brandl, H.; Knee, E.J., Jr.; Fuller, R.C.; Gross, R.A.; Lenz, R.W. *Int. J. Biol. Macromol.* **11,** 49 (1989).
31. Doi, Y.; Tamaki, A.; Kunioka, M.; Soga, K. *Appl. Microbiol Biotechnol.* **28,** 330 (1988).
32. De Smet, M.; Eggink, G.; Witholt, B.; Kingma, J.; Wynberg, H. *J. Bacteriol.* **154,** 870 (1983).
33. Lageveen, R. Ph.D. Thesis University of Groningen, Groningen, The Netherlands (1986).
34. For studies carried out using continuous two-liquid-phase cultures to evaluate PHA microbial formation by *P. oleovorans* see: Preusting, H.; Kingma, J.; Witholt, B. *Enzyme Microb. Technol.* **13,** 770 (1991).
35. Lageveen, R.G.; Huisman, G.W.; Preusting, H.; Ketelaar, P.; Eggink, G.; Witholt, B. *Appl. Environ. Microbiol.* **54,** 2924 (1988).
36. Brandl, H.; Gross, R.A.; Lenz, R.W.; Fuller, R.C. *Appl. Environ. Microbiol.* **54,** 1977 (1988).
37. Gross, R.A.; DeMello, C.; Lenz, R.W.; Brandl, H.; Fuller, R.C. *Macromolecules* **22,** 1106 (1989).
38. Ballistreri, A.; Montaudo, G.; Impallomeni, G.; Lenz, R.W.; Kim, Y.B.; Fuller, R.C. *Macromolecules* **23,** 5059 (1990).
39. Doi, Y.; Abe, C. *Macromolecules* **23,** 3705 (1990).

40. Fritzsche, K.; Lenz, R.W.; Fuller, R.C. *Int. J. Biol. Macromol.* **12,** 92 (1990).
41. Huisman, G.W.; Leeuw, O.-D.; Eggink, G.; Witholt, B. *Appl. Environ. Microbiol.* **55,** 1949 (1989).
42. Haywood, G.W.; Anderson, A.J.; Ewing, D.F.; Dawes, E.A. *Appl. Environ. Microbiol.* **56,** 3354 (1990).
43. Timm, A.; Steinbuchel, A. *Appl. Environ. Microbiol.* **56,** 3360 (1990).
44. See ref. 8, pp. 89–98 (1990).
45. Doi, Y.; Kunioka, M.; Nakamura, Y.; Soga, K. *Macromolecules* **21,** 2722 (1988).
46. Kunioka, M.; Nakamura, Y.; Doi, Y. *Polymer Commun.* **29,** 174 (1988).
47. Doi, Y.; Segawa, A.; Kunioka, M. *Polymer Commun.* **30,** 169 (1989).
48. Kunioka, M.; Kawaguchi, Y.; Doi, Y. *Appl. Microbiol Biotechnol.* **30,** 569 (1989).
49. Doi, Y.; Segawa, A.; Kunioka, M. *Int. J. Biol. Macromol.* **12,** 106 (1990).
50. Doi, Y.; Tamaki, A.; Kunioka, M.; Soga, K. *Makromol. Chem. Rapid Commun.* **8,** 631 (1987).
51. Gross, R.A.; Brandl, H.; Ulmer, H.W.; Posada, M.A.; Fuller, R.C.; Lenz, R.W. *Polym. Preprints, Am. Chem. Soc., Div. Polym. Chem* **30,** 492 (1989).
52. Lenz, R.W.; Kim, B.-W.; Ulmer, H.W.; Fritzsche, K. Functionalized poly-β-hydroxyalkanoates produced by bacteria. In *Novel Biodegradable Microbial Polymers* Dawes, E.A., ed. pp. 23–25. Kluwer, Dordrecht (1990).
53. Fritzsche, K.; Lenz, R.W.; Fuller, R.C. *Int. J. Biol. Macromol.* **12,** 85 (1990).
54. Huijberts, G.-N.M.; Eggink, G.; Waard, P.-De; Huisman, G.W.; Witholt, B. *Appl. Environ. Microbiol.* **58,** 536 (1992).
55. Fritszche, K.; Lenz, R.W.; Fuller, R.C. *Makromol. Chem.* **191,** 1957 (1990).
56. Kim, Y.B.; Lenz, R.W.; Fuller, R.C. *Macromolecules* **24,** 5256 (1991).
57. Recently, it has been demonstrated for both *P. oleovorans* and *A. eutrophus* that there are two DNA sequences present that encode for PHA polymerases. See Huisman, G.W.; Wonink, E.; Meima, R.; Kazemier, B.; Terpstra, P.; Witholt, B. *J. Biol. Chem.* **266,** 2191 (1991); Peoples, O.P.; Sinskey, A.J. *J. Biol. Chem.* **264,** 15298 (1989).
58. Kim, Y.B.; Lenz, R.W.; Fuller, R.C. *Macromolecules* **25,** 1852 (1992).
59. Gross, R.A.; Ulmer, H.W.; Lenz, R.W.; Tshudy, D.J.; Uden, P.C.; Brandl, H.; Fuller, R.C. *Int. J. Biol. Macromol.* **14,** 33 (1992).
60. Doi, Y.; Kunioka, M.; Nakamura, Y.; Soga, K. *Macromolecules* **20,** 2988 (1987).

CHAPTER 8

Biosynthetic Polysaccharides

David L. Kaplan, Bonnie J. Wiley, Jean M. Mayer, Steven Arcidiacono, Julia Keith, Stephen J. Lombardi, Derek Ball, and Alfred L. Allen

8.1 Introduction

Microorganisms produce a variety of polymers that function in growth and survival (Table 8.1). These polymers may be found associated with processes such as

David L. Kaplan, Bonnie J. Wiley, Jean M. Mayer, Steven Arcidiacono, Julia Keith, Stephen J. Lombardi, Derek Ball, and Alfred L. Allen, Biotechnology Division, U.S. Army Natick Research Development and Engineering Center, Natick, Massachusetts 01760-5020, U.S.A.

Table 8.1   Natural Functions of Microbial Biopolymers

a. Intracellular: energy storage
   Structural organization, compartmentalization

b. Extracellular: adhesion (slime layers, surface colonization)
   Collection of nutrients, metals
   Protection from antibacterial agents (capsular layer)
   Protection from adverse environmental conditions (water holding capacity)

c. Cell wall/membrane
   Structural support
   Molecular recognition
   Barrier

biofouling, flocculation, adhesion, dental plaque formation, and soil colonization, and have an impact on food storage/stability, pulp and paper processing, production of pharmaceuticals by fermentation, wastewater treatment, medical treatments, and agricultural activities.

One class of microbial polymers is the polysaccharides, which are ubiquitous in nature and the most abundant biopolymers on earth (Table 8.2). Polysaccharides are carbohydrates characterized by the presence of repeating structures in which the interunit linkages are of the O-glycosidic type [1]. The variety of saccharide monomers

Table 8.2   Some Polysaccharides Produced by Microorganisms

Polymer[a]	Organism	Structure
Fungal		
Pullulan (N,E)	*Aureobasidium pullulans*	1,4-; 1,6 α-D-Glucan
Scleroglucan (N,E)	*Sclerotium glucanicum*	1,3; 1,6 α-D-Glucan
Chitin (N,W)	Filamentous fungi	1,4 β-D-Acetyl glucosamine
Chitosan (C,W)	*Mucor rouxii*	1,4 β-D-N-Glucosamine
Elsinan (N,E)	*Elsinoe leucospila*	1,4-; 1,3 α-D-Glucan
Bacterial		
Xanthan gum (A,E)	*Xanthomonas campestris*	1,4 β-D-Glucan with D-mannose; D-glucuronic acid side groups
Curdlan (N,E)	*Alcaligenes faecalis*	1,3 β-D-Glucan (with branching)
Dextran (N,E)	*Leuconostoc* sp.	1,6 α-D-Glucan with some 1,2; 1,3; 1,4 α-linkages
Gellan (A,E)	*Pseudomonas elodea*	1,4 β-D-Glucan with rhamose, D-glucuronic acid
Levan (N,E)	*Erwinia herbicola*	2,6 β-D-Fructan with some β-2,1-branching
Emulsan (A,E)	*Acinetobacter calcoaceticus*	Lipoheteropolysaccharide
Cellulose (N,E)	*Acetobacter xylinum*	1,4 β-D-Glucan

[a] N = neutral, A = anionic, C = cationic, I = intracellular, E = extracellular, W = cell wall/membrane.

(≈ 200) and the variety of possible O-glycosidic linkages result in a diversity of polysaccharide structures and conformations [2]. Aside from microbial sources of these polymers, polysaccharides are synthesized by plants (starch, cellulose, hemicellulose), algae and seaweeds (gums), and animals (glycogen).

Microbial polysaccharides are of interest because of their unusual and useful functional properties. Some of these properties include: (1) film-forming and gel-forming capabilities; (2) stability over broad temperature ranges; (3) biocompatibility (natural products avoid the release/leaching of toxic metals, residual chemicals, catalysts, or additives); (4) unusual rheological properties; (5) biodegradability; (6) water solubility in the native state or reduced solubility if chemically derivatized; and (7) thermal processability for some of these polymers. In addition, fermentation production and control, and genetic manipulation to control product expression, molecular weight distribution, stereospecificity, and functional properties, are advantages to working with these polymers when compared with many synthetic polymers.

Many excellent books, reviews, and symposia have been published on polysaccharides [3–13]. We confine this chapter to polysaccharides that are biosynthesized and not associated with lipids, proteins, or nucleic acids in a significant way. Four microbial, predominately straight chain, glycan homopolymers that exhibit good film forming properties are discussed: chitosan, pullulan, levan, and elsinan (Figure 8.1). For each

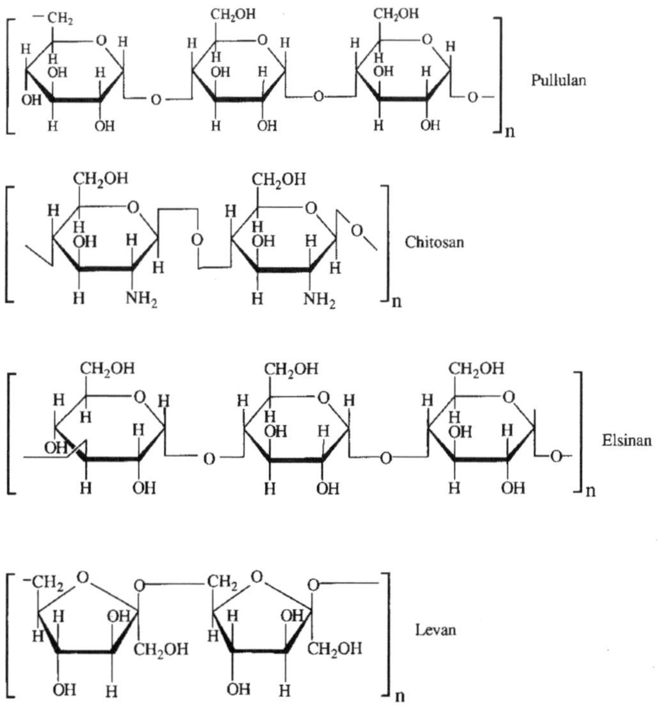

Figure 8.1   Primary structures of the four polysaccharides reviewed.

of these polysaccharides the following areas are addressed: (1) background—structure, sources, enzymatic hydrolysis, (2) biosynthesis—fermentation, (3) processing including film or fiber formation; and (4) applications, including biomedical.

Biodegradable polymers for material applications including food packaging, biomedical uses, and agricultural applications, are becoming increasingly desirable from an environmental, medical, and functional viewpoint. The four biopolymers described here were chosen for study based primarily on their potentially useful properties with regard to food coatings or packaging. Studies have included various aspects of fermentation/processing controls, film and fiber formation and evaluation of functional properties, chemical modification of the natural polymer to enhance functional properties, evaluation of rates of biodegradation in different environments, and characterization of biosynthetic enzymes toward the development of new expression systems.

Consideration should be given to the potential importance of polysaccharides for a number of applications, including: food coatings and packaging, enteric coatings for drug delivery, and encapsulation for agricultural uses such as controlled release of pesticides or fertilizers. Beneficial properties include: low rates of oxygen permeability for improved storage stability; modified water solubility through chemical derivatization of the natural polymers; stability to fats and oils; and clear, easily formed films and coatings. In addition, for food coatings and packaging, the potential for edible/nutritional properties is also a benefit. On the negative side, water solubility and some limitations in physical properties (e.g., tensile strength, tear strength, extensibility) in some cases prompt chemical or physical modifications (e.g., crosslinking, co-extrusion) to overcome these drawbacks.

8.2   Chitosan

8.2.1   Background

Chitosan is a cationic polymer of β-1,4-linked 2-amino-2-deoxy-D-glucose. This contrasts with chitin which is a polymer of N-acetylglucosamine (2-acetamido-2-deoxy-D-glucose), also linked by β-1,4 bonds.

Chitin is common in nature; it is found in crustacean exoskeletons (e.g., crab, shrimp), zooplankton, insects, marine diatoms, and filamentous fungi (cell wall). Chitosan is rare in nature; the primary occurrence is in the cell walls of some fungi, particularly the Zygomycetes, which contain both chitosan and chitin. Chitosan has been identified in the genera *Mucor* [15], *Absidia* [16], *Saccharomyces,* and *Phycomyces* [17]. Commercially, chitosan is derived from crustacean chitin by chemical conversion with strong alkali. Variability in source material can lead to variable physiochemical characteristics. The solubility of chitosan in dilute acids is a function of the degree of acetylation of the glucosamine polymer. Chitosan produced from crustacean chitin has been reported to be anywhere from 0% to 50% acetylated [18]. Cell wall chitosan isolated from *M. rouxii* has been reported to contain 5% to 10% acetyl content [19].

The chitosan content of the fungal cell wall is reported to be 33% for the mycelia and 28% for yeast-like forms, accompanied by less than 10% chitin [15].

Chitosan was first identified by Roget [20] from chitin boiled in caustic potassium hydroxide. Chitosan was first found in nature in 1954 by Kreger [17] in the cell walls and sporangiophores of *Phycomyces blakesleeanus,* and subsequently in *Mucor rouxii* by Bartnicki-Garcia and Nickerson [15]. In 1978, Fenton et al. [21] reported that each family of the Zygomycetes contained chitosan in the cell wall. Recent reports have indicated that some strains of *Absidia* produce chitosan in higher yields and with a higher molecular weight distribution than the genus *Mucor* [16]. Little of the previous work concerning chitosan from fungal sources has focused on the effect of culture conditions and processing steps on the molecular weight distribution of the polymer; the focus was primarily on yield.

Many derivatives of chitosan have been synthesized, including substitution at C-2 and C-6, *N*-acyl, *N*-carboxyalkyl, *N*-carboxyacyl, *O*-carboxyalkyl, and water-soluble chitosan acetate, among others. Some of these reactions have been reviewed [22,23]. Ogura et al. reported that concentrations of chitosan higher than 40% formed a lyotropic liquid crystal phase [24].

Chitosanase is an endo-enzyme hydrolyzing β-1,4 homoglycans of D-glucosamine with diequatorial linkages [25]. Products of the hydrolysis include di- and trisaccharides. Chitosanase has been reported in *M. rouxii* [26]. Lysozyme (EC 3.2.1.17) is unreactive toward chitosan but hydrolyzes chitin [27]. In the fungus *M. rouxii*, chitosan is synthesized by the tandem action of two enzymes, chitin synthetase and chitin deacetylase [28].

8.2.2   Biosynthesis

Chitin is formed by the repeated transfer of *N*-acetylglucosamine from uridyl diphosphate *N*-acetyl-D-glucosamine to a β-1,4-*N*-acetyl-D-glucosamine acceptor through the action of chitin synthetase [UDP-2-acetamido-2-deoxy-D-glucose: chitin-4-(2-acetamido-2-deoxy-β-D-glucosyl transferase), EC 2.4.1.16]. Chitin in its nascent state is then converted to chitosan by the enzyme chitin deacetylase (EC 3.5.1.41). Chitin synthetase is membrane bound whereas chitin deacetylase is found in the cytosol [28]. Chitin deacetylase is unreactive toward crystalline chitin.

Chitin synthetase and chitin deacetylase were initially studied by Araki and Ito [29] and subsequently partially purified from harvested mycelia of *M. rouxii* [30,31]. Chitin synthetase required protease activation and was stimulated by Mn^{2+} and Mg^{2+}. Active synthesis of polymer was at the growing hyphal tips. The chitin deacetylase was purified by ammonium sulfate precipitation, high-performance liquid chromatography, and preparative polyacrylamide gel electrophoresis [31]. Chitin deacetylase activity was quantified with radiolabelled glycol chitin as the release of tritiated acetate. Chitin synthetase activity was determined by the incorporation of radiolabelled *N*-acetylglucosamine into polymer and assayed by thin-layer chromatography. Chitin deacetylase has been purified over 70-fold and the enzyme was stable at 4°C for extended periods [31].

Work is continuing toward complete purification and sequencing of both chitin deacetylase and chitin synthetase from *M. rouxii* [30]. Once completed, genomic and cDNA libraries constructed from *M. rouxii* will be screened for appropriate clones and lead to the development of new expression systems for the biosynthesis of chitosan. This approach will avoid the requirement to extract chitosan from the cell wall of the fungus by expressing the polymer intra- or extracellularly from alternative organisms. This approach will also allow for better regulation and control of molecular weight, yield, and degree of acetylation.

Compared with the traditional chemical conversion of crustacean chitin to chitosan, a fungal source of chitosan offers many potential benefits including: avoidance of calcium interference, improved control over molecular weight, improved batch to batch consistency, reduced potential for depolymerization, and options for genetic manipulation and control. The effect of incubation time in 750-ml and 10-liter batch studies on yield and molecular weight distribution of chitosan isolated from *M. rouxii* is shown in Table 8.3. The molecular weight distribution and yield could be affected to some extent by the manipulation of incubation time. The effect of pH in two different media (nutrient rich and defined) on yield and molecular weight of chitosan isolated from the fungus is shown in Table 8.4. Although yield of chitosan is affected by pH, the molecular weight distribution changes less than twofold over the range evaluated. In all studies, molecular weight ranged from 2×10^5 to 1.4×10^6. In most

Table 8.3   Effect of Length of Incubation on Chitosan Yield and Molecular Weight from the Fungus *M. rouxii*

Time (days)	% Yield[a]	Molecular weight[b] (K)	Dispersity[c]	pH
A. 750 ml Cultures, defined medium, 125 rpm				
1	5.2	959	8.3	3.93
2	4.8	761	5.6	3.87
3	2.6	821	10.0	3.90
4	5.3	937	5.9	3.93
5	5.6	1442	18.2	4.00
6	6.0	1208	7.3	3.93
7	5.9	999	5.6	3.96
B. 10 liter, complex medium, 1.25 liter/minute aeration				
0.25	1.7	136	4.0	5.1
1.0	5.9	864	5.1	4.8
1.5	6.5	1015	6.0	4.5
2.0	5.8	1047	3.8	4.8
3.0	5.8	832	9.8	6.0
7.0	3.9	332	5.6	8.7

[a] Chitosan to biomass (w/w). To convert to chitosan yield based on cell wall (w/w) multiply by 4.5.
[b] Weight average molecular weight (Mw) by gel permeation chromatography (GPC).
[c] Mw/Mn.

Table 8.4   Effect of pH on Chitosan Yield and Molecular Weight from the Fungus M. rouxii[a]

pH	Medium	% Yield[b]	Mol wt[c] (K)	Dispersity[d]
3.0	Defined	3.07	458	4.2
	Complex	3.70	319	4.0
4.0	Defined	5.10	594	6.9
	Complex	8.00	468	4.6
5.0	Defined	6.28	471	5.4
	Complex	9.70	459	8.9
6.0	Defined	5.40	484	6.4
	Complex	8.80	521	8.4

[a] 750 ml, 125 rpm.
[b] Chitosan to biomass (w/w).
[c] Weight average molecular weight by GPC.
[d] Mw/Mn.

experiments yields of chitosan were between 5% and 10% of total biomass, depending on culture conditions. This compares favorably with the commercial conversion of crustacean chitin to chitosan, which has about a 5% yield of polymer based on total biomass. The chemical conversion of chitin from the cell wall of *Aspergillus niger* to chitosan using strong alkali has been reported to produce a 14% yield based on total biomass [32].

8.2.3   Processing

M. rouxii (ATCC 24905) was used in many of the fermentation and processing studies summarized here [30,33]. The fungus was maintained on yeast extract/peptone/glucose agar and batch and continuous studies were conducted with both complex and defined media. The extraction procedure to isolate chitosan from the cell wall was a modification of the process used by White et al. [19] and includes an autoclaving step in 1 *N* sodium hydroxide, homogenization in 5% acetic acid, and precipitation under alkaline conditions [30,33]. The weight average molecular weight of chitosan isolated from the cell wall of *M. rouxii* varied from around 200,000 to over one million, depending on culture and processing conditions [30,33]. The dispersity of the polymer ranged from 5 to 8, presumably as a result of the cell wall isolation procedure which may cause random chain cleavage. A variety of culture and processing conditions were evaluated to determine their effect on molecular weight distribution, dispersity, and yield, including nutrient availability, pH, length of incubation, and oxygen availability. Molecular weight determinations were by gel permeation chromatography. The percent acetylation of some of the fungally derived samples was determined by UV spectroscopy [34]. Values were generally between 8% and 13% and did not appear to depend on fermentation or processing conditions, whereas corresponding values for commercially available chitosan from crustacean waste chemical conversion processes ranged from 15% to 21% [33].

Films with and without plasticizers were cast in 2% or 5% acetic acid and dried overnight in an oven at 40–45°C, followed by neutralization with sodium hydroxide [35]. Films were analyzed for physical properties including tensile strength, flexibility, thickness, percent elongation, and oxygen permeability by standard methods. Oxygen permeability was extremely low under the conditions studied. Crosslinked films were synthesized using epichlorohydrin under alkaline conditions [36] by methodology adapted from work with amylose [37]. Chitosan films cast as 3.2% solutions were treated with epichlorohydrin in alkaline solution and methanol and then dried overnight as before. These films showed improved functional properties such as tensile strength and flexibility in both wet and dry states when compared with the non-crosslinked controls (Table 8.5). Processing steps used to prepare chitosan from crustacean sources have been reported to affect film quality [38].

Similar studies have been conducted to crosslink chitosan fibers using the same chemistry [39]. Chitosan fibers (5% w/v) in acetic acid were spun into a coagulating

Table 8.5   Physical Properties of Films of Chitosan, Chitosan/Pullulan, and Crosslinked Chitosan

Film composition[a]	Thickness (mm)	Flexibility (cm)	Elongation (%)	Tensile strength (MPa)[b]
Commercial chitosan[c]	0.035 ± 0.008	12.4 ± 1.7	2.9 ± 1.0	51.0 ± 11.3
Commercial chitosan wet[d]	0.038	12.1 ± 0.6	61.8 ± 15.5	0.5 ± 0.6
Commercial chitosan 10% glycerol[d]	0.034 ± 0.009	8.1 ± 1.2	8.8 ± 6.2	47.0 ± 5.2
Chitosan/pullulan (50/50 wt/wt)	0.043	N.D.[e]	1.9	48.3
Chitosan/pullulan (50/50 10% glycerol)	0.043	14.4 ± 0.8	2.7 ± 0.4	54.3 ± 5.3
Crosslinked chitosan 5% epichlorohydrin dry	0.029 ± 0.006	>18.5	1.6 ± 0.4	67.2 ± 47.3
Crosslinked chitosan 5% epichlorohydrin wet	0.023 ± 0.004	>18.5	18.2 ± 5.0	10.6 ± 3.3
Crosslinked chitosan 12.5% epichlorohydrin dry	0.037 ± 0.001	>18.5	2.7 ± 0.5	135.8 ± 36.6
Crosslinked chitosan 12.5% epichlorohydrin wet	0.035 ± 0.005	>18.5	47.1 ± 23.7	13.1 ± 7.8

[a] Chitosan around 9×10^5 and pullulan around 1.5×10^6 weight average molecular weight by GPC.
[b] Protan Laboratories, Inc., Redmond, Washington, U.S.A.
[c] Tensile strength test performed on wet films.
[d] Glycerol used as plasticizer.
[e] No data.

bath containing 1 N NaOH and then crosslinked in the wet state with epichlorohydrin. Wet tenacity improved and swelling ratio decreased with increased crosslinking reaction time. Tenacity up to 1.15 g/d, elongation up to 38.1%, and modulus up to 11.0 g/d were found for the crosslinked fibers. These values compared with 0.68, 15.0, and 7.3, respectively, for the fiber controls prior to crosslinking. In general, the increase in tensile strength due to crosslinking was greater for the films than the fibers, presumably because the fibers are more orientated initially as reflected by wet strength values of 50 MPa for films and 120 MPa for fibers.

8.2.4   Applications

The list of potential and existing applications for chitosan is extensive and includes uses as adhesives, additives in food processing, paper and textile additives, carrier or encapsulating material for pharmaceuticals, wound healing accelerants, immunological defense, enzyme immobilization medium, and wastewater treatments [22,40–51]. Gels and hybrid gels with collagen, powders, films, and fibers have been formed from chitosan. A partial list of uses for these materials include blood compatibilizers, treatments for antitumor activity, use in cosmetics as humectants, cationic flocculants for wastewater treatment, metal chelation including uranium, fibers for sutures, membranes for water purification, contact lens material, cell culture matrices, gel chromatography support matricies, and inhibitors of blood coagulant enzymes.

8.3   Pullulan

8.3.1   Background

Pullulan is a linear α-D-glucan consisting of α-1,4-linked maltotriose units connected by α-1,6-linkages [52–55]. Pullulan may be more of a generic name due to the presence of other linkages besides the maltotriose. For example, α-maltotetraose units, other neutral sugars, uronic acids, and α-1,3-linked glucosyl residues have been identified, depending on the organism and culture conditions [56]. Pullulan is released into the growth medium as a secondary metabolite by the dimorphic fungus *Aureobasidium pullulans* during the yeast-like phase of the growth cycle [57–61]. Pullulan was first characterized in 1959 [53]. The fungus is responsible for deterioration of paint and wood and also attacks plants. The natural function for the polymer may be to provide adherence to surfaces to promote the deteriorative process [56]. The production of pullulan has been commercialized by Hayashibara Biochemical Laboratories in Japan using batch culture and starch as a carbon source [59,60].

Pullulan exhibits improved water solubility when compared with amylose, which is partially crystalline. The results of conformational calculations demonstrated that the α-1,6-linkage is responsible for much of the conformational freedom in D-glucan chains and the resulting improved water solubility [62]. Solubility and rates of biodegradation can be decreased through esterification, etherification, or crosslinking.

Pullulan degrades at around 250°C. Viscosity is stable over a wide range of pH values (2–12) and salt concentrations [58,59].

Pullulan is hydrolyzed by pullulanase (pullulan 6-glucanohydrolase, EC 3.2.1.41, endo-enzyme), which cleaves α-1,6-linkages to form maltotriose [27,63]. Isoamylase (glycogen 6-glucanohydrolase, EC 3.2.1.68) also hydrolyzes α-1,6-glucosidic linkages, reacting slightly with pullulan, and α-amylase (α-1,4-D-glucan glucanohydrolase, EC 3.2.1.1) hydrolyzes α-1,4-glucan linkages in polysaccharides with three or more α-1,4-linked D-glucose units [64]. Glucoamylase (α-1,4-glucan D-glucohydrolase, EC 3.2.1.3) hydrolyzes α-1,4-glycosidic linkages successively from the nonreducing end to remove glucose monomers and also hydrolyzes some α-1,6-linkages [34]. Isopullulanase (pullulan 4-glucanohydrolase, EC 3.2.1.57) cleaves α-1,4-linkages of pullulan to form isopanose (O-α-D-glycopyranosyl-1,4-O-D-glucopyranosyl-1,6-D-glucopyranose) [27]. Neopullulanase will also hydrolyze 1,4-α-D-glucosidic linkages to form panose (O-α-D-glucopyranosyl-1,6-α-O-glucopyranosyl-1,4-D-glucopyranose) [14]. Pullulan biosynthesis is associated with the cell membrane and involves sugar nucleotide carriers [65].

8.3.2   Biosynthesis

Ueda et al. [66] studied the production of pullulan by 16 strains of A. pullulans and compared various sugars as carbon sources for yield of polymer. They found molecular weights of approximately 250,000 by light-scattering. Catley and others [65–73] carried out extensive studies on various environmental and nutritional requirements for production of the biopolymer.

The yeast form (blastospores) of A. pullulans is the major producer of pullulan. The vegetative cycle of A. pullulans was studied extensively by Ramos and Garcia Acha [74]. They found that an inoculum containing a cell concentration of at least 20 million cells per flask (150 ml) was required to maintain blastospore production. They also found that chlamydospores were readily formed in a medium containing ammonium ion as a nitrogen source. Bulmer et al. [67] studied the effect of ammonium ion as well as pH on pullulan elaboration and found that ammonium ion affected protein synthesis. They concluded that new cells produced in the presence of ammonium ion do not have the ability to elaborate polysaccharide, and that ammonium ion depletion must occur before pullulan is elaborated.

Catley [68,72] and Ono et al. [77] found that the appearance of extracellular pullulan was not concomitant with an increase in cell mass, but that there was a lag in the rate of elaboration. Catley [68] studied the role of pH and nitrogen on the production of pullulan and found that the uptake of glucose at more acid pH was diverted to the synthesis of extracellular pullulan, and that high extracellular pH inhibited its production. Catley [68,72] found that pullulan elaboration correlated with the depletion of nitrogen but not of carbon in the growth medium. Ono et al. [77] found that maintaining a constant, controlled pH resulted in lower yields of pullulan than in uncontrolled culture. Lacroix et al. [76] described the development of a bistage pH fermentation process in which the first stage was conducted at a very acidic pH (2.0) and the second stage at pH 5.5 to promote the production of pullulan. Fluoroacetate additions have been shown to improve the yield of pullulan [79].

Kato and Shiosaka [80,81] showed a marked decrease in mean molecular weight of pullulan when comparing phosphate concentrations of 11.5 mM to 22.9 mM at pH 5.5, pH 6.0, and pH 6.5. High molecular weight pullulan was produced at the lower pH level; however, molecular weight decreased with increasing phosphate concentration. They reported that hydrolyzed starch as the carbon source resulted in low molecular weight pullulan in yields as high as 75%. The production of pullulan from starch is attractive because of agriculture surpluses in some areas of the world. Most of the studies referenced used batch culture incubations of a few hours to 7 days. Only a few authors have reported the molecular weight distribution of the pullulan elaborated [70,71,80,81]. Catley [70,71] reported molecular weights decreased from 2 million down to 150,000 during fermentation studies. Kato and Shiosaka [80,81] reported molecular weights of 50,000 to 2.5 million depending on strain and culture conditions.

We have studied eight strains of fungi for pullulan production, *A. pullulans* (QM 72c), *A. pullulans* var. *melanigenum* (QM 279c = ATCC 15233), *A. pullulans* (QM 3090 = ATCC 9348 and NRRL-Y 7498), *A. pullulans* (QM 5752), *A. pullulans* (NRRL-Y 6220 = ATCC 34647), *A. pullulans* (NRRL-Y 6272 = ATCC 36276), *A. pullulans* (ATCC 12535 = NRRL-Y 7469), and *A. pullulans* var. *melanigenum* (ATCC 12536) [82]. *A. pullulans* NRRL-Y 6220 was chosen for further studies based on yield of pullulan, molecular weight distribution, and the lack of pigmentation impurities. Both batch and continuous culture studies were conducted to evaluate culture condition variables on yield, molecular weight distribution, dispersity, and pigmentation.

The molecular weight distribution of pullulan varied from under 100,000 to above 5 million, depending on culture conditions [82]. Generally the highest molecular weights were found early in culture maturity. However, this was inversely correlated with yield (Table 8.6). The dispersity (ratio of weight average molecular weight to number average molecular weight, Mw/Mn) of the polymer remained around 2.0 throughout most of the studies. The effects of different carbon sources on pullulan molecular weight distribution and yield are shown in Table 8.7. Of particular interest is the high yield of polymer produced from starch.

Many different approaches to chemically modify pullulan have been considered. The usual reasons to modify the polymer are to reduce water solubility or to enhance mechanical or physical properties. We have focused on site selective substitution to control biodegradation and solubility properties [83]. The 6-chloro-6-deoxy-pullulan, 3,6-anhydro-pullulan, 6-azido-6-deoxy-pullulan, and 6-amino-deoxy-pullulan have been synthesized, characterized, and studied for susceptibility to pullulan-degrading enzymes.

8.3.3   Processing

Processing requirements to purify pullulan from the medium are shown in Table 8.8. Pullulan films are characterized as tasteless, odorless, water soluble, transparent,

Table 8.6   Pullulan Production and Effect of Incubation Time on Molecular Weight, Yield, and Residual Sugar Using *A. pullulans* NRRL-Y 6220[a]

Time (days)	Yield[b] (%)	Mol wt[c] (K)	Dispersity[d]	Fructose (mg/ml)	Glucose (mg/ml)	Sucrose (mg/ml)	Oligomers[e]
1	5.9	1658	1.8	7.5	32.4	7.8	25.3
2	7.9	1531	12.7	13.8	31.5	4.5	20.9
3	8.8	1413	10.9	18.9	30.6	0	19.9
4	12.5	834	6.7	24.0	25.8	0	16.2
5	15.4	684	4.8	30.9	19.5	0	13.1
6	16.7	583	5.1	33.6	12.0	0	7.8
7	21.9	581	8.2	37.8	7.8	0	4.7
8	27.2	211	3.3	35.4	3.0	0	1.3

[a] 0.29 M Sucrose, pH 5.4, 26 $\pm$2°C, 125 rpm, 250 ml.
[b] Based on sucrose (w/w).
[c] Weight average molecular weight by GPC.
[d] Dispersity (Mw/Mn).
[e] Includes di-, tri-, tetra- and penta-saccharides, qualitative sums based on relative peak ratios.

biodegradable, edible, nontoxic, exhibiting low oxygen permeability, oil resistant, printable, adhesive, and sealable, and may be particularly useful for dry, powdered, or high-lipid foods [58–60].

Films formed from pullulan were cast as 2% water solutions on Teflon-coated trays and dried in an oven at 40–45°C overnight [84]. A significant loss in extensibility was observed as the molecular weight increased (Table 8.9). Brittleness of the films cast from the higher molecular weight material became a significant problem despite the addition of 10% glycerol as a plasticizer. Films cast from low (300,000 to 500,000) and medium (1 to 2 million) molecular weight pullulan were evaluated for oxygen permeability. Values (cc oxygen per square meter of film per 24 h) were 3.26 for low molecular weight pullulan (0.035 mm thickness) and 3.72 for

Table 8.7   Effect of Carbon Source on Pullulan Yield with *A. pullulans* NRRL-Y 6220[a]

Carbon source (Concn)	Yield (%)[b]	Mol wt (K)[c]	Dispersity[d]
Fructose (0.56 M)	27.0	1122	2.5
Sucrose (0.29 M)	34.4	895	2.4
Maltose (0.29 M)	25.4	881	2.1
Corn Syrup (10%)	9.3	840	3.3
Dextrose (0.56 M)	23.3	563	1.8
Lactose (0.29 M)	5.8	518	1.6
Sol. starch (10%)	70.3	137	1.4

[a] Culture conditions: pH 5.4, 26°C $\pm$ 1°C; 125 rpm, 100 ml, 4 days.
[b] Amount of carbon source converted to pullulan (w/w).
[c] Weight average molecular weight by GPC.
[d] Mw/Mn.

Table 8.8   Processing Conditions for Pullulan

1. Neutralize culture medium with 1 *N* NaOH
2. Dilute medium with quaternary detergent (1% final volume)
3. Centrifuge 23,000 *g*, 20 min
4. Dilute supernatant with acetone (1:2)
5. Stand overnight at 5°C
6. Decant water/acetone solution
7. Wash precipitate with acetone
8. Filter over vacuum
9. Dry over calcium sulfate in desiccator

medium molecular weight pullulan (0.035 mm thickness). This compared with 8.37 for polyethylene terephthalate (0.012 mm thickness) and over 200 for polypropylene (0.070 mm thickness). Previous studies indicated a range in oxygen permeability of 0.60–2.50 for various pullulan films of undefined molecular weight and thickness (compared to values of 4.70 for cellophane, 8.58 for moisture proof cellophane, and 1100 for polypropylene) [58,59]. We have also found that pullulan can be thermally processed in the presence of sufficient moisture.

8.3.4   Applications

Potential applications for pullulan have been reviewed [58,59]. In Japan, pullulan is commercialized and used as a food additive and coating. Due to the hydrophilic nature

Table 8.9   Physical Properties of Films of Pullulan and Plasticized Pullulan

Film composition[a]	Thickness (mm)	Flexibility (cm)	Elongation (%)	Tensile strength (MPa)
Low MW pullulan[a]	0.035 ± 0.007	8.7 ± 1.3	1.6 ± 0.2	40.2 ± 7.72
Low MW pullulan 10% glycerol	0.046 ± 0.002	9.6 ± 0.7	1.1 ± 0.3	30.68 ± 7.60
Med MW pullulan[b]	0.380 ± 0.009	8.5 ± 1.3	2.1 ± 0.6	58.71 ± 14.58
Med MW pullulan 10% glycerol	0.400 ± 0.004	8.7 ± 0.3	2.1 ± 0.9	45.42 ± 9.80
High MW pullulan[c]	0.033 ± 0.007	10.4 ± 1.0	0.5	9.96
High MW pullulan	N.D.[d]	N.D.	N.D.	N.D.
Commercial pullulan[e]	0.270 ± 0.001	N.D.	2.1 ± 0.2	36.50 ± 2.05
Commercial pullulan	0.045 ± 0.005	10.4 ± 0.7	1.9 ± 0.3	33.54 ± 6.15

[a] 300–500 K weight average molecular weight.
[b] 1–2 million weight average molecular weight.
[c] 3–4 million weight average molecular weight.
[d] No data.
[e] Hayashibara Biochemical Laboratories, Inc., Japan.

and biocompatibility of the polymer, controlled release drug delivery applications may be appropriate. The ability to produce transparent films with no odor, no taste, partial nutritional availability and good mechanical properties leads to applications for food coatings and packaging [35,84]. Cosmetic, adhesive, paper additive and floculant applications have also been studied.

8.4   Elsinan

8.4.1   Background

Elsinan is a linear α-D-glucan consisting of 1,4- and 1,3-linkages in molar ratios of 2.0:1 to 2.5:1; approximately one in 140 linkages is α-1,6 [85–88]. Elsinan can also be described as maltotriose and maltotetraose residues joined by α-1,3-D-glucose bonds. A related polymer is mycodextran with its alternating α-1,4- and α-1,3-linkages. Elsinan is released into the growth medium as a secondary metabolite by species of the fungus *Elsinoe*. The teleomorph fungus *Elsinoe,* and its anamorph genus, *Sphaceloma,* comprise a number of plant pathogenic species, causing spot anthracnoses of citrus, grapes, pear, apple, grasses, tea, dogwood, and other species [89–103]. The production of a viscous layer in cultures of *E. australis* was described by Bitancourt and Jenkins in 1936 [91]. Other isolates have been described as producing gummy to occasionally mucoid colonies on agar media. α-Amylase (EC 3.2.1.1) hydrolyzes elsinan at α-1,4-linkages and products include 4-O-α-nigerosyl-D-glucose, D-glucose, and 3-O-α-maltosyl-maltose [64].

Elsinan is soluble in hot water, formamide, dimethyl sulfoxide, and 90% formic acid, and insoluble in most organic solvents including methanol, ethanol, acetone, and chloroform [61,87]. Molded objects are soluble only in hot water, transparent, nontoxic, edible, and stable in long-term storage [61,87].

8.4.2   Biosynthesis

Misaki et al. [85–88] produced the polysaccharide which they named elsinan from a culture of *E. leucospila*. Yields of 2.5% to 3.0% with weight average molecular weight ranges of 10,000 to 10 million were reported when the fungus was grown on a medium containing sucrose and potato extract or corn steep liquor. They also examined the effects of various amylolytic enzymes on elsinan and isolated the degradation products which included glucose and a series of oligomers [86–88].

We have reported on elsinan production from eight strains of fungi—*E. annonae* (ATCC 15027), *E. corni* (ATCC 11189), *E. fawcettii* (ATCC 13200, 36954, and ATCC 38162), *E. heveae* (ATCC 12570), *E. lepagei* (ATCC 13008), and *E. tiliae* (ATCC 24510) [104]. More extensive studies were conducted with *E. fawcettii* (ATCC 36954 and ATCC 38162) and *E. tiliae* (ATCC 24510) based on yield and molecular weight

Table 8.10   Effect of Carbon Source on Elsinan Production by *E. fawcettii* ATCC 36954[a]

Carbon source (Conc)	Yield (%)[b]	Mol wt (K)[c]	Dispersity[d]
Sucrose (0.29 M)	14.7	1373	2.2
Fructose (0.56 M)	10.9	671	1.8
Dextrose (0.56 M)	4.6	1259	1.8
Lactose (0.29 M)	—e	—	—
Maltose (0.29 M)	11.8	2250	1.9
Sol. starch (10%)	33.3	774	11.6
Corn Syrup (10%)	6.6	2012	1.7

[a] Culture conditions: pH 6.0, 26°C ± 1°C; 125 rpm, 100 ml, 7 days.
[b] Carbon source converted to elsinan (w/w).
[c] Weight average molecular weight by GPC.
[d] Mw/Mn.
[e] Product did not contain elsinan.

distribution of the polymer during the initial studies. All cultures were maintained on Wickerham's yeast malt extract agar [105]. A salts, peptone, yeast extract, and 10% carbon source liquid medium was used during batch studies.

Tables 8.10 and 8.11 and Figure 8.2 illustrate the results from two studies on the effect of different carbon sources and length of incubation on yield and molecular weight distribution of elsinan [104]. Additional studies were reported on the effects of pH over the range of 5.0–7.0 (yields ranged from 12.4% to 18.3% and weight average molecular weights ranged from 745K to 1449K), initial sucrose concentration from

Table 8.11   Effect of Incubation Time on Elsinan Molecular Weight, Yield, and Residual Sugar Concentration Using *E. fawcettii* ATCC 36954[a]

Time (days)	Yield[b] (%)	Mol wt[c] (K)	Fructose (mg/ml)	Glucose (mg/ml)	Sucrose (mg/ml)
1	0.3	563	0	0	99.0
2	0.9	1088	1.8	3.0	78.0
3	3.1	1849	3.3	6.0	66.0
4	6.5	2150	8.4	12.9	40.5
5	8.2	1771	22.2	27.9	18.0
6	9.8	1956	37.5	31.5	2.4
7	11.1	2015	41.4	30.0	0
8	10.9	1711	42.0	30.6	0
9	12.6	1733	40.5	30.9	0
10	13.3	1906	38.1	29.1	0
11	15.0	1407	38.1	30.6	0
12	14.1	1409	32.4	26.6	0

[a] Culture conditions: 10% sucrose; 26°C, 125 rpm; 100 ml.
[b] Based on sucrose (w/w).
[c] Weight average molecular weight by GPC.

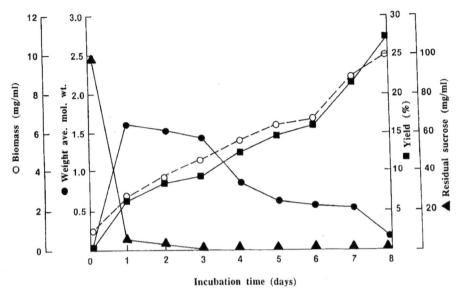

Figure 8.2   Elsinan formation in batch fermentation over 12 days. Yields of biomass and polymer, molecular weight of polymer recovered, and residual carbon source in the feed are illustrated.

0.145 M to 0.29 M (yields ranged from 5.2% to 16.4%), nitrogen sources (organic and inorganic) (yields ranged from 0.2% to 21.1% and weight average molecular weights ranged from 11K to 1803K), and phosphate type and concentration (yields ranged from 6.4% to 20.4% and weight average molecular weights ranged from 134K to 2135K) on elsinan yield and molecular weight distribution [104]. Scale-up batch and continuous fermentation studies resulted in lower yields (less than 10%) and weight average molecular weights between 100,000 and one million, depending on conditions [104]. In addition, pigmentation problems [106,107] became a significant factor under these conditions.

8.4.3   Processing

Processing of the extracellular product was similar to the procedure described earlier for pullulan. For larger volumes of media, tangential flow filtration was more efficient for purification of the product than solvent precipitation. Polymer composition was validated after acid hydrolysis using high-performance liquid chromatography against calibrated sugar monomer standards [104].

Elsinan forms gels at 5% or higher concentrations, and viscosity characteristics of elsinan solutions are affected by temperature. Misaki et al. [86] reported that elsinan forms strong films, and has physical properties similar to pullulan. Elsinan has been reported to be digestible, soluble in water, stable to salts, and stable over a pH range of 3–11 [61,86]. Films formed from elsinan are reported to be acid resistant, exhibit low oxygen permeability (1.0 and 1.3 cc per square meter for 24 h for elsinan and pullulan, respectively), water resistant, oil resistant, extensible, and heat sealable, and exhibit good physical strength properties [61,87].

8.4.4   Applications

Food and pharmaceutical coatings for controlled release would be potential applications for elsinan. The susceptibility of the polymer to digestion by salivary α-amylase would provide a potential for nutritional availability during digestion in the human digestive tract, which may be a benefit in certain applications.

8.5   Levan

8.5.1   Background

Levan is an extracellular polymer consisting of anhydro-D-fructofuranose units with predominantly β-2,6-glycosidic linkages and some β-2,1 branching (inulin type). There are an average of 10–12 monomer units between branches [108,109]. Bacterial levans have been implicated in dental diseases because of their adhesive and acidogenic properties. The fructofuranose linkages are acid labile.

Levan is produced by a wide variety of bacteria; however, it is often equated with inulin or lumped as fructosans without more definitive citation [108–111]. Bacterial sources include *Erwinia herbicola, Aerobacter lavanicum, Streptococcus salivarius, Pseudomonas prunicola, Arthrobacter acetigenum, Bacillus polymxa, and Bacillus subtilis* [111,112]. *Actinomyces* sp. also produce levan [108]. Levanase (EC 3.2.1.65) hydrolyzes β-2,6-D-linkages to produce oligosaccharides and fructose [27]. Levan is synthesized by levansucrase with sucrose as the carbon source [113]. The products of the reaction are polyfructose and glucose.

8.5.2   Biosynthesis

The effects of carbon sources and length of incubation on levan biosynthesis have been examined by Ueda et al. [114] for the bacterium *Aeromonas hydrophila*. Bodie et al. examined the effects of pH and time on the production of levan by *Arthrobacter* [115]. The effects of time, pH, temperature, sucrose concentration, and inoculum on a wide range of levan-producing bacteria were examined by Matton et al. [116] in an effort to screen for the best levan-producing system on a cost versus yield basis. Culture and processing conditions have also been explored, on a limited basis, for the bacteria *S. salivarius* and *S. bovis,* and for *Actinomyces viscous* [117,118].

Sporadic reports of molecular weights for levan have been published using viscometry [116], gel permeation chromatography [119], and light scattering [120]. The general observation has been reported that longer incubation results in a more highly branched and higher molecular weight levan [121]. Molecular weights ranging from 1×10^6 to 100×10^6, depending on organism and pH, have been reported [112,122]; however, no systematic correlation of culture conditions and molecular weight has been determined. Han and Clarke [123,124] recently reported a strain of *Bacillus polymyxa* that produced 40 g/liter of levan on a sucrose medium (50% yield based on available fructose and 25% based on available sucrose) and had a molecular weight of around 2 million by gel permeation chromatography. We studied the bacterium *E. herbicola* to better define these correlations [125].

Our initial studies involved the evaluation of yields of levan from *E. herbicola* (ATCC 15552), *A. pasteurianus* (ATCC 11142), *Microbacterium laevaniformans (ATCC 15953), and B. coagulans* QMB 1624 (*B. subtilus-pumilus*) [125]. *E. herbicola* (ATCC 15552) was chosen for further study based on yield from 5% and 10% sucrose solutions in batch and continuous cultures. Molecular weight determinations were by gel permeation chromatography and laser light scattering. After acid hydrolysis of the levan isolated from these cultures the products were fructose and its oligomers as determined by high-performance liquid chromatography.

Table 8.12 illustrates the results from a continuous fermentation study using *E. herbicola* (ATCC 15552). From the dilution rates studied, the molecular weight of the levan was relatively constant at around 1.3 million with a dispersity of 1.4–2.0 and a yield of 18% to 30% based on sucrose [125]. Using gel permeation chromatography, weight average molecular weights ranged from 1.1×10^6 to 1.6×10^6. Light scattering indicated a molecular weight of 2.97×10^7. These values were comparable when levans isolated from *E. herbicola* and *B. polymyxa* were analyzed. In addition, yields from continuous cultivation of *B. polymyxa* [126] were also comparable to the Erwinia studies. Based on a sucrose feed at 80 g/liter, steady-state yields were 23%, 22%, 20%, and 13% for dilution rates of 0.022 h^{-1}, 0.027 h^{-1}, 0.041 h^{-1}, and 0.06 h^{-1},

Table 8.12   Production of Levan by Continuous Fermentation with *E. herbicola* ATCC 15552[a]

Time (h)	Initial pH	Final pH	Volume[b] (ml)	Dilution rate (h^{-1})	Yield[c] (%)	Mol wt[d] (K)	Dispersity[e]
48	7.2	4.25	1470	0.030	20.5	1617	8.6
68	7.2	4.76	1645	0.018	29.9	1124	4.0
72	7.2	4.18	1360	0.014	19.2	1507	9.0
120	7.2	4.37	1660	0.010	18.9	1359	1.5
168	7.2	4.67	4425	0.019	29.5	1348	1.4
190	7.2	4.17	1350	—	21.7	1286	2.0

[a] Continuous culture, 10% sucrose; aeration 0.5 liter/min; 200 rpm; 25°C; culture incubated 48 h before starting media flow, total vessel volume = 1350 ml.
[b] Supernatant; total volume collected and processed.
[c] Based on sucrose.
[d] Weight average molecular weight by GPC.
[e] Mw/Mn.

respectively. Work with continuous culture using longer residence times to study changes in nutrient composition was conducted to better correlate nutritional state with molecular weight, dispersity, and yield of polymer. Results of batch studies showed that the yield of levan decreased with time, particularly after about 3 days, probably due to levanhydrolase [111,112]. In addition, the polymer became pigmented. Continuous culture resulted in higher yields of unpigmented polymer than in batch studies. Carbon, nitrogen, and phosphate sources were also studied. As would be expected, sucrose resulted in the highest yield of levan when compared with dextrose, galactose, fructose, maltose, mannose, raffinose, and mannitol [125]. A pH of between 6.0 and 7.0 was found optimal for highest yield [125].

8.5.3   Processing

The polymer was purified from batch cultures as shown in Table 8.13 and from continuous culture using tangential flow filtration.

Attempts to form films for evaluation of physical properties were only partially successful. Levan solutions with and without 10% glycerol as a plasticizer were cast and dried in a forced hot air oven under the conditions described for pullulan. The films formed were transparent but too brittle to permit physical analysis of properties. Our studies are continuing in order to overcome the brittleness encountered upon drying so that a full evaluation of functional properties can be made.

8.5.4   Applications

One of our interests in levan is for an edible food coating application. Since levan is labile to acid hydrolysis, coatings with low oxygen permeability during storage to protect the food could be rapidly depolymerized in the digestive system to release the encapsulated foods. Presumably *E. herbicola* would not be the source of choice for commercial production of levan since it is a plant parasite. As with chitosan, many medical applications with levan could also be considered.

Table 8.13   Processing Conditions for Levan

1. Add 1% (wt/vol) SDS to culture, shake 4 h, 125 rpm.
2. Dilute 1:2 with deionized water.
3. Centrifuge 23,000 *g*, 30 min.
4. Dialyze supernatant (12,000 mol wt cutoff) against deionized water.
5. Dilute 1:2 with methanol with stirring.
6. Centrifuge at 6500 *g*, 15 min.
7. Decant off water/methanol.
8. Wash precipitate with methanol.
9. Dry under vacuum over calcium sulfate with desiccator.

Acknowledgments

We would like to thank Elwyn Reese for his technical input and critical review of this manuscript. We would also like to thank Serry Sousa and Deborah Zorfass for their technical support. Appreciation is also given to Hayashibara Laboratories, Japan, for their supplies of research samples of pullulan and technical assistance.

REFERENCES

1. Aspinall, G.O. In *The Polysaccharides*, **Vol. 1,** Aspinall, G.O., ed. Academic Press, New York, pp. 1–18 (1982).
2. Atkins, E. *Int. J. Biol. Macromol.* **8,** 323–329 (1986).
3. Aspinall, G.O., ed. *The Polysaccharides,* **Vol. 1.** Academic Press, New York, pp. 1–340 (1982).
4. Aspinall, G.O., ed. *The Polysaccharides,* **Vol. 2.** Academic Press, New York, pp. 1–503 (1983).
5. Aspinall, G.O., ed. *The Polysaccharides,* **Vol. 3.** Academic Press, New York, pp. 1–470 (1985).
6. Berkeley, R.C.W.; Gooday, G.W.; Ellwood, D.C., eds. *Microbial Polysaccharides and Polysaccharases,* Academic Press, London, pp. 1–479 (1979).
7. Cottrell, I.W. In *Fungal Polysaccharides;* Sandford, P.A.; Matsuda, K., eds. *ACS Symposium Series* No. 126, American Chemical Society, Washington, D.C., pp. 249–270 (1980).
8. Harada, T. *Biochem. Soc. Symp.* **48,** 97–116 (1983).
9. Kang, K.S.; Cottrell, I.W. *Microbial Technology,* **Vol. 1.** Academic Press, New York, 1; pp. 417–481 (1979).
10. Keene, L.; Lindberg, B. In *The Polysaccharides,* **Vol. 2.** Aspinall, G.O., ed. Academic Press, New York, pp. 287–363 (1983).
11. Marchessault, R.H. *Contemp. Top. Poly. Sci.* **5,** 15–53 (1984).
12. Sandford, P.A.; Laskin, A., eds. *Extracellular Microbial Polysaccharides, ACS Symposium Series* No. 45, American Chemical Society, Washington, D.C., pp. 1–326 (1977).
13. Yalpani, M., ed. *Industrial Polysaccharides,* **Vol. 3.** Elsevier, Amsterdam, pp. 1–408 (1987).
14. Arora, D.K.; Elander, R.P.; Mukerji, K.J. *Handbook of Applied Mycology: Fungal Biotechnology,* **Vol. 4.** Marcel Dekker, New York (1992).
15. Bartnicki-Garcia, S.; Nickerson, W.J. *Biochim. Biophys. Acta* **5,** 102–119 (1962).
16. Kobayashi, T.; Takiguchi, Y.; Shimahara, K.; Sannan, T. *Nippon Nogeikagaku Kaiahi* **62,** 1463–1469 (1988).
17. Kreger, D.R. *Biochim. Biophys. Acta* **13,** 1–9 (1954).
18. Foster, A.B.; Webber, J.M. In *Advances in Carbohydrate Chemistry,* **Vol. 15.** Wolfrom, M.L.; Tipson, R.S., eds. Academic Press, New York, pp. 371–393 (1960).
19. White, S.A.; Farina, P.R.; Fulton, I. *Appl. Environ. Microbiol.* **38,** 323–328 (1979).
20. Roget, C. *Compt. Rend. Acad. Sci. Paris* **48,** 792–795 (1959).
21. Fenton, D.; Davis, B.; Everleigh, D.E. In *Proceedings of the First International Conference on Chitin/Chitosan.* Muzzarelli, R.A.; Pariser, E.R., eds. Massachusetts Institute of Technology Sea Grant 78-7, Cambridge, Massachusetts, pp. 169–181 (1978).
22. Hirano, S.; Sato, N.; Yoshida, S.; Kitagawa, S. In *Industrial Polysaccharides* Yalpani, M.; ed. Elsevier, Amsterdam, pp. 163–176 (1987).

23. Hirano, S.; Ohe, Y.; Ono, H. *Carb. Res.* **47,** 315–320 (1976).
24. Ogura, K.; Kanamoto, K.; Sarnan, T.; Tankaka, K.; Iwakura, Y. In *International Conference on Chitin and Chitosan.* Hirano, S., ed. Japanese Society of Chitin and Chitosan, Sapporo, pp. 39–44 (1982).
25. Berkeley, R.C.W. In *Microbial Polysaccharides and Polysaccharases.* Berkeley, R.C.W.; Gooday, G.W.; Ellwood, D.C., eds. Academic Press, London, pp. 203–236 (1979).
26. Reyes, F.; Lahoz, R.; Martinez, M.J.; Alfonso, C. *Mycopathology* **89,** 181–187 (1985).
27. Matheson, N.K.; McCleary, B.U. In *The Polysaccharides* **Vol. 3.** Aspinall, G.O., ed. Academic Press, London, pp. 1–105 (1985).
28. Davis, L.L.; Bartnicki-Garcia, S. *Biochem.* **23,** 1065–1073 (1984).
29. Araki, Y.; Ito, E. *Eur. J. Biochem.* **55,** 71–78 (1975).
30. Arcidiacono, S.; Lombardi, S.J.; Kaplan, D.L. In *Chitin and Chitosan.* Skjak-Braek, G.; Anthonsen, T.; Sandford, P. Elsevier Applied Science, London, pp. 319–332 (1989).
31. Lombardi, S.J. Thesis, Dartmouth College, New Hampshire, pp. 1–50 (1988).
32. Hershberger, D.F. U.S. Patent 4 806 479 (1989).
33. Arcidiacono, S.; Kaplan, D.L. *Biotech. Bioeng.* **39,** 281–286 (1992).
34. Muzzarelli, R.A.A.; Rocchetti, R. *Carb. Poly.* **6,** 461–472 (1985).
35. Kaplan, D.L.; Mayer, J.M.; Lombardi, S.J.; Wiley, B.J.; Arcidiacono, S. *Poly. Reprints* **30,** 509–510 (1989).
36. Mayer, J.M.; Kaplan, D.L. U.S. Patent 5,015,293 (1991).
37. Luby, P.; Kuniak, L. *Makromol. Chem.* **180,** 2213–2220 (1979).
38. Averbach, B.L. In *Proceedings of the First International Conference on Chitin/Chitosan.* Muzzarelli, R.A.A.; Pariser, E.R., eds. Massachusetts Institute of Technology Sea Grant 78-7, Cambridge, Massachusetts, pp. 199–209 (1978).
39. Wei, Y.C.; Hudson, S.M.; Mayer, J.M.; Kaplan, D.L. *J. Polymer Sci.* **30,** 2187–2193 (1977).
40. Muzzarelli, R.A.A. *Chitin,* Pergamon Press, New York, pp. 1–309 (1977).
41. Muzzarelli, R.A.A. *Enz. Microb. Tech.* **2,** 177–184 (1980).
42. Pangburn, S.H.; Trecossy, P.V.; Heller, J. In *Chitin, Chitosan and Related Enzymes.* Zikakis, J.P., ed. Academic Press, New York, pp. 3–19 (1984).
43. Peniston, Q.P.; Johnson, E.L. U.S. Patent 3 533 940 (1970).
44. Slagel, R.C.; Sinkovitz, G.D. U.S. Patent 3 709 780 (1973).
45. Yalpani, M.; Sandford, P. In *Industrial Polysaccharides,* **Vol. 3.** Yalpani, M., ed. Elsevier, Amsterdam, pp. 311–335 (1987).
46. Kurita, K. In *Industrial Polysaccharides,* **Vol. 3.** Yalpani, M., ed. Elsevier, Amsterdam, pp. 337–346 (1987).
47. Byrom, D. In Biomaterials: Novel Materials from Biological Sources. Byrom, D., ed. Stockton Press, New York, pp. 333–359 (1991).
48. Lohmann, D. In *Novel Biodegradable Microbial Polymers,* **Vol. 186.** Dawes, E.A., ed. Kluwer Academic Publishers, London; *NATO ASI Series E,* pp. 333–348 (1990).
49. Hirano, S. In *Chitin and Chitosan.* Skjak-Braek, G.; Anthonsen, T.; Sandford, P., eds. Elsevier, London, pp. 1–835 (1989).
50. Sandford, P.A. In *Chitin and Chitosan.* Skjak-Braek, G.; Anthonsen, T.; Sandford, P., eds. Elsevier, London, pp. 51–69 (1989).
51. East, G.C.; McIntyre, J.E.; Qin, Y. In *Chitin and Chitosan.* Skjak-Braek, G.; Anthonsen, T.; Sandford, P., eds. Elsevier, London, pp. 757–764 (1989).
52. Bender, H.; Wallenfels, K. *Biochem. Z.* **334,** 79–95 (1961).
53. Bender, H.; Lehman, J.; Wallenfels, K. *Biochim. Biophys. Acta* **36,** 309–316 (1959).

54. Catley, B.J.; Whelan, W.J. *Arch. Biochem. Biophys.* **143**, 138–142 (1971).
55. Wallenfels, K.; Keilich, G.; Bechtler, G.; Freudenberger, D. *Biochem. Z.* **341**, 433–450 (1965).
56. Jeanes, A. In *Extracellular Microbial Polysaccharides*. Sandford, P.A.; Laskin, A., eds. *ACS Symposium Series* No. 45. American Chemical Society, Washington, D.C.; pp. 284–298 (1977).
57. Sumitomo Chemical Co., Ltd.; Hayashibara Biochemical Laboratories, Inc. British Patent 1 496 017 (1977).
58. Yuen, S. *Process Biochem.* **9**, 7–22 (1974).
59. Yuen, S. Pullulan and Its New Applications (unpublished data); Hayashibara Biochemical Laboratories, Inc., Japan; Undated; Technical Literature.
60. Sugimoto, K. Pullulan: Production and Applications, Fermentation and Industry (unpublished data), Hayashibara Biochemical Laboratories, Inc., Japan; Undated; Technical Literature.
61. Yokobayashi, K.; Sugimoto, T. German Patent DE28 42 855A1 (1979).
62. Brant, D.A.; Burton, B.A. In *Solution Properties of Polysaccharides,* Brant, D.A., ed. *ACS Symposium Series* No. 150. American Chemical Society, Washington, D.C., pp. 81–99 (1981).
63. Catley, B.J.; Whelan, W.J. *Arch. Biochem. Biophys.* **143**, 138–142 (1971).
64. Guilbot, A.; Mercier, C. In *The Polysaccharides,* **Vol. 3**, Aspinall, G.O., ed. Academic Press, New York, pp. 209–282 (1985).
65. James, D.W.; Preiss, J.; Elbein, A.D. In *The Polysaccharides,* **Vol. 3**. Aspinall, G.O., ed. Academic Press, New York, pp. 107–207 (1985).
66. Ueda, S.; Fujita, K.; Komatsu, K.; Nakashima, Z. *Appl. Microbiol.* **11**, 211–215 (1963).
67. Bulmer, M.A.; Catley, B.J.; Kelly, P.J. *Appl. Microbiol. Biotech.* **25**, 362–365 (1987).
68. Catley, B.J. *Appl. Microbiol.* **22**, 650–654 (1971).
69. Catley, B.J. *J. Gen. Microbiol.* **78**, 33–38 (1973).
70. Catley, B.J. *Appl. Microbiol.* **22**, 641–649 (1971).
71. Catley, B.J. *FEBS Lett.* **10**, 190–193 (1970).
72. Catley, B.J. In *Microbial Polysaccharides and Polysaccharases*. Berkeley, R.C.W.; Gooday, G.W.; Ellwood, D.C., eds. Academic Press, New York, pp. 69–84 (1979).
73. Catley, B.J. *J. Gen. Microbiol.* **120**, 265–268 (1980).
74. Ramos, S.; Garcia Acha, I. *Trans. Br. Mycol. Soc.* **64**, 129–135 (1975).
75. Boa, J.M.; LeDuy, A. *Biotech. Bioeng.* **30**, 463–470 (1987).
76. Lacroix, C.; LeDuy, A.; Noel, G.; Choplin, L. *Biotech. Bioeng.* **27**, 202–207 (1985).
77. Ono, K.; Yasuda, N.; Ueda, S. *Agric. Biol. Chem.* **41**, 2113–2118 (1977).
78. Heald, P.J.; Kristiansen, B. *Biotech. Bioeng.* **27**, 1516–1519 (1985).
79. Finkelman, M.A.J.; Vardanis, A. *Biotech. Lett.* **4**, 393–396 (1982).
80. Kato, K.; Shiosaka, M. U.S. Patent 3 827 937 (1974).
81. Kato, K.; Shiosaka, M. U.S. Patent 3 912 591 (1975).
82. Wiley, B.J.; Arcidiacono, S.; Sousa, S.; Mayer, J.M.; Kaplan, D.L. *J. Environ. Polym. Degradation* **1**, 3-a (1993).
83. Ball, D.H.; Wiley, B.J.; Reese, E.T. *Canad. J. Microbiol.* **38**, 324–327 (1992).
84. Mayer, J.M.; Greenberger, M.; Ball, D.H.; Kaplan, D.L. American Chemical Soc. *Polymer Preprints, PMSE,* **63**, 732–735 (1991).
85. Misaki, A.; Tasumuraya, Y.; Takaya, S. *Agric. Biol. Chem.* **42**, 491–493 (1978).
86. Misaki, A.; Tasumuraya, Y. In *Fungal Polysaccharides*. Sandford, P.A.; Matsuda, K.,

eds. *ACS Symposium Series* No. 126. American Chemical Society, Washington, D.C., pp. 197–220 (1980).

87. Misaki, A.; Hyogo, N.; Takaya, S.; Yokobayashi, K.; Tsumuraya, Y. U.S. Patent 4 202 966 (1980).
88. Misaki, A.; Nishio, H.; Tsumuraya, Y. *Carb. Res.* **109**, 207–219 (1982).
89. Bitancourt, A.A.; Jenkins, A.E. *J. Agric. Res.* **54**, 1–18 (1936).
90. Bitancourt, A.A.; Jenkins, A.E. *Mycologia* **28**, 489–492 (1936).
91. Bitancourt, A.A.; Jenkins, A.E. *Phytopath.* **26**, 393–396 (1936).
92. Cunningham, H.S. *Phytopath.* **18**, 593–545 (1928).
93. Gabel, A.W.; Tifany, L.H. *Mycologia* **79**, 737–744 (1987).
94. Jenkins, A.E. *J. Agric. Res.* **42**, 545–558 (1931).
95. Jenkins, A.E. *J. Agric. Res.* **44**, 689–700 (1932).
96. Jenkins, A.E. *Mycologia* **25**, 213–223 (1933).
97. Jenkins, A.E. *Phytopathology* **23**, 538–545 (1933).
98. Jenkins, A.E.; Bitancourt, A.A. *Phytopathology* **42**, 12 (1952).
99. Miller, J.A. *Mycologia* **49**, 277–279 (1957).
100. Tiffany, L.H.; Mathre, J.H. *Mycologia* **53**, 600–604 (1961).
101. Todd, E.H. *Plant Disease Rep.* **44**, 153–165 (1960).
102. Zeigler, R.S.; Lozano, J.C. *Phytopathology* **73**, 293–300 (1983).
103. Zeller, S.M.; Dermiah, J.W. *Phytopathology* **21**, 965–972 (1931).
104. Wiley, B.J.; Arcidiacono, S.M.; Ball, D.H.; Mayer, J.M.; Kaplan, D.L. Technical Report 89/035, U.S. Army Natick Research, Development & Engineering Center, Natick, Massachusetts (1989).
105. Wickerham, L.J. Technical Bulletin No. 1029; U.S. Department of Agriculture, Washington, D.C. (1951).
106. Weiss, U.; Ziffer, H.; Batterham, T.J.; Blumer, M.; Hackeng, W.H.; Copier, H.; Salemink, C.A. *Can. J. Microbiol.* **11**, 57–66 (1965).
107. Weiss, U.; Flon, H.; Burger, W.C. *Arch. Biochem. Biophys.* **69**, 311–319 (1957).
108. Avigad, G. In *Encyclopedia of Polymer Science and Technology,* **Vol. 8.** John Wiley & Sons, New York, pp. 711–716 (1968).
109. Bell, D.J. *J. Chem. Soc.* 2866–2870 (1954).
110. Avigad, G.; Feingold, D.S. *Arch. Biochem. Biophys.* **70**, 178–180 (1957).
111. Avigad, G. In *Methods in Carbohydrate Analysis,* **Vol. 5.** Whistler, R.L., ed. Academic Press, New York, p. 161 (1965).
112. Stivala, S.S.; Zweig, J.E.; Enrlich, J. In *Solution Properties of Polysaccharides.* Brant, D.A., ed. *ACS Symposium Series* No. 150. American Chemical Society, Washington, D.C., pp. 101–110 (1981).
113. Hestrin, S.; Feingold, D.S.; Avigad, G. *Biochem. J.* **64**, 340–351 (1956).
114. Ueda, S.; Momii, F.; Osajima, K.; Ito, K. *Agric. Biol. Chem.* **45**, 1977–1981 (1981).
115. Bodie, E.; Schwartz, R.; Catena, A. *Appl. Environ. Microbiol.* **50**, 629–633 (1985).
116. Matton, J.R.; Holmlund, C.E.; Schepartz, S.A.; Vavra, J.J.; Johnson, M.S. *Appl. Microbiol.* **3**, 321–333 (1955).
117. Niven, C.; Smiley, K.; Sherman, J.M. *J. Biol. Chem.* **140**, 105–109 (1941).
118. Pabst, M.J. *Infect. Immun.* **5**, 518–526 (1977).
119. Marshall, K.; Weigel, H. *Carb. Res.* **80**, 375–377 (1980).
120. Bahary, W.S.; Stivala, S.S. *Biopoly.* **14**, 2467–2478 (1975).
121. Tanaka, T.; Oi, S.; Yamamoto, T.J. *Biochem.* **87**, 297–303 (1980).
122. Long, L.W.; Stivala, S.S.; Ehrlich, J. *Arch. Oral Biol.* **20**, 503–507 (1975).

123. Han, Y.W.; Clarke, M.A. *J. Indust. Microbiol.* **4,** 447–452 (1989).
124. Han, Y.W.; Clark, M.A. *J. Agric. Food Chem.* **38,** 393–396 (1990).
125. Keith, J.; Wiley, B.J.; Ball, D.; Arcidiacono, S.; Zorfass, D.; Mayer, J.M.; Kaplan, D.L. *Biotech. Bioeng.* **38,** 557–560 (1991).

CHAPTER 9

Chemical Modification of Proteins and Polysaccharides and Its Effect on Enzyme-Catalyzed Degradation

Waleed S.W. Shalaby and Kinam Park

Waleed S.W. Shalaby and Kinam Park, School of Pharmacy, Purdue University, West Lafayette, Indiana 47907, U.S.A.

9.1   Introduction

Proteins and polysaccharides are a unique class of natural polymers that have received considerable interest in recent years. The potential use of these systems as therapeutically active agents, drug carriers, and drug delivery systems has been the subject of many recent reviews [1–10]. Proteins and polysaccharides can be cleared from the body by means of chemically induced or enzyme-catalyzed hydrolysis mechanisms. Proteins and polysaccharides can be chemically modified to improve biological activity, to alter biodegradability, to prepare water-insoluble derivatives, or to form gels for biomedical or industrial use. The objective of this chapter is to examine how chemical modification affects the enzyme-catalyzed hydrolysis and biological activity of proteins and polysaccharides. This chapter provides information relevant to (1) methods of chemical modification, (2) factors affecting enzyme-catalyzed hydrolysis, and (3) enzymatic degradation of modified proteins and polysaccharides. In the last section, the use of modified proteins and polysaccharides in the development of biodegradable drug delivery systems, prodrugs, and drug carriers is reviewed.

9.2   Chemical Modification of Polysaccharides

The chemical modification of proteins and polysaccharides represents a vast area of research. In this section a brief overview of some general modification methods is provided. More extensive reviews are available [11–14].

9.2.1   Functional Groups Available on Polysaccharides

In biological systems, polysaccharides and their derivatives are found as components of the cell wall of bacteria and plants, and of the connective tissue, eyeball vitreous humour, and circulatory system in man [15]. Polysaccharides are either homopolymers or heteropolymers. Their repeating units or monomers may contain hydroxyl, carboxyl, amino, or sulfate side chains. Examples of monomers that contain only hydroxyl group side chains are D-glucose, D-mannose, D-galactose, L-galactose, D-xylose, and L-arabinose. Carboxyl-containing side chains include D-glucuronic acid, D-galacturonic

Table 9.1   Chemical Structure of Polysaccharide Repeating Units

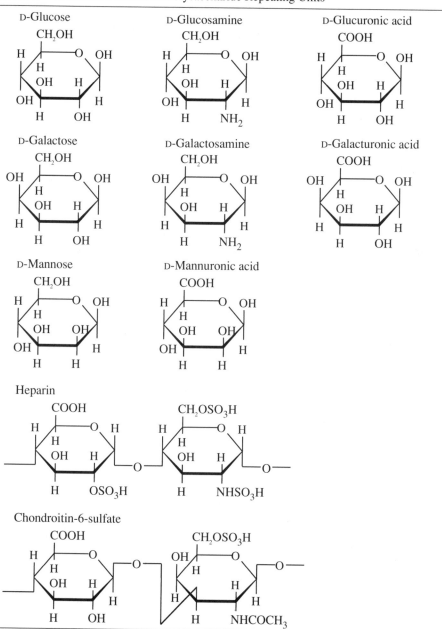

acid, and D-mannuronic acid. Polysaccharide derivatives can also be derived from sulfate-containing disaccharides such as those found in heparin and chondroitin sulfate. Table 9.1 lists the chemical structures of the various monomers found in polysaccharide chains. The chemical modification of polysaccharide chains can occur at the hydroxyl, carboxyl, amino, and sulfate moieties, or by oxidation to open up the pyranose rings. Because hydroxyl groups are by far the most abundant side chain on polysaccharides, much of the subsequent modification methods are focused on hydroxyl groups.

9.2.2   Modification of Hydroxyl Groups

9.2.2.1   Alkylation

Alkylation reactions with polysaccharides can be achieved by nucleophilic addition or substitution using the hydroxyl groups on the polysaccharide chain. Because of the abundance of hydroxyl groups, a significant degree of alkylation can be obtained with polysaccharides. Single-step alkylation of polysaccharide chains can be achieved through epoxide-based reactions, alkyl halide substitution, or alkene-based reactions (Table 9.2). Epoxide-based reactions can be carried out at room temperature under alkaline conditions where nucleophilic attack of the epoxide moiety leads to the formation of stable ether linkages. A variety of functional groups shown in Table 9.2 can be introduced onto polysaccharides via epoxide-based reactions [16–27].

Alkyl halide substitution is another type of alkylating reaction that can be used to functionalize polysaccharides (Table 9.2). Alkyl halide reagents have been especially useful for introducing amino or carboxylic pendant groups [20,28–34]. A reagent that can undergo epoxide-based addition or alkyl halide substitution is epichlorohydrin. Because of its difunctionality, this reagent is commonly used to form crosslinks between polysaccharide chains [35]. Vinylic groups also display reactivity toward hydroxyl groups under certain conditions. Two reagents that have been used for alkylation are divinyl sulfone [36] and diethylvinyl phosphate [37] (Table 9.2). In the case of divinylsulfone, crosslinking is also likely because of its tetrafunctionality.

9.2.2.2   Acylation

Esterification of hydroxyl groups on polysaccharide chains can be achieved by direct acid-catalyzed reactions using carboxylic acids or indirectly with carboxylic acid derivatives. Direct acylation of starch has been done under aqueous conditions using formic acid [38,39] and under nonaqueous conditions using formic acid, acetic acid, or citric acid [39–42]. Acylation with acetic acid and citric acid requires heat to drive the reaction whereas formic acid reacts well at room temperature. Polysaccharides have been indirectly acylated using carboxylic acid anhydrides [43–46], acyl chlorides [47–50], or vinyl esters [51–54]. It has been shown that acylation of starch with acetic anhydride depends on both the degree of alkalinity (pH 7–11) and the reaction temperature. In acylation of starch, the acid released from reactions with acyl chlorides or carboxylic anhydrides is known to degrade starch. Thus, addition of acid acceptors is often required in such reactions.

Table 9.2   Reagents Used to Alkylate Polysaccharides

Reagent	Chemical structure	References
Epoxide-based reactions		
Ethylene oxide	$H_2C \overset{O}{\triangle} CH_2$	[16]
Propylene oxide	$H_2C \overset{O}{\triangle} CH-CH_3$	[17]
2,3-Epoxypropyldiethyl amine	$H_2C \overset{O}{\triangle} CH-CH_2N(C_2H_5)_2$	[18,19]
2,3-Epoxypropyltrimethylamonium chloride		[20--26]
	$H_2C \overset{O}{\triangle} CH-CH_2\overset{+}{N}(CH_3)_3 \quad \overset{-}{Cl}$	
Glycidyl acrylate	$H_2C \overset{O}{\triangle} CH-CH_2-O-\overset{\overset{O}{\|}}{C}-CH=CH_2$	[27]
Alkyl halide substitution		
2-Chloroethylamine	$Cl-CH_2CH_2NH_2$	[28]
2-Diethylaminoethyl chloride	$Cl-CH_2CH_2N(C_2H_5)_2$	[20,31--34]
Chloroacetic acid	$Cl-CH_2COOH$	[29]
6-Bromohexanoic acid	$Br-(CH_2)_5COOH$	[30]
Epichlorohydrin	$H_2C \overset{O}{\triangle} CH-CH_2-Cl$	[35]
Others		
Divinyl sulfone	$H_2C=CHSO_2CH=CH_2$	[36]
Diethylvinyl phosphate	$H_2C=CHP\overset{\overset{O}{\|}}{<}\overset{OC_2H_5}{OC_2H_5}$	[37]

9.2.2.3   Phosphorylation

Phosphorylation of polysaccharide hydroxyl groups with organic, phosphorus-containing reagents has been used to prepare anionic polysaccharides. Three reagents developed by Tessler yield starch phosphate monoesters. The reagents used were *N*-phosphoryl-*N* '-methylimidazole [55], salicyl phosphate [56], and *N*-benzoylphosphoramidic acid [57]. Phosphorylation can also be carried out using inorganic phosphates such as sodium tripolyphosphate [58–60] or mixtures of disodium hydrogen phosphate and monosodium hydrogen phosphate [61,62].

9.2.3   Introduction of Activating Groups

Activating groups are introduced on polysaccharide chains to modify the reactivity toward a variety of nucleophilic compounds. Activating groups can be introduced either by chemical modification or by oxidation. The use of activated polysaccharides to conjugate enzymes and proteins has received considerable interest over the years. A recent article by Dumitriu et al. reviews many types of polysaccharide conjugates [63]. Some of the more common methods used in the activation of polysaccharides are listed in Table 9.3.

9.2.3.1   Cyanogen Bromide Activation

Under alkaline conditions, cyanogen bromide and other cyanogen derivatives activate hydroxyl groups of polysaccharides through the formation of cyanate ester intermediates. These intermediates can then be used to couple polysaccharide chains with various amino-containing compounds [64–70]. Because of its reactivity, the cyanate ester intermediate can undergo hydrolysis or interchain rearrangement that will lead to the formation of crosslinks or cycloimidocarbonates. The coupling of primary amine groups to the polysaccharide can occur either with the cyanate ester intermediate or with the cyclic imidocarbonate. Cyanogen bromide activation is one of the most widely used reactions for covalent linking of compounds to polysaccharides. Their particular use in the synthesis of dextran–enzyme conjugates is described below.

9.2.3.2   Carbonyldiimidazole Activation

The activation of polysaccharide chains with carbonyldiimidazole (CDI) occurs by nucleophilic attack of the CDI carbonyl group by hydroxyl groups or carboxylic anions on the polysaccharide. The extent of activation can be varied by altering the amount of CDI added to the reaction. Activated polysaccharides may be readily conjugated with other compounds by nucleophilic substitution which forms stable carbamate links. A variety of proteins and amino group-containing compounds have been coupled to CDI-activated polysaccharides [71–73]. Furthermore, CDI-activated polysaccharides may also undergo intrachain and interchain crosslinking by forming ester or diester crosslinks with hydroxyl or carboxylic groups [74].

9.2.3.3   Chloroformate Activation

The hydroxyl groups of agarose, cellulose, O-substituted celluloses, or dextrans have been activated with a variety of chloroformate derivatives [75–79]. Activated agarose can be prepared using 4-nitrophenylchloroformate, N-hydroxysuccinimidochloroformate, or trichlorophenylchloroformate. In all cases, the activated moiety can be displaced by amino groups to form N-substituted carbonate links. The activation of

Table 9.3   Reagents Used for the Preparation of Activated Polysaccharides

Reagent	Chemical structure	References
Cyanogen bromide	CNBr	[64–69]
Carbonyldiimidazole		[71–73]
4-Nitrophenylchloroformate		[75–77,80,81]
N-Hydroxysuccinimidyl-chloroformate		[75,76]
Trichlorophenylchloroformate		[75]
1-Cyclohexyl-3-(2-morpholinyl-4-ethyl)-carbodiimide		[93]
1-Ethyl-3-(3-dimethylaminopropyl)-carbodiimide		[93]
4-Toluenesulfonyl chloride		[82]
2,2,2-Trifluoroethanesulfonyl chloride		[83,84]
Cyanuric chloride		[85]

dextran by 4-nitrophenylchloroformate has been reported [77,80,81]. The aromatic carbonate can undergo nucleophilic substitution with neighboring hydroxyl groups under alkali conditions. This may lead to the formation of intrachain cyclic carbonate esters or interchain carbonate esters.

9.2.3.4   Divinylsulfone Activation

Under alkaline conditions, hydroxyl groups on polysaccharides can be alkylated with vinylsulfonyl groups by treatment with divinylsulfone [36]. The activated polysaccharide can then be derivatized with hydroxy or amine-containing compounds. Because of its tetrafunctionality, divinylsulfonyl activation can also result in crosslinking of polysaccharides.

9.2.3.5   Organosulfonyl Activation

Polysaccharide chains are frequently activated using sulfonyl chlorides [82–84]. These reagents readily react with polysaccharide hydroxyl groups to form sulfonate esters bonds. 4-Toluenesulfonyl chloride (tosyl chloride) and 2,2,2-trifluoroethanesulfonyl chloride (tresyl chloride) are two reagents commonly used for activation. Once activated, the tosyl or tresyl moieties act as good leaving groups during nucleophilic attack. Compounds containing thiol or primary amine groups will generally displace the tosyl or tresyl leaving groups.

9.2.3.6   Triazine Activation

Activation of polysaccharide chains can also be accomplished by using halogen-substituted heterocyclic rings [85]. The reaction is based on the nucleophilic substitution of halogen atoms by hydroxyl groups. Once activated, the polysaccharide can be functionalized with amine-containing compounds by displacement of a second halogen group from the ring. A common reagent that is used in triazine activation is cyanuric chloride.

9.2.3.7   Periodate Activation

Activated polysaccharides can be prepared through the oxidation of 1,2-diol groups with periodate ions [86,87]. The oxidation reaction leads to the formation of dialdehyde groups which can subsequently react with primary amine groups to form Schiff's bases [88,89]. Because of the poor stability of the Schiff's base, the imine link must be reduced with sodium borohydride or sodium cyanoborohydride to form a stable conjugate. The oxidizing efficiency of dialdehydes can be compromised to some extent due to the formation of intrahemiacetal and interhemiacetal structures [90–92].

9.2.3.8   Carbodiimide Activation

The activation of carboxyl and amino groups on polysaccharides has been achieved using water-soluble carbodiimides [85,93]. Carbodiimides react very well with carboxylate groups to form O-acyl isourea intermediates. This intermediate can then react with amino, hydroxyl, or thiol-containing compounds via nucleophilic substitution. Carbodiimide-activated polysaccharides are frequently used in coupling reactions because of their high reactivity toward different nucleophiles.

9.2.4   Crosslinking of Polysaccharides

Crosslinked polysaccharides may be prepared by the addition of difunctional or polyfunctional reagents. Reagents used for crosslinking follow the same general reaction mechanisms described above. A brief list of some common reagents used to crosslink starch is presented in Table 9.4.

9.2.5   Grafting of Polysaccharides

The grafting of synthetic polymers on natural polysaccharides is of considerable interest in biomedical, pharmaceutical, agricultural, and consumer-oriented fields. Graft copolymers are generally prepared by generating free radicals on the polysaccharide chain. The free radicals then initiate polymerization of vinylic or acrylic monomers. Free radicals can be generated either by chemical initiators or by irradiation. Some common chemical initiators include ceric salts [94], potassium permanganate [95,96], trivalent manganese [97], cupric ions [98], ammonium persulfate [99], potassium persulfate [100], and azobisisobutyronitrile [101]. Initiation by ionizing radiation can be achieved using ^{60}Co or electron beam irradiation. The effects of ionizing radiation on polysaccharides have been reviewed by Phillips [102]. With ionizing radiation, the polysaccharide chains can either be preirradiated prior to the addition of monomer or irradiated simultaneously in the presence of monomer. UV irradiation is another method to synthesize grafted polysaccharides [103,104]. Graft copolymers from both vinylic and acrylic monomers allow great versatility in altering the bulk properties of the polysaccharide chains. Some examples of monomers that have been grafted onto starch and other polysaccharides are acrylonitrile [101,105,106], acrylamide [107], acrylic acid [108], methacrylic acid [109], dimethylaminoethyl methacrylate [110], 2-hydroxy-3-methacryloyloxypropyltrimethylammonium chloride [111], methyl acrylate [99], and hydroxyethyl methacrylate [100].

9.3   Chemical Modification of Proteins

9.3.1   Chemical Introduction of Reactive Groups

Chemical modification of proteins has been used extensively in the areas of biochemistry, medicine, and pharmaceutics for the study of enzyme-catalyzed processes,

Table 9.4   Reagents Used to Crosslink Starch

Dichlorobutane

$Cl-CH_2CH_2CH_2CH_2-Cl$

Epichlorohydrin

$H_2C\overset{O}{\triangle}CH-CH_2-Cl$

Bisepoxides

$H_2C\overset{O}{\triangle}CH-R-HC\overset{O}{\triangle}CH_2$

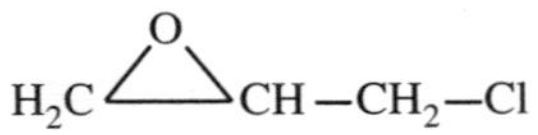

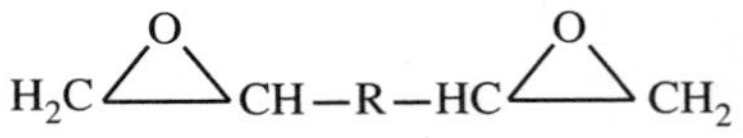

Formaldehyde

$\overset{O}{\underset{\parallel}{H-C-H}}$

Acetaldehyde

$\overset{O}{\underset{\parallel}{H_3C-C-H}}$

Divinylsulfone

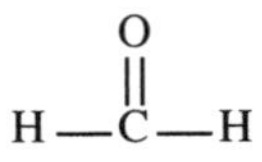

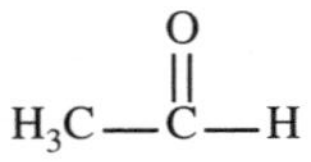

$H_2C=CHSO_2CH=CH_2$

N,N'-Methylenebisacrylamide

$H_2C=CH$ $HC=CH_2$
$O=C-NHCH_2NH-C=O$

1,3,5-Triacyl-S-triazine

Carbonyldiimidazole

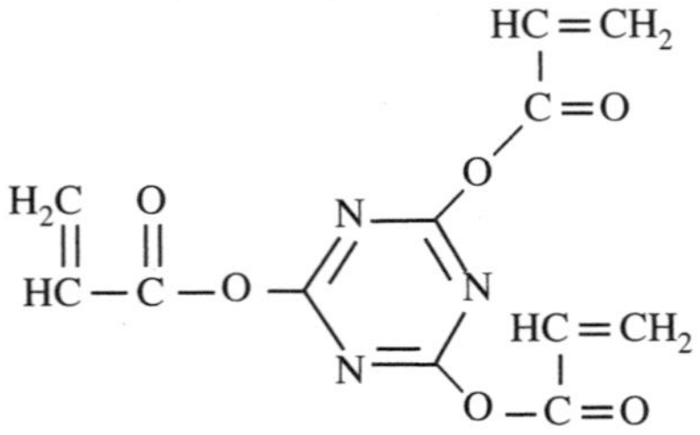

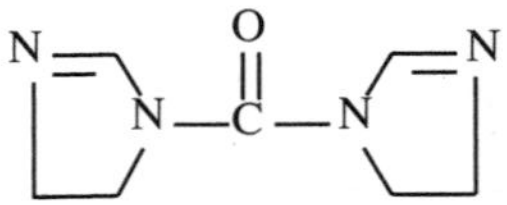

Bis-(hydroxymethyl)-urea

$\overset{O}{\underset{\parallel}{HOH_2CHN-C-NHCH_2OH}}$

Terephthaloyl chloride

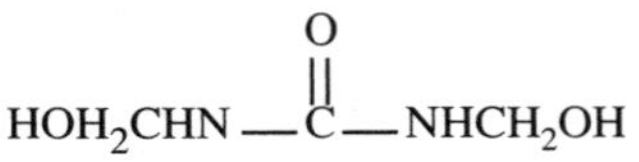

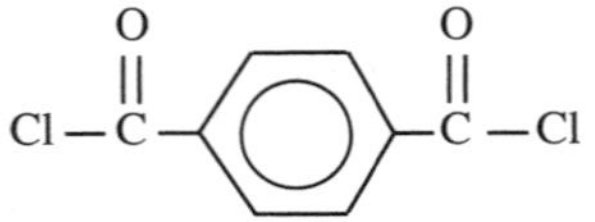

This information was obtained from ref. [13].

drug therapy, biocompatibility, and drug delivery. Because chemical modifications on proteins encompass an extraordinary amount of research, the readers are also referred to other reviews related to this topic [112–115].

Of the 20 amino acid residues, only amino, carboxylic, sulfhydryl, thioethyl, imidazolyl, guanidinyl, phenolic, and indolyl groups have been shown to be chemically reactive [116]. Thus, nine amino acid residues present along the protein backbone in addition to the N- and C-terminal amino acids are capable of chemical modification. Table 9.5 illustrates the many types of reactions that can be carried out on

Table 9.5   Types of Reactions Carried Out on Amino Acid Side Chains

Amino acid	Reactive side chain	Reactions	
Lysine	$-NH_2$	Alkylation Arylation Acylation Amidation	Diazotization
Arginine	$-NHCNH_2$ $\parallel$ NH	Amidation	
Aspartic acid and glutamic acid	$-COOH$	Acylation Esterification Amidation	
Cysteine	$-CH_2SH$	Alkylation Arylation Acylation Oxidation	Esterification Amidation Iodination
Methionine	$-SCH_3$	Alkylation Arylation Oxidation Amidation	
Tyrosine	—OH	Alkylation Arylation Amidation Oxidation	Nitration Esterification Diazotization Iodination
Histidine		Alkylation Arylation Acylation Oxidation	Iodination Diazotization
Tryptophan		Alkylation Arylation Oxidation Amidation	

This information was obtained from ref. [12].

amino acid side chains. Of the reactions listed in Table 9.5, alkylation and acylation reactions are the most common. These reactions occur by nucleophilic addition or nucleophilic substitution depending on the reagent. The reactivity of specific amino acid residues will depend on both the nucleophilicity of the side chains as well as on the microenvironment within which the reaction takes place [12]. The reactivity in the microenvironment can be affected by the neighboring side chains in the primary sequence and tertiary structure as well as by the solvent type. Neighboring side chains affect reactivity through hydrogen bonding, electrostatic interactions, and van der Waals forces. The magnitude of these interactions can vary depending on the solvent type. It has been reported that the dissociation of hydrogen atoms from the side chains becomes more difficult as the hydrophobicity of the solvent increases [117]. Furthermore, the presence of bulky side chains in the microenvironment can impart certain steric barriers that will limit a reagent's ability to react with a particular side chain. The effects of the microenvironment, however, can be reduced by modifying less reactive side chains with activating groups or spacer arms. The following section illustrates the types of side chain modifications that can be carried out on proteins.

9.3.2   Modification of Amino Groups

Reactions involving amino-containing side chains are widely used for modifying proteins. This is due to the nucleophilicity of amino groups under alkaline conditions and their overall abundance in proteins. The two main reactions with amino groups are alkylation and acylation. Table 9.6 lists the common reagents used to alkylate and acylate amino-containing side chains.

9.3.2.1   Alkylation

Alkylation reactions can be carried out using α-haloacetyl compounds [118], N-maleimide derivatives [119], aryl halides [120], and carbonyl compounds [121,122]. Reactions with haloacetates and haloacetamides are generally slow, occur only at high pH values, and can lead to dialkylation. In the reaction with N-maleimide derivatives, amino groups attack either the double bond or the carbonyl moiety under alkaline conditions. Aryl halides show good specificity for amino groups provided that the reaction is run at high pH. Specificity is achieved because the reaction products from other nucleophiles are unstable at high pH. One of the most widely used types of alkylating reagents are the carbonyl compounds. Carbonyl-based activation produces Schiff's bases. These reactions occur more readily under alkaline pH conditions. Due to the instability of the Schiff's base, however, reducing agents such as sodium borohydride or cyanoborohydride are required. Carbonyl compounds show good specificity for amino groups since reaction products from other nucleophilic groups such as sulfhydryl groups are unstable under alkaline conditions. Another approach to alkylating amino groups are epoxide-based reactions [123]. Epoxide reactivity is due

Table 9.6 Alkylation and Acylation Reactions with Amino Groups

Reagent	Chemical structure	Reaction product
α-Haloacetyl compounds	$X-CH_2\overset{\displaystyle O}{\overset{\|\|}{C}}-R$	(P)—NH—$CH_2\overset{\displaystyle O}{\overset{\|\|}{C}}$—R
N-Maleimide compounds		
Aryl halides		(P)—NH—
Carbonyl compounds	$-\overset{\displaystyle O}{\overset{\|\|}{C}}-$, reducing agent	(P)—NH—CH
Epoxide compounds	$H_2C\overset{\displaystyle O}{\diagup\!\!\diagdown}CH-R$	(P)—NH—CH_2—$\overset{\displaystyle OH}{\underset{\|}{CH}}$—R
Maleic anydride		(P)—NH—$\overset{\displaystyle O}{\overset{\|\|}{C}}$—CH=CH—$\overset{\displaystyle O}{\overset{\|\|}{C}}$—O⁻
Succinic anhydride		(P)—NH—$\overset{\displaystyle O}{\overset{\|\|}{C}}$—$CH_2$—$CH_2$—$\overset{\displaystyle O}{\overset{\|\|}{C}}$—O⁻
Isocyanate	R—N=C=O	(P)—NH—$\overset{\displaystyle O}{\overset{\|\|}{C}}$—NH—R
Isothiocyanate	R—N=C=S	(P)—NH—$\overset{\displaystyle S}{\overset{\|\|}{C}}$—NH—R
Acyl halides	$R-\overset{\displaystyle O}{\overset{\|\|}{C}}-X$	(P)—NH—$\overset{\displaystyle O}{\overset{\|\|}{C}}$—R

Continued

Table 9.6   Continued

Reagent	Chemical structure	Reaction product
Sulfonyl halides	R—S(=O)(=O)—Cl	(P)—NH—S(=O)(=O)—R
Imidoesters	R—C(=NH)—OR′	(P)—NH—C(=NH)—R
p-Nitrophenyl esters	R—C(=O)—O—C₆H₄—NO₂	(P)—NH—C(=O)—R
N-Hydroxysuccinimidyl esters	(succinimidyl)N—O—C(=O)—R	(P)—NH—C(=O)—R

to the strain on the three-membered ring which makes it energetically susceptible to cleavage during nucleophilic attack. Under mild or strongly alkaline conditions, reagents such as glycidyl acrylate and butene 2,3-oxide react with amino groups. Under strongly alkaline conditions, however, the epoxide moiety may also react with hydroxyl groups.

9.3.2.2   Acylation

Acylation reactions can occur with any nucleophilic side chains on the protein. Due to the effects from the microenvironment, steric factors, and reaction reversibility, however, acylation reactions are generally directed toward amino groups [12]. Maleic anhydride and succinic anhydride are commonly used under alkaline conditions to convert amino groups on lysine and N-terminal amino acids to carboxylic acids [124,125]. Isocyanates and isothiocyanates form unstable products with sulfhydryl, tyrosyl, and carboxyl groups, but display good stability with amino groups. Halide-containing reagents such as acyl halides and sulfonyl halides are also effective for amino-group-directed acylation. Under moderate alkaline conditions, imidoesters show the greatest specificity for amine groups. The reaction between amino groups and imidoesters leads to the formation of imidoamides. Other ester reagents that acylate amino group side chains are p-nitrophenyl esters and N-hydroxysuccinimidyl esters.

9.3.3   Modification of Carboxyl Groups

Carboxylic acid side chains and C-terminal amino acids can be modified using carbodiimides and diazoacetyl derivatives. Modification with carbodiimides is a two-step process that begins by activating the carboxylate ion to form an O-acylisourea intermediate. Once activated, nucleophilic attack by amino groups will displace the intermediate, resulting in amidation. Two water-soluble carbodiimide reagents used for amidation are 1-cyclohexyl-3-(2-morpholinyl-4-ethyl) carbodiimide and 1-ethyl-3-(3-dimethylaminopropyl) carbodiimide [93]. If amino-containing compounds are not added, carbodiimide activation will lead to crosslinking by condensing with amino-, hyroxyl-, or thiol-containing side chains of the protein. It should be noted, however, that carbodiimides also react very well with amino, sulfhydryl, and phenolic side chains. Thus, specificity for carboxyl groups may be compromised when using carbodiimides. Diazoacetyl derivatives are also used to esterify carboxylic acid side chains [126]. Two classes of diazoacetyl derivatives are diazoacetate esters and diazoacetamides. Under mildly acidic conditions, the diazo moiety acts as a good leaving group during nucleophilic attack by carboxylate ions.

9.3.4   Modification of Sulfhydryl Groups

The thiolate ion is the most reactive, nucleophilic functional group on a protein. For this reason, many of the reagents discussed thus far will also react with the thiolate ion. In general, reagents that are classified as sulfhydryl-directed react very rapidly with the thiolate ion as compared to other nucleophilic side chains. Sulfhydryl groups can be converted to carboxylic acid and amine-containing groups using α-haloacetates [127] and ethyleneimine [28] or 2-bromoethylamine [129], respectively. Under neutral or slightly acidic conditions, maleimides will also undergo nucleophilic attack by thiolate ions [130–132]. The resulting thioether bond has been shown to be quite stable under physiological conditions.

9.3.5   Crosslinking of Proteins

Many of the preceding reagents used for modifying amino acid side chains can be synthesized to form difunctional or polyfunctional compounds that can be used to chemically crosslink proteins. Bifunctional crosslinking agents are divided into three classes: homobifunctional, heterobifunctional, and zero length [12]. Examples of crosslinking agents are shown in Table 9.7. Homobifunctional crosslinking agents contain two identical functional groups that react with the same amino acid side chains. Heterobifunctional agents contain two dissimilar functional groups, and hence react with different amino acid side chains. Zero length crosslinking agents, however, link protein side chains without the addition of extrinsic compounds. Many zero-length crosslinking agents condense carboxyl groups with primary amino, hydroxyl,

Table 9.7   Reagents Used to Crosslink and Conjugate Proteins

Reagent	Chemical structures	References
Homobifunctional		
Bisimidoesters	$\overset{\overset{+}{N}H_2Cl^-}{\underset{\parallel}{}} \; \overset{\overset{+}{N}H_2Cl^-}{\underset{\parallel}{}}$ RO—C—R—C—OR	[133–139]
bis-*N*-Succinimides		[140–143]
Diisocyanates	O=C=N—R—N=C=O	[144–146]
Diisothiocyanates	S=C=N—R—N=C=S	[144,146]
Bisnitrophenylesters		[147,148]
Dialdehydes	$\underset{H-C-R-C-H}{\overset{O\quad\;\;\; O}{\parallel \quad\;\;\; \parallel}}$	[149–155]
Formaldehyde	$\underset{H-C-H}{\overset{O}{\parallel}}$	[151,156]
Diketones	$\underset{H_3C-C-R-C-CH_3}{\overset{O\quad\;\;\; O}{\parallel \quad\;\;\; \parallel}}$	[157–159]
Bisepoxides		[160]
Acydichlorides	$\underset{Cl-C-R-C-Cl}{\overset{O\quad\;\;\; O}{\parallel \quad\;\;\; \parallel}}$	[161,162]
Heterobifunctional		
Succinimides		[163–166]
Halosuccinimides		[167–169]

Continued

Table 9.7 Continued

Reagent	Chemical structure	References
Haloacetimidates	$RO-C(=\overset{+}{N}H_2Cl^-)-R-X$	[170,171]
Maleimidobenzoyl chloride	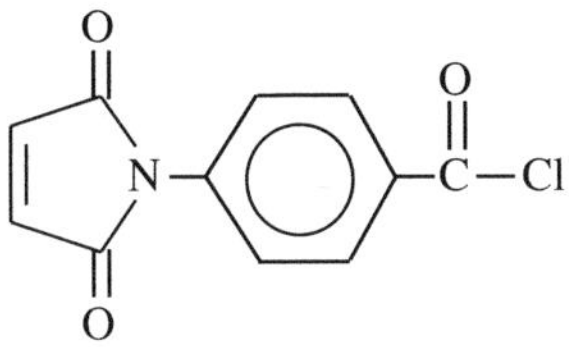	[172]
Epichlorohydrin	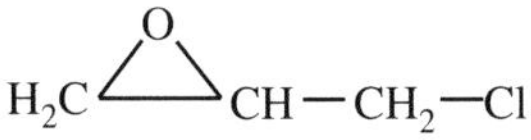	[173]
Carbodiimides	$R-N=C=N-R$	[93,174–179]
Chloroformates	$Cl-C(=O)-OR$	[180–182]
Carbonyldiimidazole	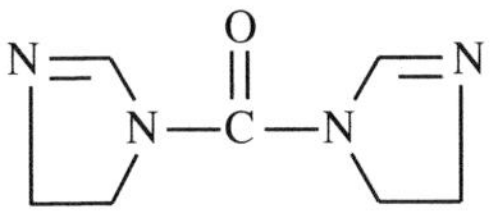	[183,184]
Others		
Glycerol polyglycidyl ether	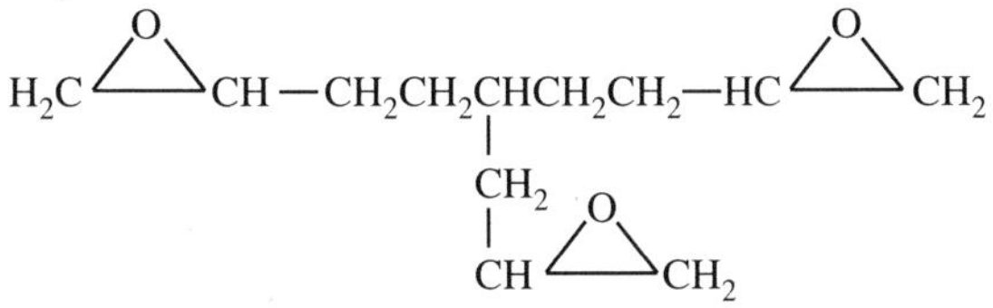	[160]
Polyglycerylpolyglycidyl ether	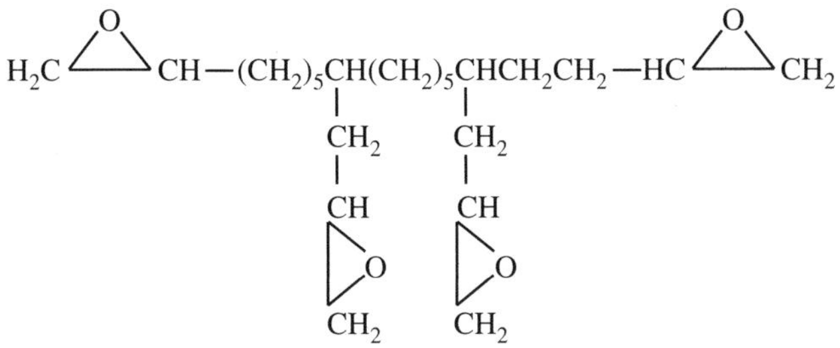	[160,185]
Polyglycerolpolyglycidyl ether	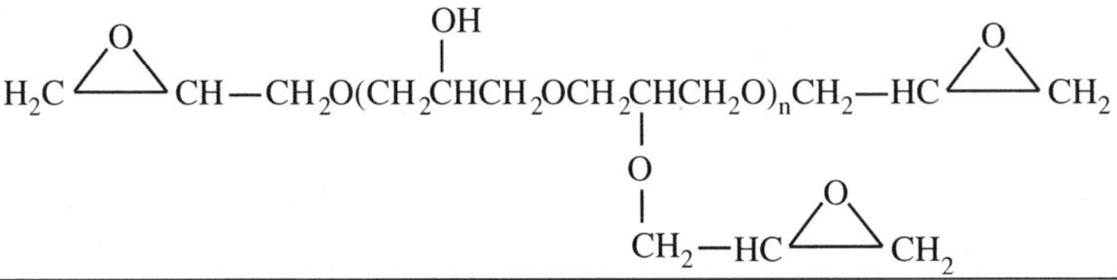	[186,187]

and thiol groups to form amide, ester, and thioester bonds, respectively. Zero-length crosslinking agents react by first activating carboxyl side chains followed by nucleophilic substitution by a neighboring side chain. Recently, polymeric crosslinking agents have been used in the preparation of valvular and vascular bioprostheses. Table 9.7 also gives a listing of polymeric reagents used for crosslinking proteins. A more extensive discussion on protein crosslinking and conjugation has been presented by Wong [12].

9.4   Enzyme-Catalyzed Degradation of Polysaccharides and Proteins

9.4.1   Properties of Enzymes

Enzymes are polypeptide macromolecules that catalyze biochemical events in living organisms. Their primary roles are in the molecular recognition and the acceleration of reaction rates. All enzymes increase reaction rates because the active site is complementary to the structure of the reaction's transition state [188]. The active site of the enzyme is made up of catalytic and noncatalytic amino acid residues. The catalytic groups induce changes in electron density on the substrate and such changes promote the formation or cleavage of chemical bonds. Enzymes are also capable of catalyzing reactions that involve different substrates providing that the electron density changes resemble that of the native substrate [189]. The noncatalytic groups in the active site lie in close proximity to the site of bond cleavage or formation. The main function of the noncatalytic groups is to interact with the substrate and lower the activation energy of the reaction [190,191].

In humans, enzyme-catalyzed hydrolysis of proteins and polysaccharides occurs throughout the body. For example, hydrolysis reactions are present in the plasma and interstitium, in the brush border membrane and lumen of the gastrointestinal tract, and in the tubular epithelium of the kidneys [192]. Other sites include tissues that comprise the reticuloendothelial system. These are the histiocytes in the skin and subcutaneous tissue; the Kupffer's cells in the liver; and tissue macrophages in the lymph nodes, alveoli, spleen, and bone marrow [193]. In many of the sites listed above, enzyme-catalyzed reactions occur intracellularly. In this case, intracellular transport of substrates occurs by specific or nonspecific receptor-mediated endocytosis.

Enzymes are divided into six classes: oxidoreductases, transferases, hydrolases, lyases, isomerases, and ligases [194]. Hydrolases are of particular relevance to proteins and polysaccharides because they catalyze the hydrolysis of C–O, C–N, and C–C bonds. Hydrolases that act on proteins are divided into peptidases (exopeptidases) and proteinases (endopeptidases) [194]. Aminopeptidases and carboxypeptidases catalyze the hydrolysis of the N-terminal and C-terminal ends of polypeptide chains, respectively. Peptidases require certain amino acid residues or transition metals to be present at the active site. As a result, there are different types of peptidases. For example, carboxypeptidases are subdivided into serine carboxypeptidases, cysteine carboxypeptidases, and metallocarboxypeptidases. Proteinases catalyze the hydrolysis of peptide bonds within the polypeptide chain. Like peptidases, proteinases are subdivided further into serine proteinases, cysteine proteinases, aspartic proteinases, and metalloproteinases depending on the requirements of the active site.

Enzymes that catalyze hydrolysis of polysaccharides are called glycosidases. There are three types of glycosidases that hydrolyze either O-glycosyl, N-glycosyl, or S-glycosyl bonds in a polysaccharide chain. The largest class of glycosidases are the O-glycosidases. Enzyme-catalyzed hydrolysis of O-glycosyl bonds is carried out by nucleophilic substitution at the anomeric carbon [195]. The reaction results in either net retention of configuration or inversion of configuration at the anomeric carbon [196]. Thus, glycosidases are classified by the overall stereochemistry of the reaction. O-Glycosidases that act on pyranosides are divided into four classes: (1) those that produce retention of equatorial configuration, (2) those that produce retention of axial configuration, (3) those that produce equatorial to axial inversion of configuration, and (4) those that produce axial to equatorial inversion of configuration. O-Glycosidases that act on furanosides are classified by the retention or inversion of configuration following enzyme catalysis. The properties of some protein and polysaccharide hydrolases are presented in Table 9.8. The information in Table 9.8 was gathered from the recommendations of the nomenclature committee of the International Union of Biochemistry regarding the nomenclature and classification of enzyme-catalyzed reactions [194].

As mentioned above, the main function of the noncatalytic groups at the active site of an enzyme is to lower the activation energy of the reaction. Their contribution to enzyme-catalyzed hydrolysis has been studied extensively [197–201]. Schechter and Berger [197] divided the active site of an enzyme into subsites (S) in which each subsite can accommodate one amino acid residue of the substrate (P) (Figure 9.1). Subsites are designated by their relative position from the cleavage site (Figure 9.1). The rate of enzyme-catalyzed hydrolysis, therefore, is dependent on the favorability of the subsite–substrate (S–P) interactions. For example, collagenase from *Clostridium histolyticum* contains six subsites in its active site [200,201]. Reactivity is determined by the S_3–P_3, S_2–P_2, S_1–P_1, S_1'–P_1', S_2'–P_2', and S_3'–P_3' interactions. It has been shown that the best substrates for collagenase have glycine at P_3 and P_1'; proline or alanine at P_2 and P_2'; hydroxyproline, arginine, or alanine at P_3'; and a large hydrophobic residue at P_1. Thus, any chemical modification near the substrate's cleavage site may alter the rate of enzyme-catalyzed hydrolysis depending on the changes it caused in subsite–substrate interactions. The importance of subsite–substrate interactions in controlling enzyme-catalyzed hydrolysis of modified proteins and polysaccharides is discussed below.

9.4.2   Factors Affecting Enzyme-Catalyzed Reactions

9.4.2.1   Conformation of Enzymes and Substrates

Changes in the conformation of enzymes or substrates will have a dramatic effect on the rate of enzyme catalysis. Conformational changes on the catalytic and noncatalytic

Table 9.8   Enzymes That Hydrolyze Proteins and Polysaccharides

Enzyme	Type	Other names	Preferential cleavage sites
Protein hydrolases			
Alanine carboxypeptidase	Metallocarboxypeptidase	—	Peptidyl-L-alanine
Arginine carboxypeptidase	Metallocarboxypeptidase	Carboxypeptidase N	Peptidyl-L-arginine
Aspartate carboxypeptidase	Metallocarboxypeptidase	—	Peptidyl-L-aspartate
Bromelain	Cysteine proteinase	—	Lys-, Ala-, Tyr, Gly-
Carboxypeptidase A	Metallocarboxypeptidase	Carboxypolypeptidase	Peptidyl-L-amino acid (aa)
Carboxypeptidase B	Metallocarboxypeptidase	Protaminase	Peptidyl-L-lysine (-L-arginine)
Cathepsin B	Cysteine proteinase	Cathepsin B$_1$	Arg-, Lys-, Phe-X-aa (carbonyl side of amino acid residue next to Phe)
Cathepsin D	Aspartic proteinase	—	Phe-, Leu-
Cathepsin G	Serine proteinase	—	Tyr-, Trp-, Phe-, Leu-
Chymosin	Aspartic proteinase	Rennin	Cleaves a single bond in casein κ
Chymotrypsin	Serine proteinase	Chymotrypsins A and B	Tyr-, Trp-, Phe-, Leu-
Coagulation factor Xa	Serine proteinase	Thrombokinase, prothrombase, prothrombinase	Arg-Ile, Arg-Gly; activates prothrombin to thrombin
Ficin	Cysteine proteinase	—	Lys-, Ala-, Tyr-, Gly-, Asn-, Leu-, Val-
Glycine carboxypeptidase	Metallocarboxypeptidase	Yeast carboxypeptidase	Peptidyl-L-glycine
Lysosomal carboxypeptidase B	Cysteine carboxypeptidase	Cathepsin B$_2$, cathepsin IV	Peptidyl-L-amino acid
Papain	Cysteine proteinase	Papaya peptidase 1	Arg-, Lys-, Phe-X-aa (carbonyl side of amino acid residue next to Phe)
Pepsin A	Aspartic proteinase	Pepsin	Phe-, Leu-
Plasmin	Serine proteinase	Fibrinase, fibrinolysin	Arg-, Lys-
Plasminogen activator	Serine proteinase	Urokinase	Arg-Val in plasminogen
Proline carboxypeptidase	Serine carboxypeptidase	Angiotensinase C, lysosomal carboxypeptidase C	Peptidylprolyl-L-amino acid
Serine carboxypeptidase	Serine carboxypeptidase		Peptidyl-L-amino acid

Enzyme	Class	Other names	Specificity
Thrombin	Serine proteinase	Fibrinogenase	Arg-; activates fibrinogen to fibrin
Trypsin	Serine proteinase	α- and β-trypsin	Arg-, Lys-
Tyrosine carboxypeptidase	Serine carboxypeptidase	Thyroid peptide carboxypeptidase	Peptidyl-L-tyrosine
Vertebrate collagenase	Metalloproteinase	—	Cleaves a single bond in native collagen leaving an N-terminal (75%) and a C-terminal (25%) fragment
Glycosidases			
Agarase	O-Glycosidase	—	Hydrolysis of 1,3-β-D-galactosidic linkages in agarose
α-Amylase	O-Glycosidase	Glycogenase	Endohydrolysis of 1,4-α-D-glucosidic linkages containing three or more 1,4-α-linked D-glucose units
β-Amylase	O-Glycosidase	Saccharogen amylase, glycogenase	Hydrolysis of 1,4-α-D-glucosidic linkages from nonreducing ends of chains
κ-Carrageenase	O-Glycosidase	—	Hydrolysis of 1,4-β-D-linkages between D-galactose-4-sulfate and 3,6-anhydro-D-galactose in carrageenans
Cellulase	O-Glycosidase	Endo-1,4-β-glucanase	Endohydrolysis of 1,4-β-D-glucosidic linkages in cellulose
Chitinase	O-Glycosidase	Chitodextrinase, 1,4-β-poly-N-acetylglucosaminidase, poly-β-glucosaminidase	Random hydrolysis of N-acetyl-β-D-glucosaminide 1,4-β-linkages in chitin and chitodextrins
Dextranase	O-Glycosidase	—	Endohydrolysis of 1,6-α-D-glucosidic linkages in dextran
α-Glucosidase	O-Glycosidase	Maltase, glucoinvertase, glucosidosucrase, maltase-glucoamylase	Hydrolysis of terminal, nonreducing 1,4-linked α-D-glucose residues

Continued

Table 9.8 Continued

	Type	Other names	Preferential cleavage sites
β-Glucosidase	O-Glycosidase	Gentiobiase, cellobiase, amygdalase	Hydrolysis of terminal, nonreducing β-D-glucose residues
Hyaluronoglucosaminidase	O-Glycosidase	Hyaluronidase	Random hydrolysis of 1,4-linkages between N-acetyl-β-D-glucosamine and D-glucuronate residues in hyaluronate
Hyaluronoglucuronidase	O-Glycosidase	Hyaluronidase	Random hydrolysis of 1,3-linkages between β-D-glucuronate and N-acetyl-D-glucosamine residues in hyaluronate
Inulinase	O-Glycosidase	Inulase	Endohydrolysis of 2,1-β-D-fructosidic linkages in inulin
Isoamylase	O-Glycosidase	Debranching enzyme	Hydrolysis of 1,6-α-D-glucosidic branch linkages in glycogen, amylopectin, and their β-limit dextrins
Lysozyme	O-Glycosidase	Muramidase	Hydrolysis of 1,4-β-linkages between N-acetylmuramic acid and N-acetyl-D-glucosamine residues in peptidoglycan and between N-acetyl-D-glucosamine residues in chitodextrin

This information was obtained from ref. [194].

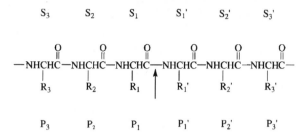

Figure 9.1   Subsite–substrate interactions between a hydrolase and a polypeptide. The arrow denotes the site of cleavage.

residues in the active site will impair substrate recognition and binding by the enzyme. Changes in enzyme conformation, and thus enzyme activity, can arise as a function of temperature, pH, ionic strength, and solvent type [202]. These factors will either impair or enhance the formation of enzyme–substrate complexes. For example, the activities of proteolytic enzymes found in the gastrointestinal tract are pH dependent. Trypsin and chymotrypsin show the highest activity in the small intestine whereas the activity of pepsin is highest in the stomach. The conformational and configurational properties of the substrate will also affect the rate of enzyme-catalyzed reactions. In some instances, the inherent three-dimensional configuration of the substrate may sterically block close encounters by enzymes [203–205]. It has been reported that denaturation of proteins by changes in temperature, pH, ionic strength, or organic solvents will make the protein more susceptible to enzyme-catalyzed hydrolysis [202].

Quite often, the kinetics of enzyme-catalyzed hydrolysis can be altered by chemically modifying the substrate. The effect of chemical modification on the formation of enzyme–substrate complexes has been studied extensively with gastrointestinal, lysosomal, and blood proteases [202,204,206–208]. Rejmanova, Duncan, Ringsdorf, Rihova, and others have developed site-directed lysosomotropic drug delivery systems that are composed of synthetic polymers containing drug-bound oligopeptide side chains [209–214]. It was reported that the amino acid sequence of the oligopeptide side chains could be manipulated to produce a structure that was resistant to hydrolysis by blood proteases yet susceptible to lysosomal proteases. Because the drug was coupled to the oligopeptide side chains, drug release was dependent on the enzyme-catalyzed hydrolysis of the side chain. It was reported that drug release rates increased when the structure of the side chain favored subsite–substrate interactions and decreased otherwise. The effects of chemical modification on human serum albumin have been studied in our laboratory. In this work, human serum albumin molecules were modified with glycidyl acrylate [215,216]. It was shown by sodium dodecyl sulfate–polyacrylamide gel electrophoresis (SDS-PAGE) that the rate of pepsin-catalyzed hydrolysis was slower with the modified albumin compared to the native albumin. After degradation by pepsin (Figure 9.2), the molecular weights from the hydrolyzed fragments of the modified albumin were larger than those from the native albumin. Increased resistance to hydrolysis was believed to result from steric hindrances produced by

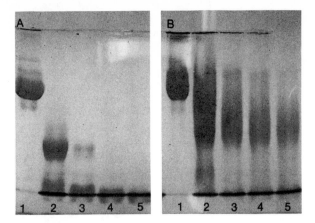

Figure 9.2   Gel electrophoresis of native albumin (A) and modified albumin (B). Column 1 represents the pepsin-free samples. Columns 2–5 represent pepsin-catalyzed hydrolysis of native albumin and modified albumin over 15 min (column 2), 30 min (column 3), 60 min (column 4), and 120 min (column 5) incubation periods. (From ref. 216, with permission.)

the vinylic pendant groups on albumin. The presence of such pendant groups may have impaired subsite–substrate interactions at the active site of pepsin. The notion of steric hindrances at the active site may be supported by the fact that in the primary structure of human serum albumin, more than 30% of the reactive lysine residues are within three amino acid residues from the major or minor sites of cleavage by pepsin [217].

9.4.2.2   Presence of Polymer Molecules

Polymers can have a significant effect on the rate of enzyme-catalyzed hydrolysis. The presence of polymer molecules in the reaction media gives rise to mutual exclusion effects, sieving effects, and steric constraint effects [218]. Mutual exclusion is based on the observation that the chemical potential of an enzyme is higher in the presence of neutral, linear polymers than in the absence of such polymers [219–221]. The magnitude of mutual exclusion is directly related to the activity coefficient of the enzyme. In the presence of polymer chains, the enzyme's activity coefficient, and thus its degree of mutual exclusion, is inversely related to the partitioning of the enzyme into the polymer medium as described by the following equation:

$$\gamma = 1/K_{av} \tag{9.1}$$

where γ represents the activity coefficient and K_{av} is the partitioning coefficient derived by Laurent [221]. Enzyme activity will increase as the enzyme partitioning into the polymer medium decreases. The exclusion effect from dextran chains increased the rate of lactate dehydrogenase-catalyzed reactions at low substrate concentrations

[218]. In addition to mutual exclusion, polymer chains will also lower the diffusion coefficients of both enzyme and substrate. As a result, the enzyme–substrate collision frequency is lowered and the overall reaction rate is reduced. This phenomenon is known as the sieving effect.

Another significant effect that polymers have on enzyme-catalyzed reactions is steric constraints. Both linear and crosslinked polymers will restrict the position, orientation, and conformation of enzymes near cleavage sites on substrates. Under such conditions, shifts in the reaction rate relative to the bulk phase will occur due to restrictions that the polymer chains impart on the formation of enzyme–substrate complexes [222,223]. Steric constraints on enzyme activity become prominent if the substrate is interpenetrated by or chemically bound to polymer chains. Polymers based on hydroxypropylmethacrylamide were synthesized with oligopeptide side chains [224,225] or oligopeptide crosslinks [226–228]. The degradability of the side chains or crosslinks by chymotrypsin increased as the length of the oligopeptide increased. A similar relationship was also observed in the presence of lysosomal enzymes [229]. The authors concluded that steric constraints from polymer chains increased as the site of cleavage moved closer to the polymer backbone. In our laboratory, albumin-crosslinked polyvinylpyrrolidone hydrogels were prepared for long-term oral drug delivery [215,216]. The degradation of hydrogels was studied in the presence of pepsin, chymotrypsin, and trypsin. It was found that as the degree of albumin modification increased, the crosslinking density of the network increased, and the rate of gel degradation decreased. It was postulated that the degree of albumin incorporation within the network controlled the size of oligopeptide sequences that exist between polymer chains. As the degree of incorporation increased, the size of oligopeptide sequences decreased, and the steric constraints from polymer chains became more prominent. As a result, the hydrogel was more resistant to enzyme-catalyzed hydrolysis.

9.5   Enzyme-Catalyzed Degradation of Modified Polysaccharides

9.5.1   Dextran

The biodegradable properties associated with modified dextrans as drug carriers were studied by Schact et al. [80,230] and Vercauteren et al. [231]. Dextran was modified by three methods: (1) periodate-activated oxidation followed by reduction with sodium borohydride (reduced aldehyde dextrans); (2) acylation with succinic anhydride (dextran monosuccinate derivative); and (3) acylation with 4-nitrophenyl chloroformate followed by subsequent reactions with amines (dextran carbamate derivatives). The objective of the work was to show how the extent of chemical modification affected dextranase-catalyzed degradation. Samples were incubated with either penicillum dextranase or lysosomal enzymes isolated from the rat liver. Dextranase-catalyzed degradation was monitored by gel permeation chromatography. In all samples, as the degree of modification increased, the size of the hydrolyzed fragments increased, that is, the rate of degradation decreased. The authors contended that the increased degree of modification impaired subsite–substrate interactions. It was noted, however, that the type of chemical modification on dextran had a minor effect on the rate of degradation. Similar relations between the extent of chemical modification and the rate of enzymatic degradation were observed by Chaves et al. [232]. In their studies, dextran was modified with ethyl and butyl chloroformate. Such modification led to the formation of cyclic carbonates and interchain carbonates (crosslinked dextran). The crosslinked, water-insoluble dextran was resistant to hydrolysis by dextranase. The hydrolysis rates of the water-soluble ethyl and butyl carbonate derivatives, however, were dependent on the degree of substitution and the concentration of dextranase. It was found that the specific viscosity of modified dextran decreased more rapidly over time as the degree of dextran substitution decreased or as the dextranase concentration increased. The more rapid decline in viscosity with lower dextran substitution indicated that the rate of enzyme-catalyzed hydrolysis increased.

The importance of chemical structure on subsite–substrate interactions between dextranase and modified dextran was studied by Crepon et al. [233]. Dextran modification was carried out by carboxymethylation followed by benzylamine coupling to a fraction of the carboxyl groups to form carboxymethylbenzylamine dextran. Additional modification was carried out by sulfonating a fraction of the benzylamine groups to form carboxymethylbenzylamine sulfonated dextran (Figure 9.3). The degree of each type of substitution was determined by potentiometric titration and elemental analysis. The extent of dextranase-catalyzed degradation was determined by the changes in molecular weight as measured by high-performance steric exclusion chromatography. The extent of degradation by dextranase after 1 h of reaction decreased as the degree of carboxylic functionality increased from 30% to 80%. Dextran derivatives were completely resistant to degradation when the carboxylic functionality was 54%

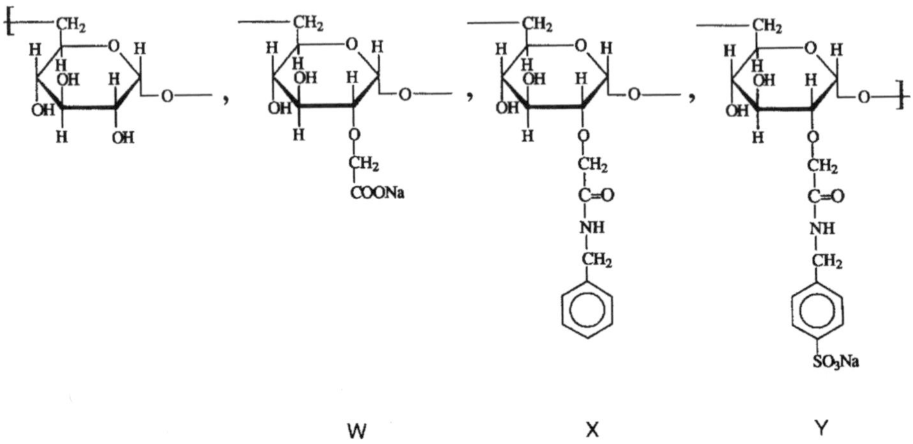

Figure 9.3 Structure of dextran derivatives. Glucosyl units were substituted with carboxyl (W), benzylamine (X), and benzylamine sulfonate (Y) groups. (From ref. 233.)

and the benzylamine-sulfonate functionality was 19.5%. When the overall degree of modification was nearly equivalent, the benzylamine and benzylamine-sulfonate groups had a stronger inhibitory effect on dextran degradation than the carboxylic group. Furthermore, it was also observed that the benzylamine group had a slightly stronger inhibitory effect than the benzylamine-sulfonate group. The results suggest that the bulky size of the benzylamine and benzylamine-sulfonate side chains was most effective in hindering the formation of the enzyme–substrate complexes.

Edman and Sjöholm have studied the use of modified dextran microspheres as lysosome-directed drug delivery systems [234–236]. The objective of their work was to deliver agents to cells of the reticuloendothelial system using microspheres that were susceptible to degradation by lysosomal enzymes. To prepare the microspheres, samples of dextran were first modified with glycidyl acrylate. The content of acrylic pendant groups was determined by halogenation.[14]C-labeled microspheres were prepared by emulsion polymerization where the degree of crosslinking was increased by using dextran derivatives with higher degrees of modification. After intravenous injection to rats, it was found that the microspheres were phagocytized predominantly by the liver, spleen, and bone marrow. The half-life of the microspheres in the liver and spleen was directly related to degree of dextran modification. It was shown that particles containing a smaller number of acrylic groups were eliminated more rapidly. In vitro degradation studies with dextranase provided proof that as the degree of dextran modification decreased, the rate of microsphere degradation was faster. The data illustrate how chemical modification and crosslinking alter substrate conformation and affect an enzyme's ability to form enzyme–substrate complexes.

9.5.2   Starch

Heller et al. developed enzyme-degradable starch hydrogels as a part of a "triggered" drug delivery system [237]. Soluble starch was functionalized with glycidyl methacrylate under alkaline conditions. The degree of modification was controlled by altering the amount of glycidyl methacrylate or the reaction time. As the degree of modification increased, the prepared hydrogels exhibited increased firmness, indicating higher crosslinking density. All hydrogels tested were susceptible to hydrolysis by α-amylase as measured by gel swelling and dissolution. The rate of hydrogel degradation decreased as the degree of starch modification increased. It was surprising, however, to see that the rate of α-amylase-catalyzed hydrolysis of the modified starch increased as the degree of modification increased. Therefore, even though the hydrolysis rate of modified starch increased with increasing chemical modification, the corresponding hydrogels were more resistant to degradation as the degree of chemical modification increased. The results suggest that the formation of a three-dimensional network imparts significantly larger steric constraints on the enzyme and such constraints increase with higher degrees of crosslinking density (i.e., increased starch modification).

Artursson et al. [238] and Laakso et al. [239,240] studied the degradation properties of modified starch microspheres prepared as lysosomotropic drug delivery systems.

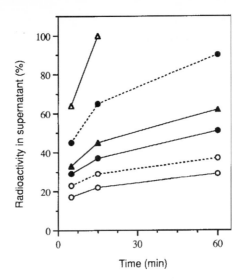

Figure 9.4   Degradation of ^{14}C-labeled starch microspheres prepared using different amounts of TEMED in mouse serum. Microspheres with a high (—) and low (- - -) degree of starch modification were prepared using 0.1 ml (○), 0.5 ml (●), 1.0 ml (▲), or 2.0 ml (Δ) of TEMED. (From ref. 240, with permission.)

Soluble starch was modified by alkylation using glycidyl acrylate or by acylation using acryloyl chloride. Through [1]H-nuclear magnetic resonance (NMR) it was found that the degree of modification increased by increasing the reaction time with the glycidyl acrylate or by increasing the amount of the added acryloyl chloride. Microspheres were prepared by free radical emulsion polymerization. As the degree of starch modification increased, the crosslinking density of the microsphere increased. The degradability of the microsphere by mouse serum, α-amylase, amyloglucosidase, or lysosomal enzymes decreased as the degree of modification increased. When the modified starch was copolymerized with N,N′–methylene–bis-acrylamide, the degradability of microspheres increased as compared to modified starch controls. It was observed, however, that the pore size in microspheres increased due to copolymerization. This resulted in increased susceptibility to degradation due to enhanced penetration of enzymes into the microsphere. The degradable properties of the starch microspheres were also altered by varying the content of the coinitiator, tetramethylethylenediamine (TEMED), in the monomer solution. The rate of microsphere degradation in mouse serum was dependent on the content of TEMED. Figure 9.4 shows that the release rate of ^{14}C-labeled starch increased as the amount of the added TEMED increased from 0.1 ml to 2.0 ml in the monomer solution. It was postulated that the rise in TEMED content increased the number of propagating units during polymerization. This produced hydrocarbon units with a lower molecular weight between crosslinks and a network with higher crosslinking density. Microsphere degradability was thought to be enhanced because the lower molecular weight hydrocarbon units were more soluble once hydrolyzed. Altering TEMED content in the monomer solution had a similar effect when acrylamide was copolymerized with the glycidyl acrylate-modified starch [241]. The molecular weight of the hydrolyzed samples of the copolymer decreased as the TEMED content increased. This lends support to the hypothesis that high levels of TEMED in the monomer solution lower the molecular weight between crosslinks. In summary, the results of these studies illustrate how structural changes in the three-dimensional network can affect the rate of degradation. The delivery of high [242] and low molecular weight [72,73] agents to tissues of the reticuloendothelial system using the glycidyl acrylate-modified starch microspheres has been shown to be quite promising.

9.5.3   Amylose

The enzyme-degradable properties associated with epichlorohydrin-crosslinked amylose was studied by Mateescu et al. [243]. The rate of enzyme-catalyzed hydrolysis was studied by an iodometric method. The iodine formed inclusion complexes with the hydrolyzed amylose chains that were released from the network. Thus, the rate of hydrolysis was determined by monitoring the change in color of the supernatant. It was found that the rate of hydrolysis by α-amylase was inversely related to the crosslinking density of the amylose. The authors postulated that the increase in crosslinking density limited penetration of enzymes into the network, and thus restricted the enzymes to the surface of the network. As the crosslinking density decreased, enzyme penetration was enhanced and more substrates were available for hydrolysis.

9.5.4   Chitosan

The effect of N-acylation on the hydrolysis of chitosan by lysozyme and chitinase was studied by Hirano et al. [244]. Chitosan was N-acylated with a mixture of acetic anhydride and D-glucosamine. The carbon/nitrogen ratio obtained from elemental analysis was used to determine the degree of substitution. The extent of N-acylation was controlled by varying the content of acetic anhydride in the reaction mixture. The hydrolytic activity of lysozyme was dependent on the degree of chitosan modification. There was little or no hydrolysis of the modified chitosan if the degree of substitution was below 0.2. The extent of hydrolysis, however, increased as the degree of substitution increased from 0.4 to 0.8. The maximum rate of hydrolysis was observed when the degree of substitution was 0.8. This was nearly four times higher than the rate of hydrolysis observed when the degree of substitution was 1.0. The same type of relationship was observed with chitinase-catalyzed hydrolysis. The maximum rate of hydrolysis was observed, however, when the degree of substitution ranged from 0.4 to 0.8. The data demonstrate how chemical modification can affect the rate of enzyme-catalyzed hydrolysis by restricting or enhancing subsite–substrate interactions. A similar effect on subsite–substrate interactions between lysozyme and partially deacetylated chitin has also been reported [245].

9.5.5   Chondroitin-6-Sulfate

Chondroitin-6-sulfate is a mucopolysaccharide found in animal connective tissue, especially cartilage, and consists of D-glucuronic acid linked to *N*-acetyl-D-galactosamine which is sulfated at C-6. Modified chondroitin-6-sulfate has been suggested as an enzyme-degradable drug delivery system [246,247]. It can be degraded by anaerobic bacteria found in the large intestine of man. Thus, Rubinstein et al. [246] prepared crosslinked chondroitin-6-sulfate as a suitable matrix for indomethacin release in the colon. Chondroitin-6-sulfate was crosslinked with diaminododecane via dicyclohexylcarbodiimide activation. The extent of crosslinking was increased by increasing the quantity of diaminododecane. The release of indomethacin from crosslinked chondroitin-6-sulfate tablets in the presence of rat cecal contents was degradation dependent. As the degree of crosslinking increased, the amount of indomethacin released within 28 h decreased. The results suggest that increasing the crosslinking density restricts the formation of enzyme–substrate complexes and slows the rate of indomethacin release. Although it appears that enzyme-catalyzed hydrolysis may correlate with indomethacin release, additional studies are needed to verify the presence of lumenal enzymes that are capable of degrading chondroitin-6-sulfate.

9.6   Enzyme-Catalyzed Degradation of Modified Proteins

Chemical modification of proteins can lead to conformational changes in the tertiary structure. Such changes can alter a protein's susceptibility to enzyme-catalyzed degradation by either enhancing or impairing subsite–substrate interactions. For example, glycoproteins are known to be resistant to trypsin-catalyzed hydrolysis. However, periodate-oxidized or acid-hydrolyzed glycoproteins were shown to be highly susceptible to degradation by trypsin. Increased hydrolysis by trypsin was due to conformational changes that arose from the destruction or loss of sialic acid from the glycoprotein [248]. Trypsin-catalyzed degradation of bovine serum albumin was enhanced following methylation with thionyl chloride and methanol as compared to native albumin [249]. Methylation produced changes in the albumin's optical rotation, viscosity, and solubility as compared to the native protein. Increased hydrolysis by trypsin was due to protein denaturation which made the protein more accessible by trypsin. Increased hydrolysis of lysozyme by trypsin was reported when lysozyme was modified by tetranitromethane and nitrophenylsulfonyl chloride [250,251]. Modification at one region of the molecule by tetranitromethane induced conformational changes at a distal region that exposed bonds accessible by trypsin. These examples clearly show that conformational changes on proteins due to chemical modification can significantly affect the rate of enzyme-catalyzed degradation. The enzyme-catalyzed degradation of some modified proteins is described below.

9.6.1   Albumin

9.6.1.1   Albumin-Crosslinked Hydrogels

Enzyme-degradable hydrogels have been prepared as long-term oral drug delivery systems that could be selectively retained in the stomach [215,216,252–257]. The hydrogels were designed to be susceptible to hydrolysis by proteolytic enzymes found in the gastrointestinal tract. The degradable component of the network was the crosslinking agent, functionalized albumin. Functionalized albumin was prepared by reacting human serum albumin with glycidyl acrylate. The degree of modification was determined by measuring the free amine groups of albumin using 2,4,6-trinitrobenzenesulfonic acid [258]. In these studies, 100% modification indicated that all of the reactive amine groups on albumin were alkylated. It was found that the degree of modification was easily controlled by varying the reaction time. Albumin-crosslinked hydrogels were prepared from the copolymerization of functionalized albumin with acrylic acid, acrylamide, or vinylpyrrolidone. The swelling, mechanical, and degradation properties of the hydrogels were controlled by varying the concentration of functionalized albumin in the monomer solution, the degree of modification on albumin, or the concentration of chemical initiator in the monomer solution. Both the extent of hydrogel swelling in pepsin-free simulated gastric fluid and the rate of hydrogel degradation by pepsin were inversely related to the concentration of functionalized albumin in the monomer solution, the degree of albumin modification, and the concentration of chemical initiator in the monomer solution. A typical example of hydrogel swelling and degradation is shown in Figure 9.5. The hydrogel swelling in pepsin-free simulated gastric fluid resulted in a dramatic increase in size. In the presence of pepsin, the hydrogel degraded and significant mechanical integrity was lost over time to the point of gel disruption. At the gel disruption point, the gel behaved like a viscoelastic solution. The mode of hydrogel degradation by pepsin was altered by controlling the degree of modification on albumin or the concentration of chemical initiator in the monomer solution. Hydrogels underwent a predominance of surface degradation when the degree of albumin modification was below 27% or when the concentration of chemical initiator in the monomer solution was below 0.10% (w/w) of the monomer. A predominance of surface degradation was characterized by the following observations: (1) at times exceeding 1 h of incubation in pepsin solution, the swelling ratio of the degrading gels became significantly lower than that of the nondegrading control samples, indicating a loss of polymer chains from the gel; (2) the swelling ratio decreased to as low as 2 before complete dissolution of the gel occurred; and (3) the integrity of surface degrading gels, while being reduced in size, was comparable to that of nondegrading control samples. It was observed that the rate of surface degradation decreased as the degree of albumin modification increased from 8% to 15% or as the concentration of chemical initiator in the monomer solution increased from 0.01% to 0.10% (w/w) of the monomer. The predominance of bulk degradation, however, was observed when the degree of alkylation was 27%

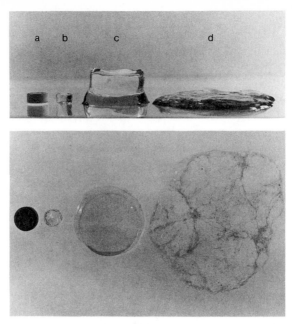

Figure 9.5   Side (A) and top (B) views of albumin-crosslinked hydrogel in various states. (a) Extra-strength Tylenol tablet as a size indicator; (b) dry glassy hydrogel; (c) hydrogel swelling in pepsin-free simulated gastric fluid; (d) hydrogel degradation by pepsin. (From ref. 215, with permission.)

or greater or when the concentration of chemical initiator was < 0.1% (w/w) of the monomer. A predominance of bulk degradation was characterized by the following observations: (1) at times exceeding 8 h of exposure to pepsin, the swelling ratio and the size of the degrading gels became significantly larger than that of the nondegrading control samples, indicating cleavage of the functionalized albumin (FA) throughout the gel; (2) the release of polyvinylpyrrolidone (PVP) from the degrading network was delayed; and (3) the integrity of the hydrogel was reduced over time, leading to complete gel disruption. It was observed that the time to reach the gel disruption point was prolonged as the degree of albumin modification increased or as the concentration of chemical initiator increased.

The rate and mode of hydrogel degradation are thought to depend largely on the degree of albumin incorporation within the network and the steric constraints imposed by the polymer chains. As the albumin incorporation increases, the size and mobility of the oligopeptide segments between crosslinks is reduced due to increased chemical crosslinking and physical entanglements within the network. Therefore, the formation of enzyme–substrate complexes is restricted by the larger conformational constraints on the substrate and by the steric constraints imposed by the polymer chains. Bulk degradation results when albumin incorporation is high enough to slow down the rate of enzymatic degradation, and thus allow swelling-controlled penetration of the enzyme into the hydrogel. Surface degradation results when the degree of albumin incorporation is very low. Because the size and mobility of oligopeptide segments between crosslinks is large, enzyme-catalyzed degradation is rapid. It was postulated that surface degradation occurred due to restricted enzyme penetration during gel swelling from the glassy dry state and the rapid hydrolysis of albumin crosslinks followed by loss of polymer chains from the rubbery phase of the gel.

9.6.1.2   Albumin Microspheres

Albumin microspheres have been prepared as injectable, biodegradable carriers for a variety of drugs [259]. It has been shown that albumin microspheres can be degraded by collagenase, papain, protease, and trypsin [259–261]. Lee et al. studied the enzyme-degradable properties associated with progesterone-loaded, glutaraldehyde-crosslinked microspheres [262]. The microspheres were prepared by crosslinking emulsified serum albumin with glutaraldehyde in the presence of progesterone. The crosslinking density of the microsphere was altered by varying the glutaraldehyde concentration. It was reported that the number of modified lysine residues increased from 21 to 47 as the concentration of glutaraldehyde increased from 1% to 4% (v/v) of the monomer solution. Microspheres prepared from 1% glutaraldehyde were the only samples susceptible to hydrolysis by chymotrypsin. Microsphere degradation following subcutaneous and intramuscular administration was also studied in rabbits. At the intramuscular site, all the microspheres had been degraded 2 months after administration.

Willmott et al. studied the release and degradation of adriamycin-loaded glutaraldehyde-crosslinked microspheres [263]. Again, the properties of the microspheres were altered by varying the glutaraldehyde concentration from 0.25% to 2% (v/v) of the monomer solution. The residence time of the intravenously injected microspheres in the lungs was studied by examining treated tissue samples with fluorescence microscopy. The residence time of microspheres entrapped in the lungs was dependent on the concentration of glutaraldehyde used for crosslinking. Two days after administration, samples crosslinked with 0.3% glutaraldehyde were nearly all degraded whereas samples crosslinked with 0.5% and 1.0% samples underwent 87% and 57% degradation, respectively. Microphotographs taken at 24 h following administration verified that degradation was taking place. The resistance to degradation was likely due to crosslinking of the network which limits penetration of enzymes into the microsphere and increases conformational constraints on the substrate that impair subsite–substrate interactions. In more recent work, Willmott et al. [260] studied the biodegradability of ^{125}I-labeled, glutaraldehyde-crosslinked albumin microspheres that were entrapped in the capillary beds of the lungs, liver, and kidneys of rats. The rate of biodegradation was measured by the loss in radioactivity in each respective tissue. It took only 2 days to lose 50% of the radioactivity in the lungs whereas it took 3.6 days in the liver. The degradation of microspheres was also observed in the kidneys. When the same microspheres were incubated in serum, however, only a slight loss in radioactivity was observed over a 9-day period. Similar findings were also reported for doxorubicin-loaded albumin microspheres in human plasma [264]. Because of the rapid degradation observed in vivo, the authors suggested that other enzymes such as those involved in the inflammatory response might be more responsible for regulating biodegradation.

9.6.2   Gelatin

9.6.2.1   Gelatin Hydrogels

The enzymatic degradation of gelatin hydrogels has been studied [265]. Gelatin was modified with glycidyl acrylate to introduce vinylic pendant groups. Hydrogels were prepared by exposing solutions of the modified gelatin to γ-irradiation. The effect of γ-irradiation dose on the swelling and degradation properties of the hydrogels was studied. In pepsin-free simulated gastric fluid, the swelling ratio (ratio of the swollen gel weight to the dry gel weight) of the γ-irradiated samples decreased by 30% as the irradiation dose increased from 0.16 Mrad to 0.48 Mrad. The rate of bulk degradation in the presence of pepsin decreased as the γ-irradiation dose was increased. The swelling ratio measured over time in the presence of pepsin passed a transient maximum. The time to reach the maximum swelling was prolonged as the γ-irradiation dose increased. Once again, it was observed that the crosslinking density of the network significantly affects the rate of enzyme-catalyzed hydrolysis.

9.6.2.2   Gelatin Microspheres

The degradation properties associated with interferon-loaded, glutaraldehyde-cross-linked gelatin microspheres were studied by Tabata et al. [266]. Gelatin microspheres were prepared by crosslinking with glutaraldehyde. The crosslinking reaction was carried in the absence and in the presence of ^{125}I-labeled interferon. It was reported that the resistance to collagenase-catalyzed hydrolysis increased as the reaction time increased and as the concentration of glutaraldehyde increased. At reaction times exceeding 4 h, however, no further changes in degradation resistance were observed. Furthermore, as the resistance to enzymatic degradation increased, a slower rate of ^{125}I-labeled interferon release was observed. The release of hydrolyzed gelatin was monitored using the ninhydrin method [267]. The phagocytosis of ^{125}I-labeled gelatin microspheres containing ^{125}I-labeled interferon was studied using mouse peritoneal macrophages. The degradation of microspheres by endosomes was monitored by measuring the radioactivity present in the cell lysates. The rate of the microsphere degradation and the subsequent release of ^{125}I-labeled interferon were reduced as the glutaraldehyde content increased. Thus, the release of ^{125}I-labeled interferon correlated well with microsphere degradation. As seen in the previous examples, the crosslinking density of the network profoundly affected the rate of enzyme-catalyzed hydrolysis presumably by limiting penetration of enzymes into the network and by restricting subsite–substrate interactions. Glutaraldehyde-crosslinked gelatin microspheres labeled with ^{131}I have been used as a carrier for mitomycin C, a chemotherapeutic agent [268]. It was shown that the gelatin microspheres achieved sufficient peripheral embolization in the hepatic artery and could be detected in the liver for as long as 28 days. The half-life of the microspheres was between 7 days and 14 days. Biodegradation was believed to be the result of lysosomal breakdown following phagocytosis from surrounding cells.

9.6.3   Epidermal Growth Factor

Chemical modification of epidermal growth factor (EGF) was recently studied by Njieha and S.W. Shalaby [269,270]. The objective of the work was to improve the polypeptide's resistance to degradation by enzymes found in the gastrointestinal tract while maintaining the biological activity. N-Acylation of the three primary amine groups on EGF was accomplished using acetic anhydride under slightly alkaline conditions. The extent of modification was determined by either fluorescence measurements with methoxydiphenyl furanone or by HPLC using a cation-exchange column. The digestion of modified and unmodified EGF by trypsin was monitored over time by HPLC. Approximately 75% of the triacylated EGF was resistant to trypsin-catalyzed hydrolysis for over 4 h. Native EGF, however, was completely digested within 2 h. The extent of enzymatic degradation was found to be dependent on the degree of acylation (i.e., mono-, di-, or triacylation), because the monoacylated and diacylated EGFs were completely digested within 4 h after incubation with trypsin. Because the major cleavage sites of trypsin are lysine and arginine residues, it was believed that N-acylation sterically hindered the subsite–substrate interactions during the formation of enzyme–substrate complexes.

9.6.4   Modified Enzymes

Recently, polymer–enzyme conjugates have been used in various applications. Polymer–enzyme conjugates have been studied as metabolite depletion agents for the treatment of acute lymphoblastic leukemia in man, canine lymphosarcoma, and also in the treatment of non-Hodgkin's lymphoma. The use of polymer–enzyme conjugates in enzyme replacement therapy, as oxygen free radical scavengers, or as fibrinolytic agents has also received considerable attention. The pharmacological properties associated with polymer–enzyme conjugates and other polymer–protein conjugates have been recently reviewed by Nucci et al. [271], Pizzo [272], Maeda [273], and Sehon [274]. The main attributes of polymer–protein conjugates is their ability to minimize immunological side effects as compared to the native proteins and to prolong the elimination half-life. The dramatic increase in elimination half-life is still not fully understood, but the steric repulsion of the water-soluble polymer is expected to prevent interactions with other proteins in the blood and thereby increase the elimination time. It is generally believed that polymer–protein conjugates undergo slow degradation by proteolytic enzymes in plasma and/or by intracellular lysosomal enzymes.

9.6.4.1   Polyethylene Glycol Conjugates

Monomethoxypolyethylene glycol (MPEG) has been most widely used in the preparation of polymer–enzyme conjugates. Polymer–enzyme conjugates based on polyvinylpyrrolidone have also been prepared [275–277]. MPEG is generally activated with either cyanuric chloride or carbonyldiimidazole [278,279]. Triazine-activated MPEG was coupled to liver catalase by Abuchowski et al. [280]. The degree of enzyme modification, as measured by the loss in primary amine groups, ranged from 13% to 37% depending on the amount of activated-MPEG added to the reaction mixture. Native catalase was completely inactivated by trypsin within 40 min, whereas MPEG–catalase retained 90% of its original activity even after 150 min. A similar yet slower rate of degradation was observed with chymotrypsin. After intravenous administration, native catalase lost nearly 80% of its activity within 10 h whereas MPEG–catalase remained active for over 50 h. Triazine-activated MPEG was also coupled to trypsin [281]. The degree of trypsin modification was varied from 24% to 59%. The 24%-modified MPEG–trypsin was completely resistant to inactivation by native trypsin. Resistance to inactivation by trypsin was thought to be due to the steric hindrances imparted by the MPEG chains and restricted subsite–substrate interactions due to the modification of lysine residues. Triazine-activated MPEG and its derivatives have also been coupled to other enzymes such as phenylalanine ammonia–lyase [282], asparaginase [283,284], superoxide dismutase [285], and acyl-plasmin–streptokinase complex [286]. MPEG conjugates of phenylalanine ammonia–lyase and asparaginase elicited a marked resistance to trypsin-catalyzed degradation as observed with other MPEG conjugates. The elimination half-lives of the native and MPEG-modified phenylalanine ammonia–lyases in rabbits were 6 h and 20 h, respectively. In rats, the elimination half-life of the MPEG–asparaginase was 56 h as compared to 2.9 h for the native asparaginase. The degree and type of MPEG modification on superoxide dismutase conjugates were shown to have significant effects on its elimination half-life in rats. As the degree of modification increased, the elimination half-life of the MPEG–superoxide dismutase was prolonged from 3 h to 25 h. At the same degree of modification, however, the elimination half-life increased from 1.5 h to 3.0 h as the molecular weight of the conjugated MPEG increased from 1900 to 5000. Increased half-lives for MPEG–superoxide dismutase conjugates have also been observed by Veronese et al. [287]. Katre et al. described a similar effect on elimination when MPEG was conjugated with recombinant interleukin-2 (MPEG–rIL-2) [288]. In these studies, MPEG was activated by acylating the MPEG with glutaric anhydride followed by succinimidation to form N-hydroxysuccinimide esters. The plasma level of MPEG–rIL-2 was nearly 14-fold higher than that of the native lymphokine when the degree of MPEG modification was low. The plasma level was increased more than 140-fold when the degree of MPEG modification was high. It was suggested that the prolonged level of MPEG–rIL-2 in the circulation was due to the steric hindrances from the MPEG chains and/or by size-dependent differences in glomerular filtration.

9.6.4.2   Dextran Conjugates

Dextran–enzyme conjugates are generally prepared with cyanogen bromide-activated dextran or periodate-activated dextran. Cyanogen bromide-activated dextran has been conjugated with lysozyme, chymotrypsin, and β-glucosidase by Vegarud et al. [289]. Dextran–lysozyme conjugates showed significant resistance to inactivation by chymotrypsin. Whereas the native lysozyme was completely inactivated within 2 days, the dextran–lysozyme conjugates retained 40% activity after the same time period. As the degree of dextran modification increased from 17% to 40% (w/w) of the lysozyme, the activity retained after 2 days of incubation with chymotrypsin increased from 40% to 52%. α-Amylase and catalase have also been conjugated with cyanogen bromide-activated dextran [290,291]. In these studies, the molecular weight of the conjugated dextran ranged from 60,000 to 90,000. The activities of both the dextran–α-amylase and dextran–catalase conjugates were prolonged as compared to the native enzymes in mice and rats. Two hours after intravenous injection, dextran–α-amylase retained 75% of its original activity whereas native α-amylase retained only 16% of its original activity. Similarly, dextran–catalase retained 70% of its original activity whereas native catalase retained only 7% of its original activity. It was postulated that the increased elimination half-life was related to shifts in the enzyme's isoelectric point which may have impaired cellular recognition and subsequent uptake. In the case of dextran–α-amylase, however, reduction in glomerular filtration may have also contributed to the prolonged elimination half-life. The authors also suggested that steric constraints from the dextran chains restricted enzyme-catalyzed inactivation in the plasma since it was reported that dextran-catalase conjugates displayed a marked resistance to trypsin-induced inactivation [292]. Conjugates of carboxypeptidase G_2 and dextran were studied by Melton et al. [293,294]. The molecular weight of cyanogen bromide-activated dextran ranged from 40,000 to 150,000. The activity of dextran–carboxypeptidase G_2 was not affected by trypsin or chymotrypsin for 18 h, whereas native carboxypeptidase G_2 was completely inactivated in less than 3 h. The elimination half-life of carboxypeptidase G_2 was prolonged from 3 h to 14 h by conjugation with dextran. The value was increased even more up to 46 h as the molecular weight of dextran increased. It was shown from tissue distribution studies that nondegraded, fully active dextran–carboxypeptidase G_2 was preferentially taken up by the liver. It was observed that the carboxypeptidase G_2 component of the conjugate degraded more rapidly in the liver than dextran did. In contrast, native carboxypeptidase G_2 showed no preferential tissue uptake and was rapidly cleared in the urine as low molecular weight fragments. The degradation site of native carboxypeptidase G_2, however, could not be determined. Apparently, conjugation with dextran alters the uptake and subsequent degradation of carboxypeptidase G_2 by parenchymal cells. Furthermore, the affinity for macrophage uptake appears to be decreased as the molecular weight of dextran increased.

L-Asparaginase has been conjugated to periodate-activated dextran for the treatment of lymphoblastic leukemia [295]. Dextran-L-asparaginase conjugates showed a marked resistance to inactivation by trypsin and chymotrypsin and were able to prolong activity in the plasma. The extent of prolongation was related to the molecular weight of dextran [296]. As the molecular weight of dextran increased from 10,000 to 70,000, the plasma half-life in rabbits increased nearly sevenfold. Increased circulation time of dextran-conjugated enzyme in blood was also reported by Benbough et al. [297]. Native L-asparaginase had a half-life of 12 h in man whereas dextran-L-asparaginase conjugate displayed a prolonged half-life of 11 days. The pharmacological properties associated with dextran–asparaginase conjugates have recently been reviewed by Wileman [298]. Although the degradation mechanism of polymer–protein conjugates requires further clarification, it is clear that the circulation half-life of therapeutically active proteins can be significantly prolonged by conjugation with water-soluble polymers. With this approach, the therapeutic efficacy of certain drugs could be profoundly enhanced.

9.7   Summary

The biodegradable properties of polymer molecules and polymeric matrices are becoming more and more important not only in biomedical and pharmaceutical applications, but also in agricultural and environmental applications such as herbicide release and waste management. In general, many synthetic polymers cannot be degraded by enzymes endogenous to eukaryotic and prokaryotic cells, whereas natural polymers such as proteins and polysaccharides are readily metabolized. One possible reason for this is that living cells have not evolved to the point where they can utilize synthetic polymers as a source of metabolic energy.

One of the main differences between enzyme-catalyzed hydrolysis and chemically induced hydrolysis is that enzyme-catalyzed hydrolysis is enzyme specific. Thus, polymer degradation requires the presence of specific enzymes. This specificity allows fabrication of proteins and polysaccharides into unique formulations that can be degraded only under predetermined conditions. As described in this chapter, the enzyme-catalyzed degradation of proteins and polysaccharides can be controlled through chemical modification. The ability to control the degradation rate will be useful in the design of new biodegradable polymers for various applications. Through judicious modification of selected natural polymers, new polymers with specific properties can be prepared.

REFERENCES

1. W. Rubas; Grass, G.M., *Adv. Drug Deliv. Rev.* **7,** 15 (1991).
2. Drobnik, J. *Adv. Drug Deliv. Rev.* **7,** 295 (1991).
3. Wearley, L.L. *Crit. Rev. Ther. Drug Carrier Syst.* **8,** 331 (1991).
4. Langer, R.; Moses, M. *J. Cell. Biochem.* **45,** 340 (1991).
5. Braybrook, J.H.; Hall, L.D. *Drug Des. Delivery* **6,** 73 (1990).
6. Amkrat, A.; Eckenhoff, J.B.; Nichols, K. *Adv. Drug Deliv. Rev.* **4,** 255 (1990).
7. Bocci, V. *Adv. Drug Deliv. Rev.* **4,** 149 (1990).
8. Lee, V.H.L.; Yamamoto, A. *Adv. Drug Deliv. Rev.* **4,** 171 (1990).

9. Larson, C. *Adv. Drug Deliv. Rev.* **3**, 103 (1989).
10. Shalaby, S.W. In *Encyclopedia of Pharmaceutical Technology, Vol. I.* Swarbrick, J.; Boylan, J.C., eds. Marcel Dekker, New York, p. 465 (1988).
11. Means, G.E.; Feeney, R.E. *Bioconj. Chem.* **1**, 2 (1990).
12. Wong, S.S. *Chemistry of Protein Conjugation and Crosslinking,* CRC Press, Boca Raton, FL (1991).
13. *Modified Starches: Properties and Uses.* Wurzburg, O.B., ed. CRC Press, Boca Raton, FL (1986).
14. *Carbohydrate Chemistry.* Kennedy, J.F., ed. Clarendon Press, Oxford (1988).
15. *Biochemistry,* Zubay, G., ed. Addison-Wesley, Reading, MA (1983).
16. Haas, S.; Geng, R.; Bleumel, G. In *Kontrolle Antithrombotika. Hamb. Symp. Blutgerrinnung. 23rd.* Marx, R.; Thies, H.A., eds. Editones Roche, Basel, Switzerland, p. 271 (1981).
17. Griffith, A.J. British Patent 1 583 006 (C1.A61K31/715), 21 Jan. 1981.
18. Antal, M.; Kuniak, L. *Die Starke* **32**, 276 (1980).
19. Antal, M.; Kuniak, L. *Starch* **35**, 94 (1983).
20. Harvey, R.D.; Klem, R.E.; Bale, M.; Hubbard, E.D. *Retention Drain.. Semin. Notes* **39** (1979).
21. Paschall, E.F., U.S. Patent 2,876,217 (1959).
22. Shildneck, P.A.; Hathaway, R.J. U.S. Patent 3,346,563 (1967).
23. Doughty, J.B.; Klem, R.E. U.S. Patent 4,066,673 (1978).
24. Fischer, W.; Langer, M.; Pohl, G. German Patent 2,949,886 (1981).
25. Tasset, E.L. U.S. Patent 4,464,528 (1984).
26. McClure, J.D.; Williams, P.H. U.S. Patent 3,475,458 (1968).
27. Edman, P.; Ekman, B.; Sjöholm, I. *J. Pharm. Sci.* **69**, 838 (1980).
28. Taniguchi, M.; Samal, R.K.; Suzuki, M.; Iwata, H.; Ikada, Y. *Graft Copolymerization of Lignocellular Fibers.* Hon, D.N.S., ed. *ACS Symposium Series* 187, American Chemical Society, p. 217 (1982).
29. Mauzac, M., Josefonvicz, J. *Biomaterials* **5**, 301 (1984).
30. Takahura, Y.; Kitajima, M.; Matsumoto, S.; Hashida, M.; Sezaki, H. *Int. J. Pharm.* 37, 135 (1987).
31. Caldwell, C.G.; Wurzburg, O.B. U.S. Patent 2,813,093 (1957).
32. Hullinger, C.H.; Uyi, N.H. U.S. Patent 2,970,140 (1961).
33. Roberts, H. *Starch Chem. Technol.* **1**, 439 (1965).
34. Streitweiser, A. *Chem. Rev.* **56**, 571 (1956).
35. Konigsburg, M. U.S. Patent 2,500,950 (1950).
36. Porath, J.; Laas, T.; Janson, J.C. *J. Chromatogr.* **103**, 49 (1975).
37. Verbanac, F.; Moser, K.B. U.S. Patent 3,553,194 (1971).
38. Roberts, H. *Starch Chem. Technol.* **2**, 293 (1967).
39. Roberts, H. *Methods Carbohydrate Chem.* **4**, 289 (1964).
40. Tranquair, J. *J. Soc. Chem. Ind.* **28**, 288 (1909).
41. Clarke, H.T.; Gillespie, H.B. *J. Am. Chem. Soc.* **54**, 2083 (1932).
42. Cross, C.F.; Bevan, E.; Tranquair, J. *Chem. Ztg.* **29**, 527 (1905). 43. Caldwell, C.G. U.S. Patent 2,461,139 (1949).
44. Wurzburg, O.B. U.S. Patent 2,935,510 (1960).
45. Wurzburg, O.B. *Methods Carbohydrate Chem.* **4**, 286 (1964).
46. Ferrutti, P.; Tanz, M.C.; VAccaroni, F. *Makromol. Chem.* **180**, 375 (1979).
47. Mullen, J.W.; Pacsu, E. *Ind. Eng. Chem.* **34**, 1209 (1942).

48. Parmerter, S.M. U.S. Patent 3,620,913 (1971).
49. Gunasingham, H.; Teo, P.Y.T.; Lai, Y.H.; Tan, S.G. *Biosensors* **4**, 349 (1989).
50. Frautschi, J.R.; Munro, M.S.; Llyod, D.R.; Eberhart, R.C. *Trans. Am. Soc. Intern. Organs* **29**, 242 (1983).
51. Smith, C.E.; Tuschhoff, J.V. U.S. Patent 2,928,828 (1960).
52. Tuschhoff, J.V. U.S. Patent 3,022,289 (1962).
53. Evans, R.B.; Kunze, W.G. U.S. Patent 3,318,868 (1976).
54. Eastman, J.E. U.S. Patent 3,959,514 (1976).
55. Tessler, M.M. U.S. Patent 3,838,149 (1974).
56. Tessler, M.M. U.S. Patent 3,719,662 (1973).
57. Tessler, M.M. U.S. Patent 3,824,071 (1974).
58. Paschall, E.F. *Methods Carbohydrate Chem.* **4**, 294 (1964).
59. Wurzburg, O.B.; Jarowenko, W.; Rubens, R.W.; Patel, J.K. U.S. Patent 4,166,173 (1979).
60. Wurzburg, O.B.; Jarowenko, W.; Rubens, R.W.; Patel, J.K. U.S. Patent 4,216,310 (1980).
61. Hamilton, R.M.; Paschall, E.F. *Starch Chem. Technol.* **2**, 351 (1967).
62. Radley, J.A. *Starch Production Technology*, Applied Science Publishers, London, p. 543 (1976).
63. Dumitriu, S.; Popa, M.; Dumitriu, M. *J. Bioact. Compatible Polym.* **3**, 243 (1988).
64. Kohn, J.; Wilchek, M. *Anal. Biochem.* **115**, 375 (1981).
65. Kohn, J.; Wilchek, M. *FEBS Lett.* **154**, 209 (1983).
66. Wilchek, M.; Miron, T. *Methods Enzymol.* **104**, 3 (1984).
67. Nagaoka, S.; Kurumatani, H.; MOri, Y.; Tanazana, H. *J. Bioact. Compatible Polym.* **4**, 323 (1989).
68. Hashida, M.; Kojima, T.; Takahashi, Y.; Muranishi, S.; Sezaki, H. *Chem. Pharm. Bull.* **25**, 2456 (1977).
69. Kojima, T.; Hashida, M.; Muranishi, S.; Sezaki, H. *J. Pharm. Pharmacol.* **32**, 30 (1980).
70. Kohn, J.; Wilchek, M. *Appl. Biochem. Biotechnol.* **9**, 285 (1984).
71. Degling, L.; Stjärnkvist, P.; Sjöholm, I. *Proceed. Intern. Symp. Control. Rel. Bioact. Mater.* **18**, 391 (1991).
72. Laakso, T.; Stjärnkvist, P.; Sjöholm, I. *J. Pharm. Sci.* **76**, 134 (1987).
73. Stjärnkvist, P.; Degling, L.; Sjöholm, I. *J. Pharm. Sci.* **80**, 436 (1991).
74. Wong, S.S. *Chemistry of Protein Conjugation and Crosslinking*, CRC Press, Boca Raton, FL, p. 195 (1991).
75. Wilchek, M.; Miron, T. *Biochem. Int.* **4**, 629 (1982).
76. Drobnik, J.; Labsky, J.; Kudlvasrove, H.; Sandek, H.; Svec, F. *Biotechnol. Bioeng.* **24**, 487 (1982).
77. Vandoorne, F.; Vercauteren, R.; Permentier, D.; Schacht, E. *Macromol. Chem.* **186**, 2455 (1985).
78. Kennedy, J.F.; Burke, S.A.; Rosevear, A. *J. Chem. Soc. Perkin Trans.* **1**, 2293 (1973).
79. Kennedy, J.F.; Rosevear, A. *J. Chem. Soc. Perkin Trans.* **1**, 757 (1974).
80. Schacht, E.; Vandoorne, F.; Vermeersch, J.; Duncan, R. In *Controlled-Release Technology, Pharmaceutical Applications*. Lee, P.I.; Good, W.R., eds. *ACS* Symposium Series 348, American Chemical Society, Washington, D.C., p. 188 (1987).
81. Vansteenkiste, S.; DeMarre, A.; Schacht, E. *J. Bioact. Compatible Polym.* **7**, 4 (1992).
82. Nilsson, K.; Norrloew, O.; Mosbach, K. *Acta Chem. Scand. Ser. B* **B35**, 19 (1981).
83. Nilsson, K.; Mosbach, K. *Biochem. Biophys. Res. Commun.* **102**, 449 (1981).
84. Tay, S.W.; Merril, E.W.; Salzman, E.W.; Lindon, J. *Biomaterials* **10**, 11 (1989).

85. Sturgeon, C.M. *Carbohydrate Chemistry* Kennedy, J.F., ed. Clarendon Press, Oxford, p. 560 (1988).
86. Wilson, M.B.; Nakane, P.K. *J. Immunol. Methods* **12,** 171 (1976).
87. Wright, J.F.; Hunter, W.M. *J. Immunol. Methods* **48,** 311 (1982).
88. Singh, M.; Vasudevan, P.; Sinha, T.J.M.; Ray, A.R.; Misro, M.M.; Guha, K. *J. Biomed. Mater. Res.* **15,** 655 (1981).
89. Schacht, E.; Vermeersch, J.; Vandoorne, F.; Vercauteren, R.; Remon, J.P. *J. Controll. Rel.* **2,** 245 (1985).
90. Ishak, M.F.; Painter, T. *Acta Chem. Scand.* **25,** 3875 (1971).
91. Painter, T.; Larson, B. *Acta Chem. Scand.* **24,** 2366 (1970).
92. Aalmo, K.M.; Painter, T. *Carbohydr. Res.* **89,** 73 (1981).
93. Kurzer, F.; Douraghi-Zadeh, K. *Chem. Rev.* **67,** 107 (1967).
94. Mino, G.; Kaizeman, S. *J. Polym. Sci.* **32,** 242 (1958).
95. Carlsohn, H.; Hartman, H. *Acta Polym.* **33,** 640 (1982).
96. Hebeish, A.; Abdel-Thalouth, I.; El-Kashouti, M.A.; Abdel-Fattah, S.H. *Angew, Makromol. Chem.* **78,** 101 (1979).
97. Mehrotra, R.; Randy, B. *J. Appl. Polym. Schi.* **21,** 1647 (1977).
98. Imoto, M.; Morita, E.; Ouchi, T. *J. Polym. Sci. Polym. Symp.* **68,** 1 (1980).
99. Nishioka, N.; MInami, K.; Kosai, K. *Polym. J.* **15,** 591 (1983).
100. Nishioka, N.; Matsumoto, Y.; Yumen, T.; Monmae, K.; Kosai, K. *Polym. J.* **18,** 323 (1986).
101. Nishioka, N.; Kosai, K. *Polym. J.* **13,** 1125 (1981).
102. Phillips, G.O. *The Carbohydrates, Chemistry and Biochemistry, Vol. 1B.* Pigman, W.W.; Horton, D, eds. Academic Press, New York, p. 1217 (1980).
103. Merlin, A.; Fouassier, J.P. *Makromol. Chem.* **182,** 3053 (1981).
104. Trimnel, D.; Stout, E.I. *J. Appl. Polym. Sci.* **25,** 2431 (1980).
105. Fanta, G.F.; Burr, R.C.; Russel, C.R.; Rist, C.E. *J. Macromol. Sci. Chem.* **A4,** 331 (1970).
106. Mehrotra, R.; Randy, B. *J. Appl. Polym. Sci.* **22,** 2991 (1978).
107. Reyes, Z.; Clark, C.F.; Comas, M.; Russel, C.R.; Rist, C.E. *Nucl. Appl.* **6,** 509 (1969).
108. Reyes, Z.; Syz, M.G.; Huggins, M.L.; Russel, C.R. *J. Polym. Sci. C* **23,** 401 (1968).
109. Gruber, E.; Alloush, S.; John, K.; Schurz, J. *Die Starke* **25,** 325 (1973).
110. Fanta, G.F.; Burr, R.C.; Doane, W.M.; Russel, C.R. *J. Appl. Polym. Sci.* **16,** 2835 (1972).
111. Fanta, G.F.; Burr, R.C.; Doane, W.M.; Russel, C.R. *Polym. Sci. Technol.* **2,** 275 (1973).
112. Lunblad, R.L. *Chemical Reagents for Protein Modifiction,* CRC Press, Boca Raton, FL (1991).
113. Alber, T. *Annu. Rev. Biochem.* **58,** 765 (1989).
114. Meares, C.J. In *Protein Tailoring for Food and Medical Uses.* Feeney, R.E.; Whitaker, J.R., eds. Marcel Dekker, New York, p. 339 (1986).
115. Neville, D.M. *Crit. Rev. Ther. Drug Carrier Syst.* **2,** 329 (1986).
116. Means, G.E.; Feeney, R.E. *Chemical Modification of Proteins,* Holden-Day, San Francisco (1971).
117. Lunblad, R.L.; Noyes, C.M. *Chemical Reagents for Protein Modification,* CRC Press, Boca Raton, FL p. 16 (1984).
118. Baker, B.R. *Design of Active-Site-Directed Irreversible Enzyme Inhibitors: The Organic Chemistry of the Enzymic Active Site,* John Wiley & Sons, New York (1967).
119. Brewer, C.F. *Anal. Biochem.* **18,** 248 (1967).
120. Eiser, H.N.; Belman, S.; Cartsen, M.E. *J. Am. Chem. Soc.* **75,** 4583 (1953).

121. Means, G.E.; Feeney, R.E. *Biochemistry* **7,** 2192 (1968).

122. Klein, S.M.; Sagers, R.D. *J. Biol. Chem.* **242,** 301 (1967).

123. Jaworek, D. In *Insolubilized Enzymes.* Salmona, M.; Saronia, C.; Garattini, S., eds. Raven Press, New York, p. 65 (1974).

124. Butler, P.J.G.; Harris, J.I.; Hartley, B.S.; Leberman, R. *Biochem. J.* **112,** 679 (1969).

125. Gounaris, A.D.; Perlman, G.E. *J. Biol. Chem.* **242,** 2739 (1967).

126. Riehm, J.P.; Scheraga, H.A. *Biochemistry* **4,** 772 (1965).

127. Gurd, F.R.N. *Methods Enzymol.* **11,** 532 (1967).

128. Raftery, M.A.; Cole, R.D. *Biochem. Biophys. Res. Commun.* **10,** 967 (1963).

129. Lindley, H. *Nature* **178,** 647 (1956).

130. Gorin, G.; Martic, P.A.; Doughty, G. *Arch. Biochem. Biophys.* **115,** 593 (1966).

131. Gregory, J.D. *J. Am. Chem. Soc.* **77,** 3922 (1955).

132. Smyth, D.G.; Blumfeld, O.O.; Konigsburg, W. *Biochem. J.* **91,** 589 (1964).

133. Wang, T.W.; Kassel, B. *Biochemistry* **13,** 698 (1974).

134. Bartholeyns, J.; Moore, S. *Science* **186,** 444 (1974).

135. Dutton, A.; Adam, M.; Singer, S.J. *Biochem. Biophys. Res. Commun.* **23,** 730 (1966).

136. Niehaus, W.G.; Wold, F. *Biochem. Biophys. Acta* B196, **170** (1970).

137. Hartman, F.C.; Wold, F. *J. Am. Chem. Soc.* **88,** 3890 (1966).

138. Siezen, R.J.; Bindels, J.G.; Hoender, H.J. *Eur. J. Biochem.* **107,** 243 (1980).

139. Staros, J.V.; Morgan, D.G.; Appling, D.R. *J. Biol. Chem.* **256,** 5890 (1981).

140. Pilch, P.F.; Czech, M.P. *J. Biol. Chem.* **254,** 3375 (1979).

141. Staros, J.V. *Biochemistry* **21,** 3950 (1982).

142. Giedroc, D.P.; Puett, D.; Ling, N.; Staros, J.V. *J. Biol. Chem.* **258,** 16 (1983).

143. Lindsay, D.G. *FEBS Lett.* **21,** 105 (1972).

144. Ozana, H. *J. Biochem.* **62,** 419 (1967).

145. Snyder, P.D.; Wold, F.; Bernlohr, R.W.; Dullum, C.; Desnick, R.J.; Krivit, W.; Condie, R.M. *Biochem. Biophys. Acta* **350,** 432 (1974).

146. Schick, A.F.; Singer, S.J. *J. Biol. Chem.* **236,** 2477 (1961).

147. Brandenberg, D. *Hoppe Seylers Z. Physiol. Chem.* **353,** 869 (1972).

148. Plotz, P.H. *Methods Enzymol.* **46,** 505 (1977).

149. Richard, F.M.; Knowles, J.R. *J. Mol. Biol.* **37,** 231 (1968).

150. Joseph, R.; Eisenberg, H.; Reisler, E. *Biochemistry* **12,** 4060 (1973).

151. Hopwood, D. *Histochemie* **17,** 151 (1969).

152. Raymond, G.; Degenera, M.; Mikeal, R. *Drug Dev. Ind. Pharm.* **16,** 1025 (1990).

153. Burgess, D.J., Davis, S.S.; Tomlinson, E. *Int. J. Pharm.* **39,** 129 (1987).

154. Willmot, N.; Chen, Y.; Florence, A.T. *J. Control. Rel.* **8,** 103 (1988).

155. Benesch, R.; Kwong, S. *J. Protein Chem.* **10,** 503 (1991).

156. Chiao, C.S.L.; Price, J.C. *Pharm. Res.* **6,** 517 (1989).

157. Vural, I.; Kas, H.S.; Erlan, M.T.; Hincal, A.A. *Drug Dev. Ind. Pharm.* **16,** 1781 (1990).

158. Sager, P.R. *Toxicol. Appl. Pharmacol.* **97,** 141 (1989).

159. Robinson, I.D. *J. Appl. Polym. Sci.* **8,** 1903 (1964).

160. Imamura, E.; Sawatani, O.; Kouanagi, H.; Noishiki, Y.; Miyata, T. *J. Cardiac Surg.* **4,** 50 (1989).

161. Levy, M.C.; Rambourg, P.; Levy, J.; Potron, G. *J. Pharm. Sci.* **71,** 759 (1982).

162. Rambourg, P.; Levy, J.; Levy, M.C. *J. Pharm. Sci.* **71,** 753 (1982).

163. Ji, T.H. *Biochim. Biophys. Acta* **559,** 39 (1979).

164. Kitagawa, T.; Shimozana, T.; Aikana, T.; Yoshida, T.; Nishimura, H. *Chem. Pharm. Bull.* **29,** 1130 (1981).

165. Keller, O.; Rundinger, J. *Helv. Chim. Acta* **58,** 531 (1975).
166. Yoshitake, S.; Yamada, Y.; Ishikana, E.; Masseyeff, R. *Eur. J. Biochem.* **101,** 395 (1979).
167. Rector, E.S.; Schwenk, R.J.; Tse, K.S.; Sehon, A.H. *J. Immunol. Methods* **24,** 321 (1978).
168. Cuatrecasas, P.; Wilchek, M.; Anfinsen, C.B. *J. Biol. Chem.* **244,** 4316 (1969).
169. Fasold, H.; Baumert, H.; Fink, G. In *Protein Crosslinking: Biochemical and Molecular Aspects,* Plenum Press, New York, p. 207 (1976). 170. Olomucki, M.; Diopoh, J. *Biochim. Biophys. Acta* **263,** 312 (1972).
171. Diopoh, J.; Olomucki, M. *Hoppe Seylers Z. Physiol. Chem.* **360,** 1257 (1979).
172. Hermentin, P.; Doenges, R.; Granski, P.; Bosslet, K.; Kraemer, H.; Hoffman, D.; Zilag, H.; Steinstraesser, A.; Schwartz, A.; Kulhlmann, L.; Luben, G.; Seiler, F.R. *Bioconjugate Chem.* **1,** 100 (1990).
173. Baumert, H.G.; Fasold, H. *Methods Enzymol.* **172,** 584 (1989).
174. Carraway, K.L.; Koshland, D.E. *Methods Enzymol.* **25,** 616 (1972).
175. Sheehan, J.C.; Cruickshank, P.A.; Boshart, G.L. *J. Organ. Chem.* **26,** 2525 (1961).
176. Zot, H.G.; Puett, D. *J. Biol. Chem.* **264,** 15552 (1989).
177. Hennink, W.E.; Feijen, J.; Ebert, C.D.; Kim, S.W. *Thromb. Res.* **29,** 1 (1983).
178. Cremers, H.F.M.; Feijen, J.; Kwon, G.; Bae, Y.H.; Kim, S.W.; Noteborn, H.P.J.M.; McVie, J.G. *J. Controll. Re.* **11,** 167 (1990).
179. Magee, G.; Willmot, N.; Halbert, G. *Proceed. Int. Symp. Control. Rel. Bioact. Mater.* **18,** 363 (1991).
180. Patramani, I.; Katsiri, K.; Pisteuou, T.; Kalogerakos, M.; Pawlatos, M.; Evangelopoulos, A.E. *Eur. J. Biochem.* **11,** 28 (1969).
181. Avrameas, S.; Ternynck, T. *J. Biol. Chem.* **242,** 1651 (1967).
182. Aleix, J.A.; Swaminathan, B.; Minnich, S.A.; Wallshein, V.A. *J. Immunol.* **6,** 391 (1985).
183. Chang, S.I.; Hammes, G.G. *Biochemistry* **25,** 4661 (1986).
184. Barthing, G.J.; Chattopadhyay, S.K.; Barker, C.W.; Farrester, L.J.; Brown, H.D. *Int. J. Peptide Protein Res.* **6,** 287 (1974).
185. Lee, J.M.; Gratzer, P.F.; Pereira, C.A. *17th Annual Meeting of the Society for Biomaterials,* p. 161 (1991).
186. Nojiri, C.; Noishiki, Y.; Koyanagi, H. *J. Thorac. Cardiovasc. Surg.* **93,** 867 (1987).
187. Tu, R.; Lu, C-L.; Thyagarajan, K.; Wang, E.; Nguyen, H.; Quijano, R.C. *17th Annual Meeting of the Society for Biomaterials,* p. 163 (1991).
188. Pauling, L. *Chem. Eng. News* **24,** 1375 (1946).
189. Spratt, T.E.; Kaiser, E.T. *J. Am. Chem. Soc.* **106,** 6440 (1984).
190. White, H.; Solomon, F.; Jencks, W.P. *J. Biol. Chem.* **251,** 1688 (1976).
191. White, H.; Solomon, F.; Jencks, W.P. *J. Biol. Chem.* **251,** 1799 (1976).
192. Carone, F.A.; Peterson, P.R.; Flouret, G. *J. Lab. Clin. Med.* **100,** 1 (1982).
193. Guyton, A.C. *Textbook of Medical Physiology 7th Edn.,* W.B. Saunders, Philadelphia (1986).
194. *Enzyme Nomenclature, Recommendations of the Nomenclature Committee of the International Union of Biochemistry on the Nomenclature and Classification of Enzyme-Catalyzed Reactions.* Webb, E.C., ed. Academic Press, Orlando, FL (1984).
195. Sinnot, M.L. *New Compr. Biochem.* **6** (*Chem. Enzyme Action*), 389 (1984).
196. Sinnot, M.L. In *Enzyme Mechanisms.* Page, M.I.; Williams, A. eds. The Royal Society of Chemistry, Northern Ireland, p. 259 (1987).
197. Schechter, I.; Berger, A. *Biochem. Biophys. Res. Commun.* **27,** 157 (1967).
198. Schechter, I.; Berger, A. *Biochem. Biophys. Res. Commun.* **32,** 898 (1968).
199. Fruton, J.S. *Proc. Am. Philos. Soc.* **121,** 309 (1977).

200. Steinbrink, D.R.; Bond, M.D.; VanWart, H.E. *J. Biol. Chem.* **260,** 2771 (1985).
201. VanWart, H.E.; Steinbrink, D.R. *Biochemistry* **23,** 6520 (1985).
202. Kopeček, J.; Rejmanova, P. *Controlled Drug Delivery Vol. I.* Bruck, S.D., ed. CRC Press, Boca Raton, FL, p. 81 (1983).
203. Taylor, J.W.; Miller, R.J.; Kaiser, E.T. *Mol. Pharmacol.* **22,** 657 (1982).
204. Pytela, J.; Saudek, V.; Drobnik, J.; Rypacek, F. *J. Controll. Rel.* **10,** 17 (1989).
205. Hayashi, T.; Tabata, Y.; Nakajima, A. *Rep. Prog. Pol. Phys. Jpn.* **27,** 517 (1984).
206. Chandy, T.; Sharma, C.P. *Biomaterials* **12,** 677 (1991).
207. Hayashi, T.; Ikada, Y. *Biomaterials* **11,** 409 (1990).
208. Kopeček, J. *Biomaterials* **5,** 19 (1984).
209. Rejmanová, P.; Kopeček, J. *Makromol. Chem.* **184,** 2009 (1983).
210. Duncan, R.; Cable, H.; Lloyd, J.B.; Rejmanová, P.; Kopeček, J. *Makromol. Chem.* **184,** 1997 (1983).
211. Rejmanová, P.; Kopeček, J.; Duncan, R.; Lloyd, J.B. *Biomaterials* **6,** 45 (1985).
212. Ringsdorf, H.; Schmidt, B.; Ulbrich, K. *Makromol. Chem.* **188,** 257 (1987).
213. Říhová, B.; Vetuicka, V.; Strohalm, J.; Ulbrich, K.; Kopeček, J. *J. Controll. Rel.* **9,** 21 (1989).
214. Říhová, B.; Kopeček, J. *J. Controll. Rel.* **2,** 289 (1985).
215. Park, K. *Biomaterials* **9,** 435 (1988).
216. Shalaby, W.S.W.; Park, K. *Pharm. Res.* **7,** 816 (1990).
217. *Atlas of Protein Sequence and Structure, Vol. 5, Suppl. 3.* Dayhoff, M.O., ed. The National Biomedical Research Foundation, Silver Springs, Maryland, p. 306 (1978).
218. Laurent, T.C. *Eur. J. Biochem.* **21,** 498 (1971).
219. Laurent, T.C.; Ogston, A.G. *Biochem. J.* **89,** 249 (1963).
220. Edmon, E.; Ogston, A.G. *Biochem. J.* **109,** 569 (1968).
221. Laurent, T.C.; Killander, J. *J. Chromatog.* **14,** 317 (1964).
222. Giddings, J.C. *J. Phys. Chem.* **74,** 1368 (1970).
223. Giddings, J.C.; Kucera, E.; Russel, C.P.; Myers, M.N. *J. Phys. Chem.* **72,** 4397 (1968).
224. Kopeček, J.; Rejmanová, P.; Chytry, V. *Makromol. Chem.* **182,** 799 (1981).
225. Kopeček, J. *Makromol. Chem.* **178,** 2169 (1977).
226. Rejmanová, P.; Obereigner, B.; Kopeček, J. *Makromol. Chem.* **182,** 1899 (1981).
227. Ulbrich, K.; Strohalm, J.; Kopeček, J. *Makromol. Chem.* **182,** 1917 (1981).
228. Ulbrich, K.; Zakhariena, E.I.; Obereigner, B.; Kopeček, J. *Biomaterials* **1,** 199 (1980).
229. Duncan, R.; Lloyd, J.B.; Kopeček, J. *Biochem. Biophys. Res. Commun.* **94,** 284 (1980).
230. Schacht, E.; Vercauteren, R.; Vansteenkiste, S. *J. Bioact. Compat. Polym.* **3,** 72 (1988).
231. Vercauteren, R.; Brunnel, D.; Schacht, E. *J. Bioact. Compat. Polym.* **5,** 4 (1990).
232. Chaves, M.S.; Arranz, F. *Makromol. Chem.* **186,** 17 (1985).
233. Crepon, B.; Jozefonvicz, J.; Chytrý, V.; Říhová, B.; Kopeček, J. *Biomaterials* **12,** 550 (1991).
234. Edman, P.; Sjöholm, I. *J. Pharm. Sci.* **72,** 796 (1983).
235. Edman, P.; Sjöholm, I.; Brunk, U. *J. Pharm. Sci.* **72,** 658 (1983).
236. Edman, P.; Sjöholm, I. *Life Sci.* **30,** 327 (1982).
237. Heller, J.; Pangburn, S.H.; Roskos, K.V. *Biomaterials* **11,** 345 (1990).
238. Artursson, P.; Edman, P.; Laakso, T.; Sjöholm, I. *J. Pharm. Sci.* **73,** 1507 (1984).
239. Laakso, T.; Sjöholm, I. *J. Pharm. Sci.* **76,** 935 (1987).
240. Laakso, T.; Artursson, P.; Sjöholm, I. *J. Pharm. Sci.* **75,** 962 (1986).
241. Stjärnkvist, P.; Laakso, T.; Sjöholm, I. *J. Pharm. Sci.* **78,** 52 (1989).
242. Artursson, P.; Edman, P.; Sjöholm, I. *J. Pharm. Exp. Ther.* **231,** 705 (1984).

243. Mateescu, M.A.; Schell, H.D. *Carbohydr. Res.* **124,** 319 (1983).
244. Hirano, S.; Tsuchida, H.; Nagao, N. *Biomaterials* **10,** 574 (1989).
245. Pangubrn, S.H.; Trescony, P.V.; Heller, J. *Biomaterials* **3,** 105 (1982).
246. Rubinstein, A.; Nakar, D.; Sintov, A. *Pharm. Res.* **9,** 276 (1992).
247. Sintov, A.; Nakar, D.; Rubinstein, A. *Proceed. Intern. Symp. Control. Rel. Bioact. Mater.* **18,** 381 (1991).
248. Anatha Samy, T.S. *Arch. biochem. Biophys.* **121,** 703 (1967).
249. Sri Ram, J.; Maurer, P. *Arch. Biochem. Biophys.* **85,** 512 (1959).
250. Atassi, M.Z.; Habeeb, A.F.S.A. *Biochemistry* **8,** 1385 (1969).
251. Habeeb, A.F.S.A.; Atassi, M.Z. *Immunochemistry* **6,** 555 (1969).
252. Shalaby, W.S.W.; Peck, G.E.; Park, K. *Controll. Rel.* **16,** 355 (1991).
253. Shalaby, W.S.W.; Blevins, W.E.; Park, K. *Water-Soluble Polymers, Synthesis, Solution Properties, and Application, ACS Symposium Series* 467. Shalaby, S.W.; McCormick, C.L.; Butler, G.B. eds. American Chemical Society, p. 484 (1991).
254. Shalaby, W.S.W.; Blevins, W.E.; Park, K. *Polymeric Drugs and Drug Delivery Systems, ACS Symposium Series* 469. Dunn, R.; Ottenbrite, R.M., eds. American Chemical Society, p. 237 (1991).
255. Shalaby, W.S.W.; Blevins, W.E.; Park, K. *Biomaterials* **13,** 289 (1992).
256. Shalaby, W.S.W.; Blevins, W.E.; Park, K. *J Controll. Rel.* **19,** 131 (1992).
257. Shalaby, W.S.W.; Chen, M.; Park, K. *J. Bioact. Compat. Polym.* **7,** 257 (1992).
258. Snyder, S.L.; Sobocinski, P.Z. *Anal. Biochem.* **64,** 284 (1975).
259. Morimoto, Y.; Fujimoto, S. *Crit. Rev. Ther. Drug Carrier Syst.* **2,** 19 (1985).
260. Willmott, N.; Chen, Y.; Goldberg, J.; Meardle, C.; Florence, A.T. *J. Pharm. Pharmacol.* **41,** 433 (1989).
261. Mahato, R.I.; Willmott, N.; Vezin, W.R. *Proceed. Intern. Symp. Control. Rel. Bioact. Mater.* **18,** 375 (1991).
262. Lee, T.K.; Sokoloski, T.D.; Royer, G.P. *Science* **213,** 233 (1981).
263. Willmott, N.; Cummings, J.; Studart, J.F.B.; Florence, A.T. *Biopharm. Drug Dispos.* **6,** 91 (1985).
264. Jones, C., Burton, M.A.; Gray, B.N. *J. Pharm. Pharmacol.* **41,** 813 (1989).
265. Kamath, K.R.; Park, K. *Proc. Int. Symp. Control. Rel. Bioact. Mater.* **19,** 42 (1992).
266. Tabata, Y.; Ikada, Y. *Pharm. Res.* **6,** 422 (1989).
267. McGrath, R. *Anal. Biochem.* **49,** 95 (1972). 268. Yan, C.; Li, X.; Chen, X.; Wang, D.; Xhang, D.; Tan, T.; Kitano, H. *Biomaterials* **12,** 640 (1991).
269. Nzieha, F.K.; Shalaby, S.W. *Polym Prepr.*` **32,** 233 (1991).
270. Nzieha, F.K.; Shalaby, S.W. *J. Bioact. Compatible Polym.* 7, 288 (1992).
271. Nucci, M.L.; Shorr, R.; Abuchowski, A. *Adv. Drug. Deliv. Rev.* **6,** 133 (1991).
272. Pizzo, S.V. *Adv Drug Deliv. Rev.* **6,** 153 (1991).
273. Maeda, H. *Adv Drug Deliv. Rev.* **6,** 181 (1991).
274. Sehon, A.H. *Adv Drug Deliv. Rev.* **6,** 203 (1991).
275. Geiger, B.; Von Specht, B.; Arnon, R. *Eur. J. Biochem.* **73,** 141 (1977).
276. Von Specht, B.; Brendel, W. *Biochim. Biophys. Acta* **484,** 109 (1977).
277. Veronese, F.M.; Sartore, L.; Caliceti, P.; Schiavon, O. *J. Bioact. Compatible Polym.* **5,** 167 (1990).
278. Brucato, F.H.; Pizzo, S.V. *Blood* **76,** 73 (1990).
279. Berger, H.; Pizzo, S.V. *Blood* **71,** 1641 (1988).

280. Abuchowski, A.; McCoy, J.R.; Palczuk, N.C.; Van Es, T.; Davis, F.F. *J. Biol. Chem.* **252,** 3582 (1977).
281. Abuchowski, A.; David, F.F. *Biochim. Biophys. Acta* **578,** 41 (1979).
282. Wieder, K.J.; Palczuk, N.C.; Van Es, T.; Davis, F.F. *J. Biol. Chem.* **254,** 12579 (1979).
283. Matsushima, A.; Nishimura, H.; Ashihara, Y.; Yokota, Y.; Inada, Y. *Chem. Lett.* **7,** 773 (1980).
284. Kamisaki, Y.; Wada, H.; Yagura, T.; Matsushima, A.; Inada, Y. *J. Pharmacol. Exp. Ther.* **216,** 410 (1981).
285. Boccu, E.; Velo, G.P.; Veronese, F.M. *Pharmacol. Res. Commun.* **14,** 113 (1982).
286. Tomiya, N.; Watanabe, K.; Awaya, J.; Kurono, M.; Fujii, S. *FEBS Lett.* **193,** 44 (1985).
287. Veronese, F.M.; Caliceti, P.; Pastorino, A.; Schiavon, O.; Sartore, L. *J. Controll. Rel.* **10,** 145 (1989).
288. Katre, N.V.; Knauf, M.J.; Laird, W.J. *Proc. Natl. Acad. Sci.* **84,** 1487 (1987).
289. Vegarud, G.; Christensen, T.B. *Biotechnol. Bioeng.* **17,** 1391 (1975).
290. Marshall, J.J.; Humphreys, J.D.; Abramson, S.L. *FEBS Lett.* **83,** 249 (1977).
291. Marshall, J.J. *Trends Biochem. Sci.* **3,** 79 (1978).
292. Marshall, J.J.; Rabinowitz, M.L. *Biotechnol. Bioeng.* **18,** 1325 (1976). 293. Melton, R.G.; Wiblin, C.N.; Foster, R.L.; Sherwood, R.F. *Biochem. Pharmacol.* **36,** 105 (1987).
294. Melton, R.G.; Wiblin, C.N.; Baskerville, A.; Foster, R.L.; Sherwood, R.F. *Biochem. Pharmacol.* **36,** 113 (1987).
295. Wileman, T.E.; Bennett, M.; Lilleymann, J. *J. Pharm. Pharmacol.* **35,** 762 (1983).
296. Wileman, T.E.; Foster, R.L.; Elliot, P.N.C. *J. Pharm. Pharmacol.* **38,** 264 (1986).
297. Benbough, J.E.; Wiblin, C.N.; Rafter, T.N.A.; Lee, J. *Biochem. Pharmacol.* **28,** 833 (1979).
298. Wileman, T.E. *Adv. Drug Deliv. Rev.* **6,** 167 (1991).

Index